The Earth and Its History

RICHARD FOSTER FLINT

YALE UNIVERSITY

The Earth and Its History

W · W · NORTON & COMPANY · INC ·
NEW YORK

The photograph on the title page is the Front Range, Rocky Mountains, Colorado. *(T. S. Lovering, U.S. Geol. Survey.)*

FIRST EDITION

Library of Congress Cataloging in Publication Data

Flint, Richard Foster, 1902–
The earth and its history.

Includes bibliographies.
1. Earth. 2. Paleontology. I. Title.
QE501.F57 560 72–10879
ISBN 0–393–09377–8

Published simultaneously in Canada
by George J. McLeod Limited, Toronto

PRINTED IN THE UNITED STATES OF AMERICA

1 2 3 4 5 6 7 8 9 0

For Peggy,

who showed me what was less than good
and made me change it.

Contents

Preface

With the recent spectacular growth of populations and their increasing concentration in cities, a student's familiarity with the natural world can no longer be taken for granted. Because of the strains created by people living together in cities, current student taste understandably runs to social studies of various kinds, political science, psychology, and philosophy. In this stampede in recent years toward studies of human relations, natural science has been neglected. Yet its study should be encouraged, to help provide a less one-sided curriculum.

Earth science is the natural science best adapted for study in a liberal-arts context, for at least two reasons. The first is that a student can see the materials of the Earth and the forces that act on them in their natural environments rather than isolated in a laboratory. He can see mountains, valleys, and coasts being shaped, soil being created, and rocks of many kinds being made and then destroyed so that new rocks can be made with their debris. By witnessing such things in books and in the field, he soon comes to understand the harmonious interplay of the natural forces and feels himself in some degree a participant. Placed in the natural world, in contrast to an artificial city world, a student feels more understanding and less tension.

This result is heightened by the second reason, which is that the data of much Earth science are less exact than those in other natural sciences and are more commonly expressed in terms of probability. Thus a student who lacks a rigorous basic training in science and mathematics can absorb Earth science material with comparative ease, and can cover a broad spectrum in a single course.

Earth science can be studied in two different ways. The first is basically analytical, the other basically descriptive. A successful analytical study demands at least some previous experience with fundamental physics, chemistry, and biology, and some skill with the simpler mathematics. It constitutes the initial step in the training of a research Earth scientist.

The second way demands no prerequisites other than willingness to learn, nor does it entail rigorous analysis. It consists of intelligent observation and discussion through which the Earth's materials and the working of the Earth's natural cycles can be broadly appreciated, especially as to the manner in which those cycles create differing environments. Because it does not make use of close analysis, this approach does not itself train Earth scientists. It can, however, remove the air of mystery that often seems to surround natural processes, and so promote a sense of confidence, a sense of being at home in the world of nature.

This second method makes Earth science essentially a liberal art through which the student develops a nonprofessional understanding of his natural environment in the broadest sense. More than this, it enables him to regard time in a new way, to appreciate the meaning of ten million or even a billion years. This geologic view of time has two important consequences. It links the present with the past as parts of an endless continuum, and so emphasizes the extremely modest place occupied by any single human individual in that continuum.

Implicit, then, in the liberal-arts approach to Earth science is a dual study: (1) the natural world as a dynamic theater today, and (2) the long history of the world in the past. In college curricula these two aspects usually constitute separate courses, taught usually in consecutive terms. In practice many students do not take both courses. They complete a first course in "physical geology" (essentially the Earth's natural processes as we see them today) and thus satisfy an administrative requirement that they pass one course in science. These students never take the second course in "historical geology" (essentially Earth history presented in chronologic sequence). They remain ignorant of three basic aspects of Earth science: the significance of geologic time; the subjection of environments to slow, continual change; and the vital process of organic evolution.

There is no good reason why the physical and historical aspects of Earth science, skillfully woven together, cannot be included in a single, one-term, descriptive course, less detailed but more useful than a course in physical processes alone. This book is a text for such a course. It is the fruit of twenty years' experience in teaching a unique and successful one-term course in which the physical and the historical were combined. The treatment followed here was tested and refined by classroom discussion and examinations. It presents history as a continuum, the evolution of a primitive Earth to the Earth possessing today's characteristics. This is paralleled by the evolution of organisms, with emphasis on the threads that resulted in the skeletal and other characteristics of Modern Man. Throughout, the narrative stresses environment.

The book is addressed to liberal-arts college students at freshman level, who have little or no background in secondary-school science or mathematics. For such students the big picture—the outline of the Earth's history, emphasizing the development of life—has deep and lasting value. To achieve the big picture several topics in physical geology have been omitted because they do not make an essential contribution to the nontechnical understanding of Earth history.

Physical changes in the Earth's surface discussed by geological periods are far less interesting to liberal-arts students than is physical change as related to the evolution of living things. This book takes the latter approach and goes into progressively more detail as geologic history unfolds. It devotes more space to the Cenozoic than to earlier major brackets of time because the reader can relate more readily to the environments described and because to him mammals are inherently more interesting than brachiopods.

The book is written in plain English. It includes no single word merely because geologists or biologists use it. It employs only those technical terms believed essential to communicate the book's substance to nonprofessional people. There is a glossary, but it contains barely sixty terms, some of them simplified. The technical words appearing in the book thus average only three per chapter. Although self-contained and self-explanatory, the book includes supplementary reading references for each principal topic.

The text is adaptable to a course consisting of classroom discussions or to one consisting of lectures accompanied by supplementary reading. It is built to accept a program of laboratory exercises as well as one of local field trips. These programs are recommended. Some personal contact with a few minerals, a few rocks, and a few fossils, and two or three local field trips regardless of the kinds of rock, regolith, and structures they display, will illuminate the text and give its readers a greater feeling for the documents of geology and for the way in which inferences can be drawn from them.

I acknowledge with thanks the generous help of R. L. Armstrong, Theodore Delevoryas, P. M. Orville, and J. N. Ostrom, who read parts of the text and gave excellent critical advice. My thanks go also to H. W. Banks, Jr., Alphonso Coleman, D. R. Pilbeam, and E. L. Simons, who freely supplied data or provided illustrations not acknowledged in their captions.

R. F. F.
New Haven, Connecticut
July 15, 1972

The Earth and Its History

Figure 1 - 1. Astronauts' view of Planet Earth. Approximate positions of continental coasts are outlined. (Photograph by ATS-III Satellite, NASA.)

1

The Earth as a Planet

Planet Earth

On November 18, 1967, the photograph on the opposite page (Fig. 1-1) was made from a space vehicle as it reached a distance of about 22,000 miles from the Earth. The photo shows the whole Earth, our home, in one single view. Until vehicles first left our planet and began the space age, the globe on which we live had never been seen in this way, the whole world in one picture.

CHAPTER 1 THE EARTH AS A PLANET

Looking at the Earth from space is not like looking at the Moon from the Earth. The Earth's features do not appear sharp, clear, and fully visible, as the Moon's features do. They tend to look somewhat blurred and many of them are hidden behind clouds. This is an indication that the Earth has an atmosphere. The presence of an atmosphere makes life possible and also, less obviously but still importantly, makes possible the breakdown by decay of the rocks that form the Earth's solid surface. This process of decay does not occur at the surface of the atmosphereless Moon.

Another feature visible in the picture is water, in the form of an ocean. Water is a substance not found on the Moon. As we well know, water, like air, is essential for life. Also, because of its motions in streams, lakes, and the sea, water carries particles of rock from one place to another and deposits them as layers of sediment. In time the layers of sediment become layers of rock.

A viewer in space would find that the atmosphere with its prevalent clouds would prevent him from getting an immediate idea of the Earth's largest surface features. He might have to compare many photos made at different times before he could piece together enough of the Earth's solid surface to make clear the shapes of continents and oceans. Our photo shows enough coastline to reveal what part of the surface was exposed to the space camera. At left center is the southern part of South America, and at the right the great bulge of West Africa faces its neighbor continent across the Atlantic Ocean. The two coasts, now separated by a vast expanse of water, show a good fit, like two pieces of a jigsaw puzzle waiting to be pressed into place. And indeed there is good reason for believing that earlier in the Earth's history the two continents were joined together. Since then they have broken apart and, each planted on its own base, have been moving slowly away from each other, creating an ever-widening Atlantic Ocean.

However, the discovery that the Earth's continents are afloat and moving, somewhat like ice floating in water, has little to do with the conquest of space. It is the fruit of many years of difficult research by many scientists working at the surface of the Earth itself. It is the latest of a long series of discoveries about the Earth's dynamics. These discoveries form the basis of the Theory of the Earth. That theory is the statement of the principles that underlie all the activities, some physical, some chemical, and some biologic, that cause continual change. The Theory of the Earth embraces not only the floating and drifting of the continents, but also all the other activities on our planet. Some of these are the transport of water from ocean to land by the atmosphere and back to the ocean again by rivers, the gradual decay of rock, and the movement of rock particles by rivers, by waves and currents, by glaciers, and by the winds that continually sweep the Earth's surface. The energy that powers such activities is the energy radiated by the Sun. Of course that energy, in the form of light reflected from the Earth's surface, is what made possible our space photograph.

The interior

The center of our planet is only 6370 kilometers * (3958 miles) beneath our feet—less than 2 percent of the distance to the Moon above us—yet the greatest depth to which man *in person* has penetrated into the Earth is 3.6 km in a South African gold mine. In a search for oil in western Texas the solid Earth has been penetrated by drilling tools to a depth of more than 9 km and samples of rock from that depth have been brought to the surface.

Although, literally, we have barely scratched the surface of the planet, still we already have a rough model that shows what its interior is like. This picture, improved year by year, is largely the result of vast, coordinated, worldwide studies of the paths and travel times of earthquake waves, reliable messengers that travel much farther and faster than drilling tools. The earthquake studies are supported by research of other kinds, such as investigation of the properties of rocks at the Earth's surface and at greater temperatures and pressures. Our model is certainly not accurate in detail, but the probability that it is correct in very broad outline is good. Figure 1-2, the model in simplified form, shows a planet built of

* Scientists always use the metric system to state measurements related to the Earth's interior, ocean depths, heights of continents, and similar features. We shall do likewise. But for measurements of smaller features, distances and heights on the ground and the like, we shall use wherever possible the more familiar English system employing miles, feet, and inches. Meanwhile it may be useful to remember that one kilometer (abbreviated *km*) is a little less than ⅔ mile, and one meter (abbreviated *m*) is a little more than one yard.

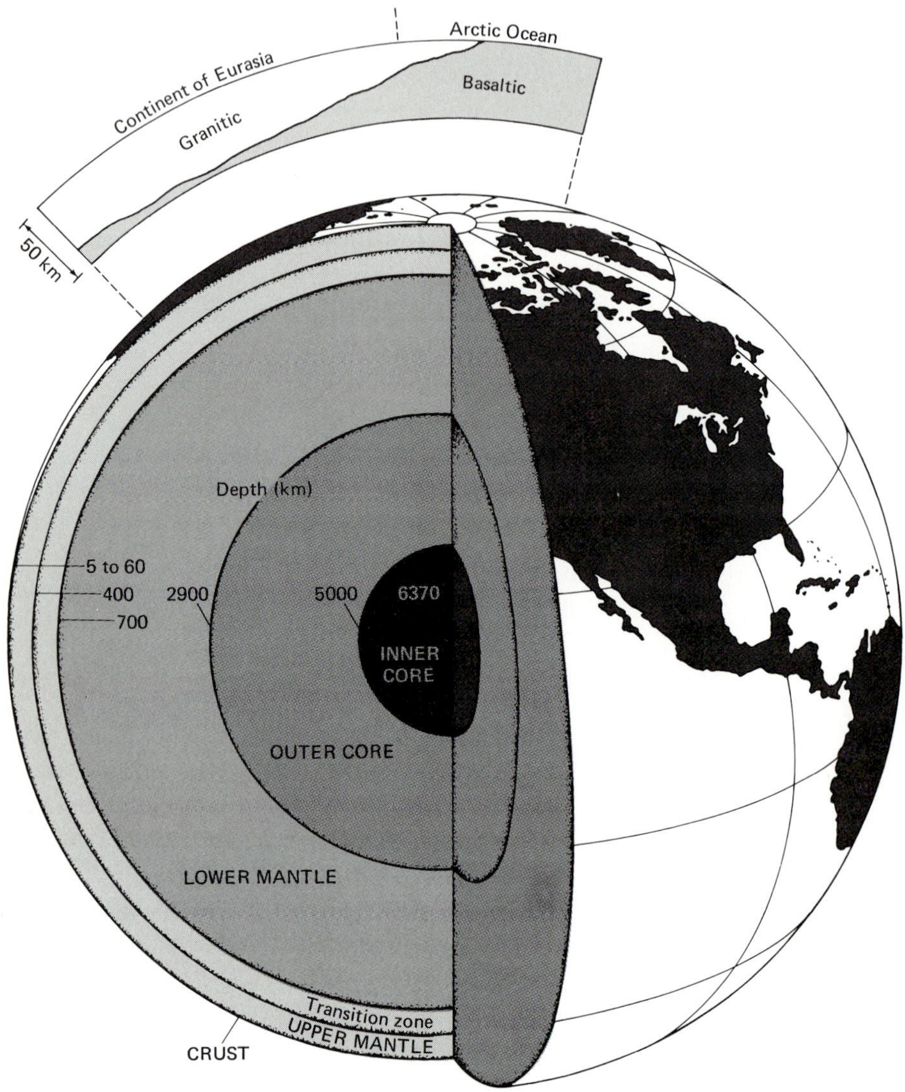

Figure 1 - 2. If the Earth could be sliced into like an apple, it would reveal a two-part core of metallic material surrounded by layers of rocky material. The Earth's crust is extraordinarily thin. Above, a piece of the crust on a much larger vertical scale, showing oceanic and continental parts.

concentric layers, like the skins of an onion but fewer and thicker. Each layer is heavier than the one above it.

At the surface there is a *crust* consisting of rock of various kinds. The crust is thinner than an eggshell in proportion to the diameter of the egg. But its thickness is less uniform than that of an eggshell. Beneath oceans it is only about 5 km; in continental areas it reaches 40 km to 60 km. The *mantle* (the "white" of the egg) consists of material heavier than the rocks of the crust. It comprises about 80 percent of the whole Earth by volume. It is underlain abruptly by the *core* (the yolk of the egg), which is metallic, very heavy, and probably an alloy of iron, nickel, and perhaps sulfur or silicon. Very likely the core, rich in iron, is the source of the Earth's magnetic field, which controls our compass needles.

Compared with temperatures at the Earth's surface, the interior is very hot indeed. Even a deep mine, at only a tiny fraction of the distance to the Earth's center, is a hot place; some are so hot that they must be ventilated artificially with cooled air. This illustrates the general fact that temperature increases with depth, at least through the crust and for an unknown distance below it. The rate of increase, called the *geothermal gradient,* varies from place to place but within the crust averages between 10° and 50° (Celsius) * per kilometer. If the average gradient were 30°, the base of the crust would be about 1800°, literally red hot. Yet within the core the gradient is believed actually to decrease, because metallic materials are efficient conductors of heat. If this is the case, the temperature at the center may be no more than a few thousand degrees. About half of this great store of heat has its origin in radioactivity. Radioactive chemical elements distributed through the Earth's body are decaying, and heat energy is a product of their decay. The other half is thought to be heat left over from the time when the Earth came into being.

Whatever the geothermal gradient, we can be sure that the outer part of the Earth's body is acting as a thermal insulating jacket. It permits heat from the inner region to pass outward only very slowly

* In most scientific literature temperatures are given on the Celsius scale, and we will use that scale too. To convert a Celsius temperature to a Fahrenheit temperature, use this formula: $F° = \frac{9}{5}(C° + 32°)$.

toward the surface, from which it radiates into the atmosphere at a slow rate—only $\frac{1}{20{,}000}$ of that at which the Sun's heat reaches the surface from above.

Another property that increases with depth below the surface is pressure. This is a direct result of the weight of the overlying material, which of course increases with increasing depth. Hence there is a pressure gradient as well as a geothermal gradient. On the Earth's surface, at sea level, the pressure caused by the weight of the atmosphere is 15 pounds per square inch. But the pressure that bears down from all directions on a point at the center of the Earth is calculated at more than 20,000 tons per square inch. Even the pressures at such comparatively shallow depths as ten or twenty kilometers are great enough to make important changes in the character of the rocks.

Continents and ocean basins

Our space photo (Fig. 1-1) shows us what we know already—from our own experience—that the surface of the solid Earth is irregular, consisting of highs and lows. The largest and broadest highs are the continents; the even broader lows are the ocean basins (Fig. 1-3). Seawater covers 71 percent of the Earth's surface, but the ocean basins taken together include less—only about 60 percent of the Earth's surface. The difference, 11 percent, is represented by continental shelves, where the water is shallow. The shelves, which surround their continents discontinuously, belong to the continents rather than to the basins. The continents are not only smaller in area than the basins; they also reach less high above sea level than the basin floors descend below it:

	AVERAGE ALTITUDE	MAXIMUM ALTITUDE (APPROX.)
Continents	+0.8 km	+8.84 km (Mt. Everest)
Ocean basins	−4.0 km	−10.85 km (Mariana Trench)

It is hardly surprising that the continents, standing on the average about 5 km above the ocean floors, turn out to consist of materials different from those that underlie the basins (Fig. 1-4). Beneath the basins is *basalt*, a heavy rock almost black in color and contain-

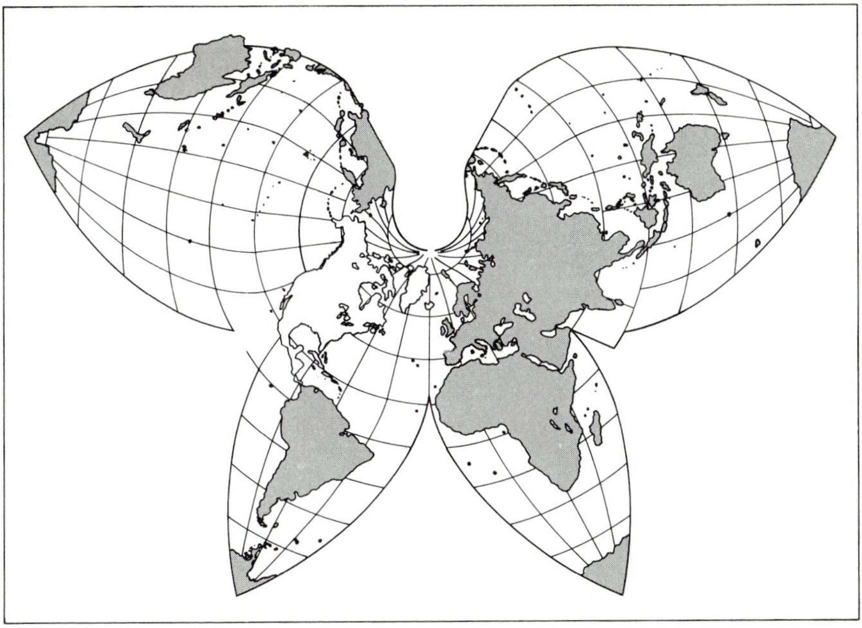

Figure 1 - 3. Continents and oceans, drawn on a grid designed to show all parts of the Earth's surface on a single scale instead of exaggerating the areas of the polar regions. Continental shelves, invisible on this map, can be seen in Figure 5 - 3. (Projection by courtesy of John Bartholomew & Son Ltd., Edinburgh.)

ing much iron and magnesium. Beneath the basalt is a layer of even heavier though otherwise rather similar material. In many places basalt wells up from below ocean floors as *lava* and, solidifying, forms volcanic cones, many of them now islands. In contrast, the rocks that form the continents are lighter in weight and color and are rich in silicon and aluminum. The continental rocks are of many kinds, but the bulk composition of the whole group is that of *granite.*

Figure 1 - 4. Continents, including their shelves, are lighter than the basaltic material beneath them, and hence float in the basalt. Most of the wide ocean area, with no lightweight rock, has been omitted (between zigzag lines).

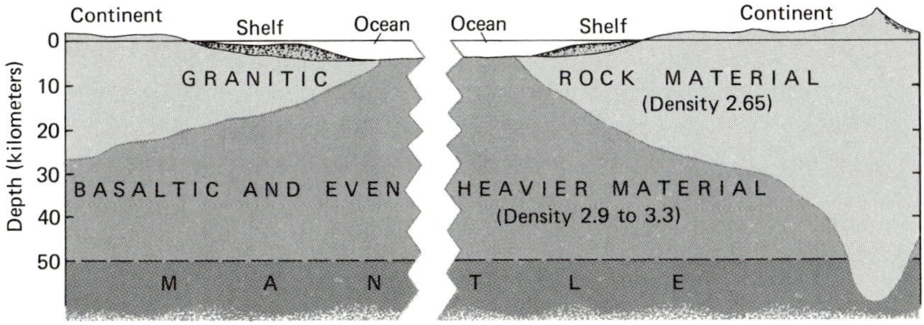

So we say that continental rocks are granitic, ocean-floor rocks basaltic. However, this basaltic material not only underlies the ocean basins but extends right around the Earth, underneath the continents as well. The granitic rocks, on the other hand, are confined to the continents and their shelves (Fig. 1-3).

The difference of weight between these two kinds of rocks explains why continents stand high above ocean basins. Just as ice, having a specific gravity of about 0.9, floats in water, so a pancake-like continent made of granite (specific gravity 2.65) can float in basalt (2.9) and in the still heavier material (3.3) that underlies it. A continent therefore is a sort of granitic raft floating in a sea of dark, heavy material. At first thought, the idea of rock floating in rock is hard to visualize because we are accustomed to think of rock as we see it at the Earth's surface—a brittle substance that is usually cut by cracks and can be broken into rough pieces by a hammer blow. But laboratory experiments have shown that rock becomes pliable when subjected to sufficient heat or pressure. At depths of 10 or 20 km or even more, where continental granitic rocks overlie basaltic rocks, both heat and pressure are so great that rock has ceased to be rock in the usual sense. It cannot break. Instead it flows very slowly. So in a sense the granitic continents really are afloat, and at those great depths both the granitic and basaltic material, if pushed from one direction, will flow in another direction somewhat as toothpaste flows through the open mouth of its tube.

The four spheres

It is more realistic to think of the Earth not as a single sphere but as consisting of four distinct, closely related spheres. Parts of three of them can be seen in our space photograph: (1) the solid land, consisting primarily of rocks; (2) the liquid ocean, consisting of water; and (3) the turbulent clouds, consisting of part of the atmosphere with its included water vapor. Rocks, water, air. We see parts of a rock sphere 12,700 km in diameter with two thin, fluid envelopes, one liquid and one gaseous, surrounding it and penetrating both it and each other. These are the *lithosphere* (its upper surface identical with the upper surface of the crust) surrounded by the *hydrosphere* (water) and the *atmosphere* (nitrogen and oxygen).

The surface of the lithosphere is the surface at which all three spheres are present, and therefore a zone of great and varied activity. It is likewise the zone in which the fourth sphere is most concentrated. That sphere (4) is the *biosphere,* the aggregate mass of everything alive. Of the animals and plants that are alive today only one and a half million *kinds* have been studied and described. But probably the total number of kinds amounts to several millions. Each kind consists of a number of individuals. In some kinds that number is enormous. We can hardly doubt, then, that the total number of living individuals is truly countless.

Chemically the biosphere is constructed from only a few of the Earth's chemical elements. Basically it consists of carbon, hydrogen, and oxygen, with much smaller amounts of phosphorus, nitrogen, sulfur, and iron. With energy drawn from sunlight, the biosphere exchanges chemically within itself and also with the other three spheres. The exchanges are of various kinds, but oxygen is the currency that most obviously keeps changing hands. With all this change and movement it is no wonder that the Earth's solid surface is where the action is.

A boundary between two different environments generally is the scene of exceptional activity, in nature as in human affairs. The coastline of a sea or a large lake is such a boundary, well defined by a belt of surf that tears at coastal rocks, grinds them up, and builds a beach with the ground-up particles. Another such boundary is a river bank, which the current of the stream attacks and crumbles. Both the crumbling bank and the sea beach are the work of "boundary conditions" where conflict occurs between two sets of forces or between an active force and a mass that stubbornly resists it. The Earth's solid surface is a realm that harbors many different environments,* each involving different combinations of the four spheres as well as local variations within each of the spheres. So it is rich in boundaries and in conflicts.

* The word *environment* simply means *surroundings.* We shall use it frequently when we discuss the activities that occur upon and within the Earth today, as well as ways in which living things are influenced by their surroundings. Recently *environment* has come to be used more narrowly, in connection with pollution, overcrowding, and other effects brought about by large numbers of people living in highly industrial societies. In this book the word is used in its broad meaning, *surroundings.*

Conflicts between internal and external processes

The boundaries and the conflicts that accompany them are of many kinds. The surface of the lithosphere is the most obvious boundary because it is the zone of conflict between dynamic processes that act within the crust (*internal* from the human point of view) and the quite different kinds of processes that act within the hydrosphere and the atmosphere. Most of the latter operate at and above the solid surface and so we call them *external.* The changes caused by all the internal and external processes as a group, and by the many conflicts among them, are responsible for most of the diversity we see in landscapes—for much of what makes the natural world interesting to its human inhabitants.

Internal processes. The big activities that take place within the crust are of three general kinds: movement of bodies of solid rock, movement of hot liquid rock material, and transformation of rock far below the surface by great pressure at high temperature. We know that all these things happen, either because we can watch them in action (lava flowing out of a volcano) or because we can see the result after the action (two mountain-size blocks of rock, one of which has moved up, down, or sidewise past the other).

In Figure 1-5 A can be seen two common ways in which the movement of bodies of solid rock can affect the surface. Scale is not important; these blocks might be one kilometer or scores of kilometers in width. What is important is that in places internal forces have moved bodies of rock upward, at least in relative terms, in one case by bending and in the other along a clean break. In so doing they have altered the surface by steepening local slopes, in one case to the point of creating a nearly vertical cliff.

Figure 1-5 B shows two ways in which molten rock commonly moves upward and makes changes in the surface. On the left, a blob of molten rock has pushed slowly upward from many kilometers down within the crust. It deformed the rock layers above it, but cooled and solidified before it reached the surface. On the right, molten rock material squeezed upward through a tube-like opening and spread out on the surface in successive layers of lava, each of which solidified before the next layer formed and covered it up.

Figure 1 - 5. Examples of the action of internal processes.

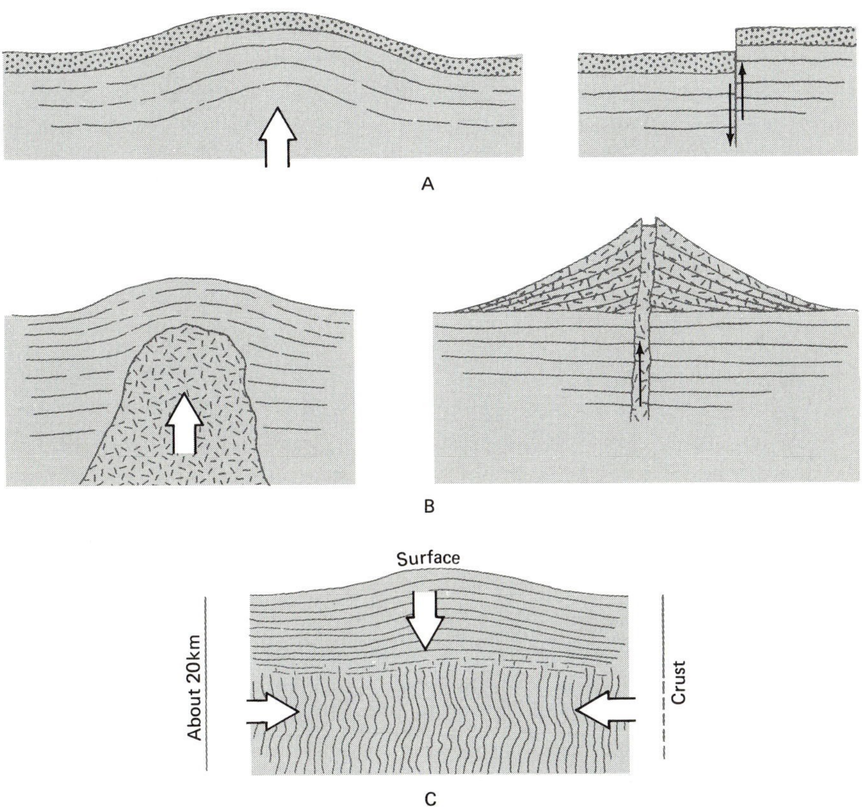

During the process the preexisting rocks were not deformed, but a new, conical edifice, a layered volcanic cone, was created. Some such cones are more than 6 km in height and have steep slopes. The sea floor is dotted with them.

Finally, Figure 1 - 5 c shows what can happen to rock that sagged downward rather than having been moved upward. In some places rock layers that formed originally at the surface are carried downward by broad bending in the crust. When they reach a depth of 10 to 20 km the rock layers are within an environment where pressure, applied from above and in some cases from the sides also, is great enough to make them flow, thus transforming them into rock material very different in appearance from the rock layers that existed earlier.

The energy that powers all three sorts of internal processes is the heat energy within the Earth. It activates internal processes on a large scale and is great enough to counteract the force of gravity.

Hence, as the illustrations show, internal processes push up or build up the Earth's surface in one place or another.

External processes. The fluid envelopes we call hydrosphere and atmosphere are the agencies responsible for most (though not quite all) of the external processes. Even where they are not moving, air and water (particularly in the form of air containing water vapor) provide an environment in which chemical exchange between water and rock sets up decay of the rock, loosening the bonds that hold the rock together and causing it to crumble. This chemical conflict goes on wherever rock is in contact with damp air. But more than that, water and air are in continual motion, and wherever they move over the surface they drag at the crumbled rock, carrying particles of it with them. Air moving as wind builds sand dunes with the rock particles it carries; water moving as a river deposits sediment on its bed, and moving as ocean waves along a shore builds beaches with the rock material it washes from the land. Ice moving as a glacier (also a part of the hydrosphere even though it is a solid) carries rock fragments and deposits them where it melts. Even water moving underground, soaking slowly through tiny openings in rocks, carries away mineral matter dissolved from the rocks.

With only a few local exceptions, the external processes act under the influence of the force of gravity. In general those processes break down solid rock into small particles and carry the rock waste away. Since their activities are controlled by gravity, they carry waste *downward* from higher areas to lower ones. Because of this loss of rock matter, the higher areas gradually waste away and become lower and lower. Sooner or later the waste is deposited in the areas that were low originally. It tends to be spread out in thin sheets, forming layers that accumulate one above another. In this way low areas are built up while the high areas are being cut down. Across the terrain in between, rock waste is continually being transferred down every slope.

The external processes draw their energy from heat radiated by the Sun and received at the Earth's surface. The mechanisms by which solar heat drives currents of air, currents of water, and other dynamic agents are mentioned in Chapter 2.

Interaction among the processes. There are, then, two groups of processes: the internal, powered by the Earth's interior heat and working mainly outward against gravity; and the external, powered by solar heat and working under the direct influence of gravity. The two groups are in endless conflict at the boundary that is the Earth's solid surface. Imagine a solid Earth smooth as a billiard ball. The pull of gravity measured along the smooth surface would be the same everywhere. Water could not flow from one place to another. The surface of the real Earth, however, is not smooth. On it internal processes create high places and low places. That means there must be slopes. Water flows down the slopes from the high places to the low, and the water carries particles of rock. The flow of water with its load of rock particles is one of the external processes.

In time this and other external processes would smooth the surfaces of the lands and would reduce them very nearly to sea level. This smoothing does indeed happen, but not to all the Earth's land at the same time. Something interferes with it: one or another of the internal processes. Now here and now there, these lift up areas of land, and by creating new heights spoil the leveling work of the flowing water. Highlands, whether steep mountains or broad smooth uplands, are always forming somewhere—not necessarily at a steady rate, of course, and not always in a regular pattern. This creation of highlands keeps things going. It continually gives the external processes something new to work on. It makes slopes, but it also makes water flow faster. It gives the flowing water something new to smooth out and renewed energy with which to do the job. It keeps sediment moving from higher places to lower ones, to be spread out in layers built, literally, from the wreckage of rock that existed earlier somewhere uphill. Thus the internal/external conflict goes on and on. Seemingly it never ends. Nor is it likely to end as long as the Earth's interior is hot enough to power the internal processes, and as long as the Sun radiates heat to the Earth's surface, powering the external processes.

This glimpse of the globe as seen from a distance shows that the Earth is a lively planet and that its liveliest and most varied zone lies at its solid surface, where complex interaction sets up conflict. The surface zone is also where we live and where our history has been enacted, shaped by the environments the zone has afforded us. Because we live in the midst of the action, we cannot avoid taking an interest in what is going on around us. The next chapter of this book focuses on the surface zone. It shows that the big processes and the rock materials they continually move and reshape are not haphazard, but constitute an orderly and understandable system that operates in accordance with natural laws.

References

EARTH'S INTERIOR:

Bott, M. H. P., 1971, The interior of the Earth: St. Martin's Press, New York.

OCEANS:

Hill, M. N., ed., 1962–1963: The sea: Interscience Publishers, New York, 3 vols.

BIOSPHERE:

Hutchinson, G. E., and others, 1970. [A group of articles related to the biosphere.] Scientific American, v. 223, p. 45–208.

2

Rocks and Geologic Cycles

The importance of rock

At this point in our discussion it would be a very good idea to look for a few moments at a piece of rock. In many cities and towns the rock of the Earth's solid crust is not wholly concealed, but is exposed to view here and there. The outer surfaces of many buildings are built of rock. In some cities the curbstones along the sides of the streets consist of rock. Many museums contain collections of rock of different kinds.

If rock is within your reach in these or other situations, look at it. Look at it closely. What impression does it make on you? Does it seem uncompromising, hard, motionless, unchanging, lifeless? All these words have been used to describe rock. During the few minutes it takes to look thoughtfully at a sample of rock, all the words may be true. But instead of merely looking at the rock, suppose we examine it closely, "take it apart" to see what it is made of and how it has been put together, see if it has weak points as well as strong points. Above all, suppose we think of what might happen to it, how it might change, not in five minutes but through the course of years, centuries, and even millions of years. If we look at the rock in these ways we shall find that not one of those words applies to it, because no rock, however hard and firm, lasts forever. It breaks down; its component parts move away, sometimes traveling long distances, and are shuffled and rebuilt into new and different rocks.

So any rock, no matter what kind it may be, carries within it a long history of destruction, of movement, and of rebuilding. The story it tells is far from lifeless if we think of life as continuous, organized activity. On the contrary, it is full of movement and of changing scenes. It is the story not only of the making, unmaking, and remaking of rock in every kind of environment, but also of the origin and the changing history of the biosphere, the living things that have inhabited and still do inhabit the Earth. Without the rocks that story would be unknown to us. So rocks are worth not only a close look but real study, because in them lies the story of all our ancestors, human and pre-human.

This chapter and the first few of those that follow it explain how rocks are made and unmade, and how they come to contain the record of past life. If they are read with care and close attention, omitting no details, it will be possible to follow in an understanding way the dramatic history that forms the second half of this book.

Rock, of course, is a material, a substance that remains passive until it is moved. What moves it? The processes, both internal and external, that we described in Chapter 1. These are activities. Although they may seem quite distinct from each other, the activities (processes) and materials (rock) are closely intertwined. The fine particles of rock material that constitute the yellow, brown, or reddish mud in a river are not mere freight in transport; they are an essential part of the river itself. The quantity and the fineness of the suspended mud and also the number and sizes of the usually invisible larger particles of rock that move along the stream bed strongly affect the characteristics of the river and of the deposits it makes along its course.

Because processes and materials intertwine, we shall have to discuss them in a not-quite-logical order. Let us begin with the water cycle, a system of processes and a single material, water. Then we shall continue with the nature and dimensions of rock materials. After that it will be possible to show how materials combine with processes to create an active rock cycle, another great system of processes.

The water cycle

A graphic sketch of the *water cycle* (Fig. 2-1) tells its own story. With heat energy from the Sun, huge amounts of water are continually evaporated from the ocean, lifted into the atmosphere, and transported from one region to another. Sooner or later the vapor condenses and is precipitated on land or sea as rain and snow. The part that falls on land returns to the sea by various paths. Some returns via streams; some soaks into the ground and percolates slowly seaward. Some evaporates from lakes and streams or is transpired by plants, and thus returns to the atmosphere by a shortcut. Another shortcut is taken by water that returns from atmosphere to ocean directly, by precipitation.

Figure 2 - 1. The water cycle.

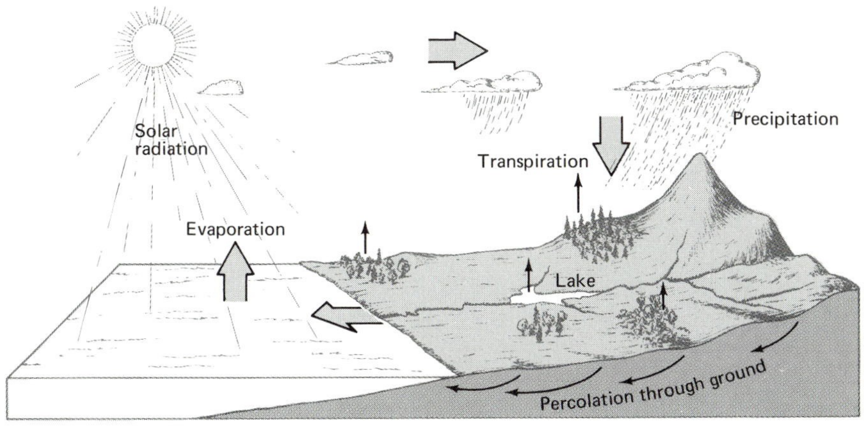

The water cycle is a big system. Enough falls in one year on the United States alone to cover the country with a layer of water averaging 30 inches in depth. About one-third of it gets back to the sea through streams. The other two-thirds is evaporated and transpired directly back to the atmosphere.

The biosphere is an essential participant in the water cycle. On any continent, probably a good deal more water substance is transpired from the surfaces of green tissues of plants and exhaled by and evaporated from animals, than is carried through rivers to the sea during the same period of time. Of course the biosphere exists by virtue of the presence of water, but it also plays an integral part in the water cycle, somewhat as rocky sediment is part of the stream that is transporting it. In this chapter we are concerned with the part played by the water cycle in the destruction of land by *erosion,* the breaking down and transport of rock material.

The three classes of rocks

We can hardly continue to mention rocks without explaining what they are. All plays and some novels as well begin with a cast of characters, often with an explanation of their relation to each other. The different rocks are our characters; if we are to understand the dramatic history of the crust, it will help a great deal to know what they are.

The matter that constitutes the crust consists of more than 100 chemical elements in varying amounts, of which oxygen, silicon, aluminum, and iron together make up nearly 87 percent by weight. These and the less common elements are combined in various proportions into *minerals,* most of which are crystalline solids having three-dimensional rigid frameworks like the steel framework of a big building. Let us take a closer look at the way these frameworks are put together.

Many of the minerals that are common in rocks are silicates, combinations of silica (SiO_2) with other chemical elements. The frameworks of silicate minerals consist of geometric combinations of a single tiny unit called the SiO_4 tetrahedron. We can liken the unit to a four-sided pyramid made up of three identical balls just touching each other in a horizontal plane. On them is placed a fourth ball of similar size, creating a pyramid. In the space at the center of the

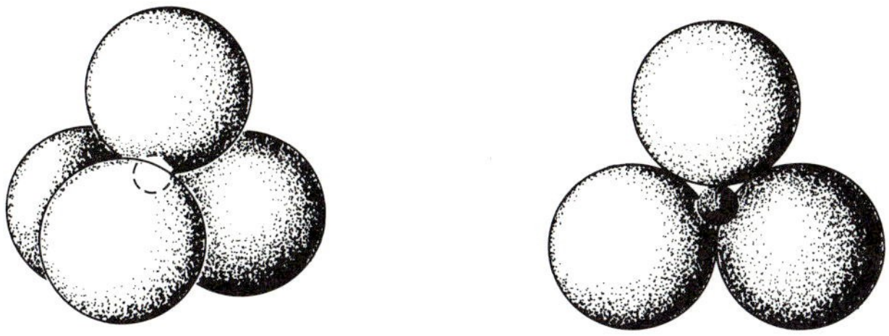

Figure 2 - 2. *Left:* Model of the SiO_4 tetrahedron, the fundamental building unit in the structure of the silicate minerals in rocks. In it a single small silicon ion is surrounded by four large oxygen ions.

Right: Same model with one of the oxygen ions removed to reveal the silicon ion inside the pyramid.

The ions are very, very small. In this illustration they are magnified more than 57 million times.

pyramid and just touching all four balls is a fifth, a small ball about a quarter the diameter of the other four (Fig. 2 - 2). Each of the large balls is an oxygen ion; the small one is a silicon ion. To fasten our model pyramid together, we would have to use glue at each point of contact. In the real pyramid the five ions are cemented firmly together by their own electric charges.

As building blocks, SiO_4 tetrahedrons, with their pyramidal shape, are very adaptable. They fit together nicely in all sorts of symmetrical patterns, varieties of layers, chains, and rings, which themselves can be built up into many different three-dimensional frameworks. Each silicate mineral has its own characteristic framework, its own geometric pattern. But the units with which each framework is constructed are the same. They are SiO_4 tetrahedra. The shape of the framework of any crystalline mineral determines the form its crystals assume. Crystals of the mineral quartz (Fig. 2 - 3 D) reflect an internal geometry that is present in each tiny fragment, no matter how small, of a broken crystal. Partly because of the geometry of their frameworks, minerals possess specific physical properties by which each can be recognized (Table 2 - A). More than 2000 minerals exist, but only a few are important constituents of common rocks.

Rocks are combinations of two or more (in some cases many) minerals, and a few consist of a single mineral only. Hundreds of kinds of rocks are known, but only a few are common. A piece of ordinary granite (Fig. 2 - 3 A), small enough to be held easily in one's hand, might consist of hundreds of crystal grains of only five

Table 2 - A. Eleven selected minerals, and properties that help identify them.

Mineral	Chemical Composition	Luster	Form	Cleavage	Hardness	Specific Gravity	Other Properties; Remarks
Carbonates							
Calcite	$CaCO_3$	Vitreous to dull	In tapering crystals or granular aggregates	Three perfect, at oblique angles	3	2.71	Colorless or white; effervesces in dilute HCl
Dolomite	$CaMg(CO_3)_2$	Vitreous to pearly	In crystals with rhomb-shaped faces; also in granular masses	Perfect in three directions, as in calcite.	3.5 to 4	2.85	White, gray, or flesh-colored; some crystals have curved faces; must be scratched or powdered to effervesce in cold dilute HCl
Feldspars							
Potassium feldspar	$KAlSi_3O_8$	Vitreous	In prismatic crystals or grains with cleavage	Two good cleavages, at right angles	6	2.56 to 2.59	Commonly flesh colored, pink, or gray; one variety green
Plagioclase	$NaAlSi_3O_8$ (albite) to $CaAl_2Si_2O_8$ (anorthite)	Vitreous to pearly	Commonly as irregular grains or cleavable masses; some varieties in thin plates	Two good cleavages, not quite at right angles	6 to 6.5	2.62 to 2.76	White to dark gray; also other colors; cleavage planes may show fine parallel striations; play of colors in some varieties
Hornblende	$Ca_2(Mg,Fe)_5(OH_2)(Al, Si)_8O_{22}$	Vitreous on cleavage surfaces	In long, 6-sided crystals; also in fibers and irregular grains	Two, intersecting at 56 and 124°	5 to 6	2.9 to 3.8	Commonly black, dark and light green, rarely white

Iron oxides							
Hematite	Fe_2O_3	Metallic or earthy	Varied: massive, granular, micaceous, earthy	None, uneven fracture	5 to 6	4.9 to 5.3	Red-brown, gray to black; red-brown streak
Limonite (Impure oxide)	Mixture of several hydrous oxides of iron	Dull to vitreous	Compact to earthy masses; irregular nodules	None, irregular fracture	1 to 5.5	3.5 to 4	Yellow, brown, black; streak yellow-brown
Micas							
Muscovite (white mica)	$KAl_2(OH,F)_2$ $AlSi_3O_{10}$	Vitreous to pearly	In uniform thin flakes; rarely in 6-sided crystals	One cleavage direction; perfect flakes or sheets	2 to 2.5	2.77 to 2.88	Colorless and transparent when pure but commonly greenish and mottled; flakes are flexible and elastic
Biotite (black mica)	$K(Mg,Fe)_3(OH,F)_2$ $AlSi_3O_{10}$	Pearly to nearly vitreous	In perfect thin flakes; 6-sided crystals	One cleavage direction; uniform flakes or sheets	2.5 to 3	2.7 to 3.3	Black, dark brown, or green; nearly or quite opaque; flakes are both flexible and elastic
Other minerals							
Pyroxene group (augite)	$Ca(Mg,Fe,Al)$ $(Al,Si)_2O_6$	Vitreous	In 8-sided, stubby crystals; also in granular masses	Two cleavages, nearly at right angles	5 to 6	3.2 to 3.9	Light to dark green, or black; alternate crystal faces at right angles (fit into corner of a box)
Quartz	SiO_2	Vitreous to greasy	Six-sided crystals, pyramids at ends, also in irregular grains and masses	None; conchoidal fracture	7	2.65	Varies from colorless and transparent to opaque with wide range of colors

For identifying minerals and rocks, a simple lens (at least 3x), a pocket knife, a small bottle of dilute hydrochloric acid, and a hardness-test set of minerals are desirable.

An elaborate practical reference on *minerals* is Vanders, Iris, & Kerr, P.F., 1967. Mineral recognition: John Wiley & Sons, Inc., New York. Abbreviated manuals to aid in identification of common rocks are included in most textbooks of physical geology, such as Longwell, C.R., Flint, R.F., & Sanders, J.E., 1969. Physical geology: John Wiley & Sons, Inc., New York. Similar manuals, extending to varying degrees of elaboration, are included in the laboratory manuals intended for use in college courses in physical geology.

Granite A · Diorite B · Gabbro C

Figure 2 - 3. Igneous rocks and a mineral, quartz.

A, B, C. Coarse-grained igneous rocks. Three members of a series that grades imperceptibly from light-colored, relatively lightweight granite through an intermediate kind, diorite, into a dark-colored, relatively heavy gabbro. The differences among them depend on the minerals of which they consist; these in turn depend on the chemical character of the magma from which they solidified.

Crystals of quartz D · Basalt with gas bubbles E · Obsidian F

D. The mineral quartz (SiO_2), an important constituent of granite. This group of nearly symmetrical crystals formed near the surface of the Earth, in a small cavern-like opening filled with a water solution of SiO_2. The crystals began to form at the cavern wall and grew generally inward toward the cavern center. Their form, a hexagonal prism ending in a pyramid, is typical of most quartz. But this perfect form is uncommon. It can develop only when the growing mineral is surrounded entirely by liquid and can grow in all directions without interference.

Although many of the grains visible in the granite specimen (A) are quartz crystals, their shapes completely lack the symmetry seen in D, because their growth was interfered with by the presence of other crystals in the cooling magma.

E. Basalt, the rapidly cooled equivalent of gabbro. Its component grains are so small as a result of rapid cooling that they are invisible except through a microscope. The holes, like the holes in Swiss cheese, are caused by imprisoned bubbles of gas that did not escape before the substance solidified. Such bubbles occur near the upper surfaces of lava flows and form only at the Earth's surface. Not all basalt, of course, contains them.

F. Obsidian (volcanic glass), in which no crystals formed because of ultra-rapid chilling of a lava flow.

Microphoto
G

G. Detail of A. A paper-thin slice of the specimen of granite (A) was made with a diamond saw and was glued between two protective thin-glass plates, forming a sandwich. The thin slice is seen through a microscope and is magnified 22 times. It reveals a pattern of interlocking crystals of feldspar (both potassium feldspar, PF, and plagioclase, P), quartz, and the mica biotite, B. The granite also contains hornblende, but our slice did not happen to cut through a crystal of that mineral. (Photo P. M. Orville.)

In this much-enlarged photograph it is even more evident than in A that the crystals of quartz do not look at all like the nearly perfect ones in D. In fact none of the mineral grains in the slice even approaches its ideal form. During solidification of the granite from a molten state each growing crystal interfered so much with its neighbors that none could be perfect.

or six common minerals, plus tiny amounts of uncommon minerals that are not significant for our very general discussion.

By examining a score or two of samples one can see that rocks fall into three broad classes. The first class consists of *igneous* ("fire-made") *rocks*, those which have solidified from a hot, liquid state. Next there are *metamorphic* ("changed-form") *rocks*, those resulting from the alteration, in the solid state, of preexisting rocks by pressure, heat, or both. Finally we have *sedimentary* ("made from sediment") *rocks*, those derived from deposition of broken particles and other waste of preexisting rocks.

These three kinds of rocks have their origins in different parts of the crust, the sedimentary rocks at the surface and the other two kinds mostly at considerable depths. But each kind can be moved relatively—slowly of course—into zones other than that of its birth. As a result of all their slow travels, rocks have become greatly mixed, so that today we see all three kinds at the Earth's surface. We must examine them more closely in order to trace, even sketchily, the history of the crust.

Igneous rocks. With a geothermal gradient of 30° per kilometer, a point 40 km down in the crust beneath a continent would be at a temperature of 1200°. Laboratory experiments in "cooking" rocks tell us that at such a temperature, some rocks melt under the lesser pressure that prevails at the Earth's surface. Far down in the crust, however, the enormous weight of the overlying rock raises melting points enough to prevent most of the material there from becoming liquid, despite very high temperature. But a small proportion of the material present there can melt. This results in the creation of pockets of molten rock (material we call *magma* as long as it remains below the surface). Such pockets are believed to exist at various places down within the crust.

Magma, being a fluid, is mobile. When it moves, it tends to move toward the surface (in other words, toward the mouth of the toothpaste tube). In part it melts its way up—melts a hole in the crust as a blowtorch melts a path through a piece of steel; its upward thrust is aided by the force of expansion of gases dissolved in it. As it moves upward—and the process is generally very slow—both pressure and temperature decrease, and at some point the process of solidification begins. Molecules of solid minerals begin to form, acting as nuclei to which similar molecules are attracted. The getting-together of enough similar molecules results in the growth of crystals, tiny solids dotted through the magma. The solids increase in size and number, gradually converting the magma into a hot mush. Eventually the mush solidifies completely, becoming an igneous rock, although still very hot. A newly formed body of granite might have a temperature of 700° and might have required ten million years or more to solidify, so slowly does the heat escape through the insulating regions of overlying rock.

If upward travel were slow and if solidification occurred well before the magma reached the surface, a very long time would be required. During so long a period, the mineral molecules assembling to form crystals would have time to build the crystals to large size. Each crystal might be several millimeters in diameter (Fig. 2-3 G), and so the resulting igneous rock would be coarse grained. But if travel were faster and the magma reached the surface and there poured out as *lava*, like water overflowing from a well, the crystals would be small, because the time would have been too short to permit their leisurely growth to large size. The crystalline grains might be only 1/10 or 1/20 inch in diameter or even less. Indeed, in the extreme case of magma that reached the surface before any mineral nuclei had had time to form, quick chilling in contact with the atmosphere could cause solidification of the jelly-like mass without formation of any crystals at all (Fig. 2 - 3 F). A rock formed in that way is not crystalline; it is a natural glass.

The interest for us in this sequence is that when we find a large body of coarse-grained granite or other igneous rock exposed at the Earth's surface we can be sure it solidified down within the crust. That being so, we then know that after the rock solidified, deep and long-continued erosion, probably requiring millions of years, must have occurred in order to unroof it and lay it bare. Instead of the magma moving up to the surface, the surface was gradually worn down to the deep-lying solidified rock. So the exposed rock body is like the foundation of an ancient building, all the rest of which has fallen away. Not only grain size, but also the kinds of minerals that formed the first crystals in the magma reveal the depth at which they formed. The making of artificial igneous rocks under controlled pressures in the laboratory reveals how much pressure is required for magma to form the proper nuclei from which any particular kind of mineral is built. The depths corresponding to those pressures can then be easily calculated.

Texture → / Composition ↓	Coarse grained	Fine grained	Glassy (no grains)
Rich in silicon and oxygen	*Granite*	*Rhyolite*	*Obsidian*
↕	*Diorite*	*Andesite*	
Rich in iron and magnesium	*Gabbro*	*Basalt*	*Basalt glass*

Table 2 - B. Eight selected igneous rocks (see footnote to Table 2 - A), forming gradational series both horizontally and vertically. Many intermediate kinds exist; also kinds even richer in iron and magnesium.

The many varieties of igneous rock (Table 2 - B; compare Fig. 2 - 3) depend mainly (though not entirely) on the chemical composition of the crust at the places where the magma originated. But as we have said, in and on the continents the *average* composition of all rocks, despite local extreme variations, approximates that of granite, while the oceanic rocks are almost wholly basalt.

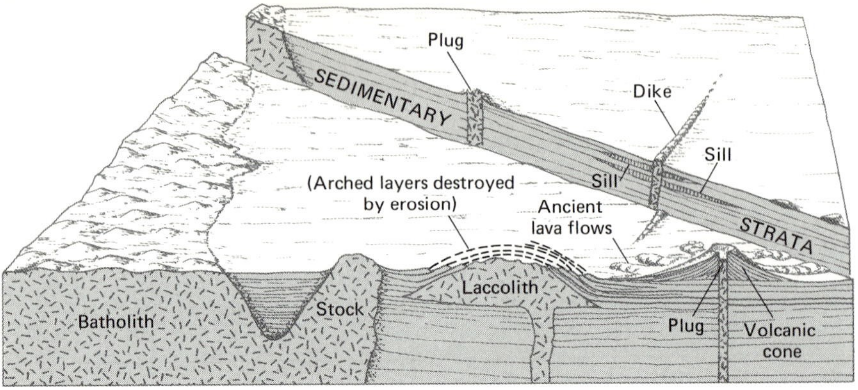

Figure 2 - 4. Bodies of igneous rock in their geometric relationships. Batholiths, with their thumblike projections called *stocks,* are by far the largest bodies. Coarse-crystal pattern indicates solidification *at depth*; fine-crystal pattern, *near surface*; black layers, *at surface*. Not to scale.

Bodies of igneous rock are of many shapes and sizes. Figure 2 - 4 illustrates typical examples, in which it is evident that after solidification of the huge batholith with its projecting stock, an enormous thickness of rock was carried away by erosion before the volcanic cone with its lava flows began to be built. Hence at least two generations of igneous rocks are represented in the diagram.

Metamorphic rocks. Although they resemble igneous rocks in that they consist of interlocking crystals, metamorphic rocks (Table 2 - C) nevertheless differ because they did not solidify from liquid

Table 2 - C. Six selected metamorphic rocks (see footnote to Table 2 - A). All are of generally granitic composition. Not included are darker, heavier rocks such as amphibolite.

Foliated rocks		
Coarse grained	*Gneiss*	(Feldspars, quartz, micas, and hornblende in irregular layers of alternately light- and dark-colored minerals.)
↓	*Schist*	(Quartz and micas predominate, with the flaky minerals conspicuous.)
↓	*Phyllite*	(Grayish to greenish, fine grained, lustrous, in thin layers; flaky minerals not conspicuous.)
Fine grained	*Slate*	(Commonly dark bluish gray, fine grained; in smooth, parallel layers that split apart readily; some luster.)
Nonfoliated rocks		
	Quartzite	(Former sandstone, made of quartz, that has been recrystallized so that it breaks through, rather than around, its component grains.)
	Marble	(Former limestone or dolostone that has been recrystallized.)

Figure 2 - 5. Foliated metamorphic rocks.
A. Gneiss, with coarse grain, consisting mostly of quartz, feldspars, and biotite.
B. Schist, with fine grain, in which muscovite is conspicuous.

melts. Most of them possess *foliation* (Fig. 2 - 5), caused by the segregation of minerals into layers as the result of chemical exchange that builds new minerals in a high-pressure environment. The building—or, really, rebuilding—process proceeds in the solid state without melting. The pressures necessary to develop foliation correspond to depths of 10 to 30 km or so, depending on what new minerals were built into the rock during the process. Foliated metamorphic rocks therefore tell us that some 10 to 30 km of overlying rock must have been stripped away by erosion in order to expose the foliated rocks to view.

Other metamorphic rocks, such as quartzite and marble, are not foliated. They are predominantly the result of the recrystallization of sedimentary rocks, buried to considerable depth and soaked in hot water, but not subjected to extraordinary pressure. Hence they have an interlocking crystalline texture but lack foliation.

Sedimentary rocks. The essential origin of sedimentary rocks (Table 2 - D) is embodied in Figure 2 - 6, the wall of a canyon in the Absaroka Mountains in western Wyoming. There a mass of igneous rock is cut by steep, narrow valleys created by runoff from rain. Most of the rock particles eroded from the valleys, including the small tributaries, have been deposited at valley mouths. The deposits form two principal heaps that spread out fanwise on the flat floor of a still larger valley, a small bit of which is visible in the left foreground. These two conical *alluvial fans* built of gravel and sand represent an early phase of the erosion and redeposition of rock material. Bedrock at and below the land surface has decayed and

Table 2 - D. Eight principal sedimentary rocks, classified according to grain size and composition (see footnote to Table 2 - A). (Adapted from Longwell and Flint, *Physical Geology*, New York, John Wiley & Sons, 1962, p. 325.)

Grain size ↓ / Composition →		Fragments of, or minerals derived from, pre-existing rock	Carbonate minerals (Mostly calcite & dolomite)
Range of diameters of dominant particles	Name of sediment	Name of rock	
More than 2 mm	Gravel, rounded	*Conglomerate*	*Limestone, dolostone*
	Gravel, angular	*Breccia*	
2mm to 1/16 mm	Sand	*Sandstone*	
1/16 mm to 1/256 mm	Silt	*Siltstone*	
Less than 1/256 mm	Clay	*Claystone* (*Shale*, if finely layered)	

Figure 2 - 6. Cone-like alluvial fans at the base of a mountain slope in western Wyoming. Runoff from the latest rainstorm has cut a channel into the upstream part of one of the cones, but the waste from the cut was redeposited in an elongate blob a little farther down the slope. (T. A. Jaggar, U.S. Geol. Survey.)

been otherwise broken down into loose particles. These have then been moved downward and deposited. In this case the deposits are localized at places where the currents of two wet-weather streams are checked by sudden flattening of their paths as they emerge onto a flat surface.

The gravel and sand of which the fans are built constitute *sediment.* Unless it is removed and scattered by erosion, the sediment will in time be compacted. Its particles will be cemented together by mineral substances in solution in the water that percolates through it. The sediment will then have become *sedimentary rock* (Fig. 2 - 7). Such rock can be likened to a body of man-made concrete in a sidewalk, or to the foundation of a building, because it consists of pieces of rock held together firmly by cement. Indeed, concrete itself is an artificial sedimentary rock.

In terms of geologic time most fans are short lived, because they occupy places that are vulnerable to erosion. Most fans are the temporary resting places of sediment that will eventually end up in a larger, more distant basin, perhaps even a basin of the sea. In a sense the movements of sediment are like those of the balls in an endless game of pocket billiards in which the vast billiard table is

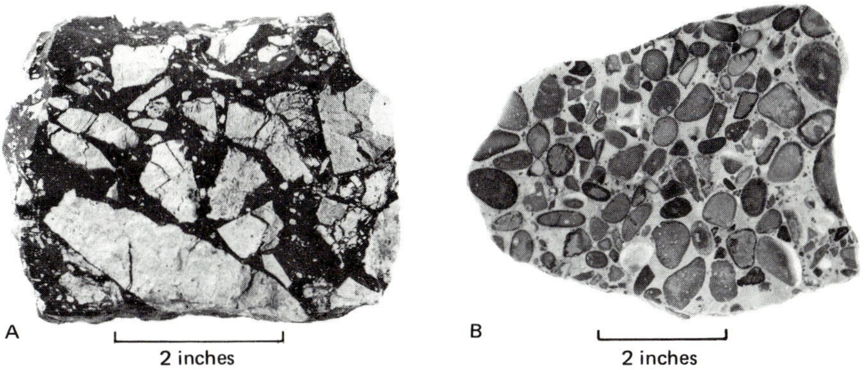

Figure 2 - 7. Common kinds of sedimentary rock.

A. Breccia, consisting of angular pieces of broken-up light-colored older rock embedded in a mass of dark-colored sandy material. Very likely the breccia accumulated at the base of a cliff consisting of the light-colored rock. The unworn shapes of the pieces indicate that they were not carried in the turbulent water of a stream or in surf.

B. Conglomerate consisting of small pebbles of various kinds of rock. Rounding of the pebbles must have been accomplished in surf along a beach or in a stream.

C 2 inches

D 1 inch

E 2 inches

C. Conglomerate consisting entirely of shells of clams and other marine animals. Many of the shells are slightly worn, indicating that they were tumbled about by surf before being deposited. Later they became stuck together by limy cement deposited from water, and thus were converted into rock. East coast of Florida.

D. Sandstone consisting of rounded grains of quartz sand together with a little silt and clay.

E. Shale consisting of well-stratified clay particles together with a quantity of silt-size particles of quartz.

(A, B, C courtesy of the National Museum of Natural History.)

not flat. The balls move about, standing still between periods of movement. In every movement some of the balls drop into pockets and out of the game—temporarily. But in nature's game the table is deformable. Sooner or later some part of it is pushed up, pockets and all, high enough so that the balls start moving again, and so the game goes on forever. The scenes of greatest action keep changing, but the same rules of the game always remain in force.

The rock cycle

Although all sediments accumulate at the Earth's solid surface and are converted into rocks at only moderate depth, nevertheless sedimentary rocks are identified in boreholes and by geophysical sensing at positions deep within the crust. This implies that at times various segments of the crust have subsided, carrying sedimentary rocks far downward. Such segments are big, deep pockets in our billiard table, filled with balls. Conversely, in many parts of the continents igneous and metamorphic rocks, some of them formed at great depth, are at the surface, even in very high mountains. This implies that parts of the crust have been lifted up to great heights, inducing much erosion. Clearly, then, parts of the crust go up and down, albeit very slowly, and their rocks, carried upward or downward into new environments that are foreign to them, must undergo changes.

What happens to the rocks is shown in outline in Figure 2 - 8. Newly uplifted rocks are acted upon by external processes and are eroded, the waste being carried downslope to some basin where it is deposited in layers as sediment. Many subsiding basins are filled with sediment about as rapidly as room is provided by subsidence; meanwhile the sediment near the bottom of the pile is being converted to sedimentary rock. If we assume that after a long time the relative movements are reversed, the area of sedimentary rocks is pushed up and in turn is eroded. Some or all of the rock waste from it would be carried into an adjacent new basin, formerly a highland. Sediments would accumulate in the basin and the whole process would be repeated.

Sometimes events are more complicated. If the basin in Figure 2 - 8 B continued to subside and went on being filled with layers of sediment for a long enough time, the lower part of the filling (now converted into sedimentary rocks) would enter a deeper zone of the crust. The pressures in a deep enough zone could convert sedimentary rocks into metamorphic rocks. With continued subsidence to still greater depths, the rocks could become hot enough to melt. If they did melt they would form a new generation of igneous rock.

Looking at these statements, one can see that they fit together into a single continuous scheme. In it (Fig. 2 - 9) are all the *phases* through which rock material can pass, and also the *processes* that

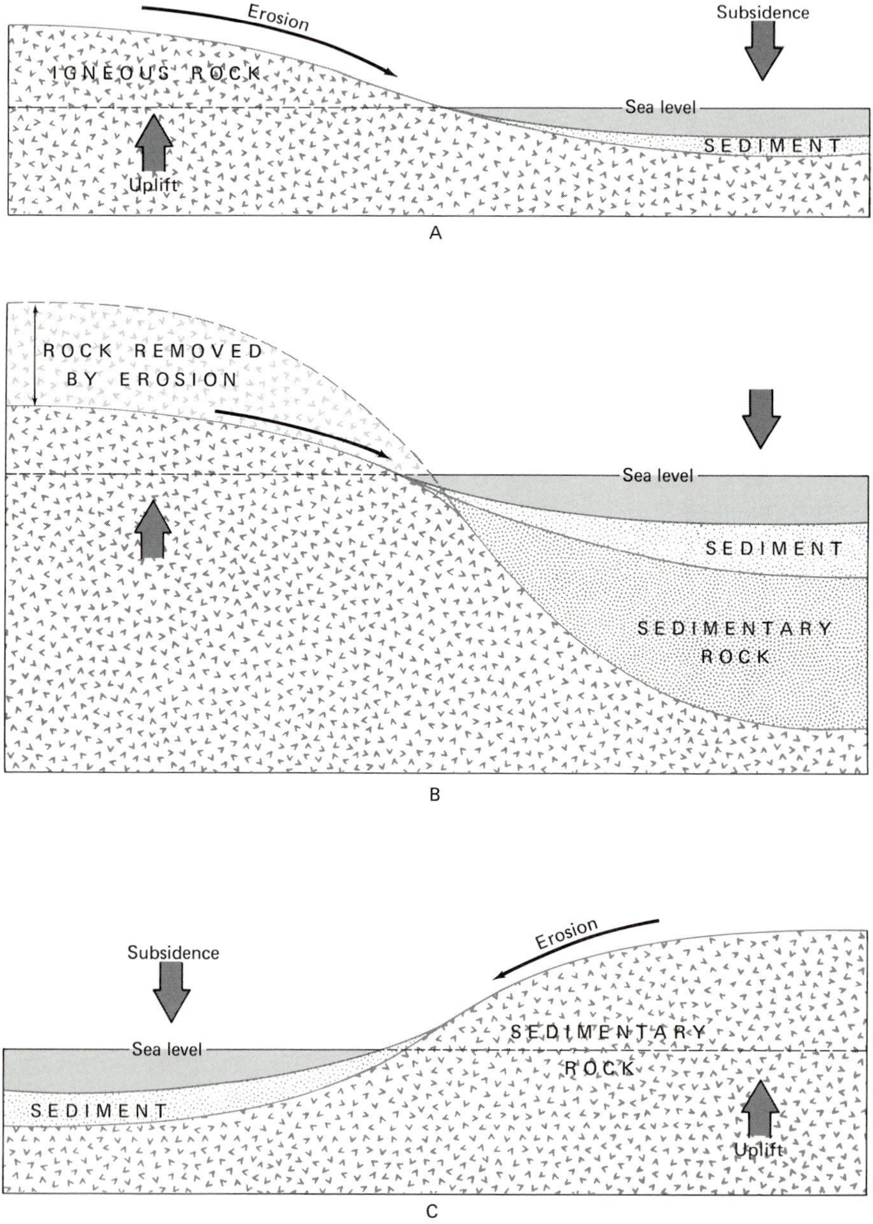

Figure 2 - 8. Erosion and deposition in adjacent regions controlled through time by uplift and subsidence of the crust. (Not to scale.) This illustrates one part of the rock cycle shown in Figure 2 - 9.

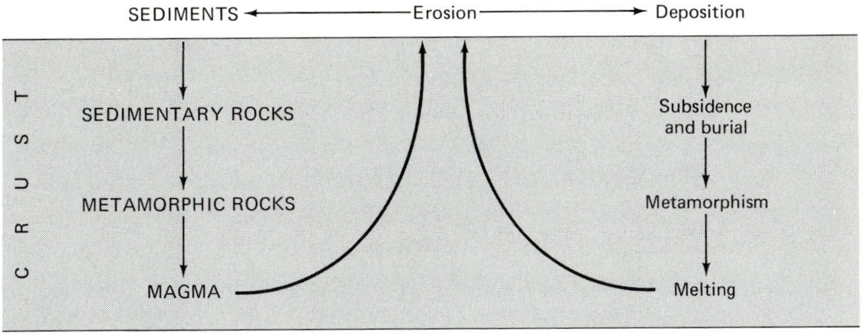

Figure 2-9. Simple diagram of the rock cycle, a continuous system. Right: A sequence of events that follow each other in directions of arrows. Left: A sequence of corresponding phases through which rock material passes. These also follow each other. Short circuits can occur at any point in the cycle.

affect rock material. We call the scheme the *rock cycle* because it embraces a continuous, unending flux of activity of natural processes to which matter responds by undergoing transformations. Although it may look as though igneous rock constituted the "beginning" of the rock cycle, this is not really true. The cycle has no beginning any more than a circle has a beginning. Rather it is an apparently endless series of transformations, repeated over and over again in generally similar fashion, although with many short circuits and with the greatest activity centered in different places at different times. Not all rock matter gets the full treatment each time around.

While these activities and transformations are going on continually within and upon the crust, the total quantity of matter in the vast system remains unchanged except for a minute amount of meteoritic material that falls in from outer space and adds to the Earth's total. Also, since 1969 an even more minute amount of equipment has been rocketed to the Moon and left there, decreasing the Earth's matter by a tiny fraction.

Although the quantity remains unchanged, particles of matter are continually shuffled and reshuffled, forming part of the substance of one mineral now and of another mineral later, forming part of a liquid mass at one time and part of a solid body at a different time. For some particles the rock cycle is a long journey full of variety. For others it means tediously occupying one environment with very little change for seemingly endless time. It is possible, though hardly probable, that some particles have never experienced all the transformations, but most have probably gone through the whole routine many times over. The crust is old enough to make this very likely.

The Principle of Uniformity

Why do we believe the rock cycle is a reality rather than just an interesting idea? The proof that it is real is this. On any continent we care to examine we find rocks that demonstrate every phase of the cycle; also we find rocks that represent all the gradations between the phases. Still further, we find examples of every possible short circuit from one phase to another. Finally, no one has ever found a body of rock that does not fit into some part of the cycle. This proof, based on the hard facts of the rocks themselves, was first perceived as the basis of a worldwide system in 1795. A Scotsman named James Hutton thought it all out and wrote a book which first set forth the idea of a rock cycle. The book is a classic in Earth history and in its essentials it has never been found defective. Hutton perceived that the rocky crust is like an old-fashioned patchwork quilt consisting of pieces of many different kinds of cloth stitched together. The various masses of rock, of all shapes and sizes and of many different ages, are chiefly remnants, like the patches in the quilt, of bodies that were once much more extensive. They fit together in irregular patterns that reflect the uplifts, subsidences, invasions of magma, and attacks by erosion to which they have been subjected. Each body consists of substance derived from older rocks either by erosion, sorting, and spreading out at the surface, or by cooking or kneading deep down within the crust.

If we believe that the rock cycle is real, we must also believe that through a time as long as the age of the oldest rocks yet discovered—a time measured in thousands of millions of years—the internal and external processes have been making and destroying rocks in just the same ways in which they are making and destroying rocks today. If during that long time there had been appreciable change in the way the processes work, that is, change in the physical/chemical laws that govern their activities, then the oldest rocks should be correspondingly different from the youngest ones. Yet they are the same.

From this fact a principle was developed early in the 19th Century. The principle says two things. *First,* the laws of nature have remained unchanged through the greater part of geologic time if not through all of it. *Second,* events in Earth history have involved the same processes, acting within the same range of rates, as those that are occurring today.

Therefore, to understand the past we study what is happening now. By closely examining rocks of any age, one can identify the particular processes that formed them and thus reason out the environments that existed and the events that occurred at the times and places of the formation of those rocks. This is the *Principle of Uniformity.* It unlocks the door to Earth history, for what is history if not our picture of the changing environments and events of the past? Before this principle and the cycle concept on which it is based were understood, men did not understand Earth history. They tended to accept supernatural legends as an explanation of what had happened. Many people in the Western world believed that the Earth was created exactly as told in the Old Testament. Others in the Eastern world accepted other legends. Not until the 19th Century was it realized that the Earth's crust is literally a book of history and that layers of rock are its pages. And not until the 20th Century was a calendar fitted to the book, giving actual dates for many events in the development of the Earth and its living things. The calendar was made possible by the application of radioactivity to the dating of rocks.

We can now begin to turn the rocky pages of this book of Earth history, looking more closely at details. After learning how the pages are dated, we shall examine the external processes and the rock characteristics that tell about environment. With the help of those characteristics we can then trace through time the evolution of the lithosphere and the biosphere.

References

Pearl, R. M., 1965, How to know the minerals and rocks: New American Library of World Literature, Inc., New York.

Zim, H. S., and Shaffer, P. R., 1957, Rocks and minerals: Golden Press, New York. (Paperback.)

3

Strata and Geologic Time

Strata and their fossils
The geologic column
Geologic time
    Radiometric dating
    Radiocarbon

Much has been learned from igneous and metamorphic rocks about temperatures, pressures, and other conditions deep within the crust. But for our purpose sedimentary rocks are more significant. They are more significant because the structures and the fossils in sedimentary rocks provide critical information about former life and its environment, and the history of living things is to many people the most absorbing aspect of the history of the Earth. With few exceptions fossils are restricted to sedimentary rocks.

Let us then look next at sedimentary rocks as a sequence of layers, and at the fossils they contain as a historical sequence of living things.

Strata and their fossils

As Figure 2 - 7 showed, most sediments washed from the lands eventually come to rest in basins, some of them on land but most of them beneath the sea. The sediments are spread out widely, in nearly horizontal layers that gradually become sedimentary rock. At length the basin is lifted up and sliced into by streams. Then the layers (universally called *strata*) appear as in a slice of layer cake. When a valley is deep and not mantled with vegetation, the pile of strata exposed in its side can be very impressive (Fig. 3 - 1). Here, seen edgewise, are pages of a kind from which the greater part of Earth history has been read. It seems hardly necessary to point out

Figure 3 - 1. Looking northwest, downstream, within the Grand Canyon, near the mouth of Shinumo Creek, at the layers of sedimentary rock exposed in the side of the canyon. The vertical distance from the brink of the canyon to the Colorado River (foreground) is just under 1 mile. (From GEOLOGY ILLUSTRATED by John S. Shelton. W. H. Freeman and Company. Copyright © 1966.)

that the lowest layer must be the oldest because it was deposited first, and that the layers are increasingly younger as they are followed up to the top. The sediments that constitute most of these strata were deposited in shallow seas that repeatedly crept in over this part of North America and as often shrank back toward the ocean. The migrations of the seacoast, so slow that they would have been imperceptible through a man's lifetime, probably were controlled by very broad gentle warping of the crust. Marine animals lived on the shallow sea floor while the sediments were being deposited, and during the times of emergence of this region plants and other land life occupied the broad lowland surface. The organisms died and some of them were buried and preserved as *fossils* (Chapter 9), embedded in the accumulating sediments and thus incorporated in the strata.

Most of the world's strata, therefore, contain fossils (Fig. 3-2), most of them the remains of sea animals. This fact tells us that most

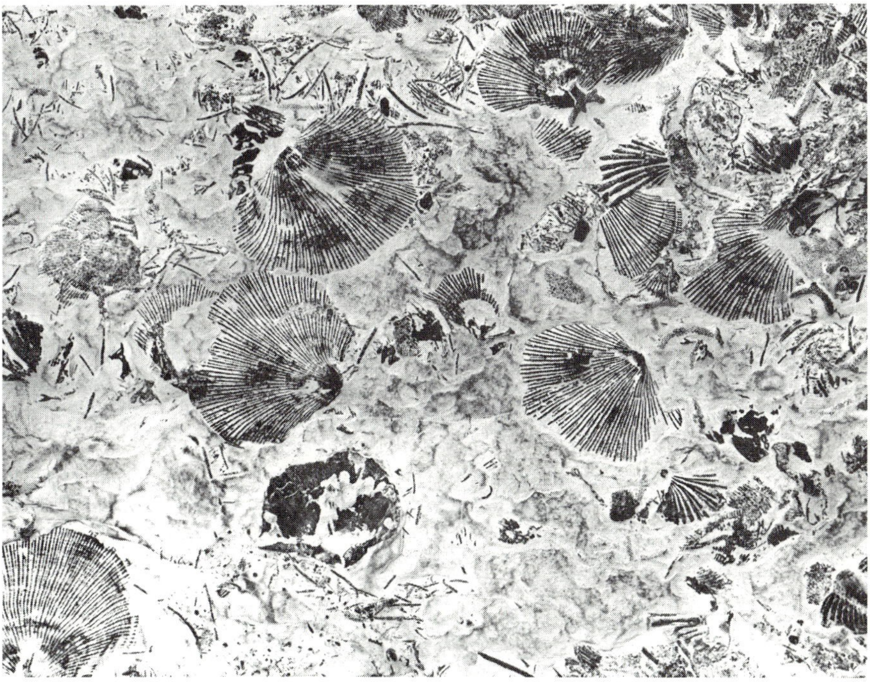

Figure 3-2. Invertebrate fossils exposed at the surface of a layer of limestone. This layer formed the bed of a shallow sea that flooded part of the North American continent hundreds of millions of years ago. (Bill Ratcliffe.)

(actually at least three-quarters) of the known strata are of marine origin. This is hardly surprising when we remember that seas are the pockets in which most sediment comes to rest, and that sediment deposited on the land stands a much better chance of being destroyed than does sediment buried beneath the sea.

When the fossil animals, mainly shellfish, are compared with creatures of similar kinds living today, it is clear that they lived not in the deep sea but in shallow water less than 200 meters or so in depth, like the water that covers continental shelves today. Many kinds lived in extremely shallow water immediately offshore. Here we come to a crucial point. Although most of the fossils in the strata resemble the shellfish living today, very few of them are precisely the same kinds. They are different. They are most different in the lowest and oldest strata and least in the highest and youngest layers. In other words, the fossils change very gradually as the strata are followed upward from bottom to top of the pile, and upward in the strata means onward in time. These changes mean that the organisms changed form through time and left in the strata a record of fossils by which the changes can be traced. Of course the changes were very slow, as we can appreciate when we consider that the difference in age between the lowest and highest flat-lying strata visible in Figure 3 - 1 is more than 300 million years.

As will appear later, this process of evolutionary change is not primarily something automatic that is inherent in the organisms themselves. Primarily it consists of changes organisms undergo in indirect response to the environments in which they live. As shallow seas spread across the lands or receded from them, impelled by gentle movements of the crust and other influences, the populations of sea animals spread and shifted accordingly and in the process slowly evolved; that is, changed form. The rates at which organisms spread were much faster than the rates at which they changed form. Therefore the changes were easily spread far and wide, so that today they show up in the form of identical or very similar fossils in strata of about the same age in two or more continents.

A modern example of the fast rate at which a population can spread is the snail *Littorina littorea,* common in the very shallow water along shorelines. Until 1852 this snail was confined to European coasts. Between 1852 and 1857 it was found established in the harbor of Halifax, Nova Scotia, where it had probably been brought by some ship arriving from Europe. It spread rapidly southward and is known to have reached as far as the harbor of New Haven, Connecticut by 1880, presumably through its own efforts without hitchhiking on other ships. This implies an average rate of spread of 30 miles per year. At such a rate (if a continuous path were available, which of course it is not) the snail could spread around the circumference of the globe in less than a thousand years.

The similarity in general degree of evolutionary progress around the world has an important meaning. We can say that two strata that occur in two continents but contain similar suites of fossils must be of similar age. The essential truth of this inference became apparent a century before measurements based on radioactivity provided dates for the strata and thus confirmed the inference. Because of this basic similarity, strata in different parts of the same continent or in two continents can be matched or *correlated,* even though there is today no physical continuity between them.

The geologic column

Correlation by means of fossils has made possible a worldwide standard catalog of strata in the form of a diagram called the *geologic column.* This column is the central core of *stratigraphy,* the study of strata. It is the pillar to which we try to attach most of our accumulating information about the Earth. In it all the known strata are brought together in their proper relative positions to create a layered, columnar diagram. Table 3 - A shows it in a simple form, including the largest units but without all the many recognized subdivisions. The units of the column were named for layers of rock that contain characteristic fossils. The units are grouped into the systems, series, and so forth seen in the figure. But their names are also the names of the corresponding units of time. Thus we can say that the strata constituting the Devonian System (or simply the Devonian strata) were deposited during the Devonian Period (or simply during Devonian time). That period, as the table shows, lasted from about 415 to about 360 million years ago, or approximately 55 million years.

The geologic column in all its detail is a system like the system by which books are shelved in a great library. New books coming into the library are catalogued and then fitted into the shelves in their logical positions. In this way the library is continually im-

Subdivisions based on **strata**/*time*				Radiometric Dates (millions of years ago)	Outstanding Events in Evolution of Living Things
		Systems/*Periods*	**Series**/*Epochs*		
PHANEROZOIC	CENOZOIC	Quaternary	Recent or Holocene Pleistocene		*Homo sapiens*
				2?	
		Tertiary	Pliocene		Later hominids
				6	
			Miocene		
				22	
			Oligocene		Primitive hominids Grasses; grazing mammals
				36	
			Eocene		Primitive horses
				58	
			Paleocene		Spreading of mammals Dinosaurs extinct
				63	
	MESOZOIC	Cretaceous	Many		Flowering plants
				145	
		Jurassic			Climax of dinosaurs Birds
				210	
		Triassic			Conifers, cycads, primitive mammals Dinosaurs
				255	
	PALEOZOIC	Permian			Mammal-like reptiles
				280	
		Pennsylvanian (upper Carboniferous)			Coal forests, insects, amphibians, reptiles
				320	
		Mississippian (Lower Carboniferous)			
				360	
		Devonian			Amphibians
				415	
		Silurian			Land plants and land animals
				465	
		Ordovician			Primitive fishes
				520	
		Cambrian			Marine animals abundant
				580	Primitive marine animals
PRECAMBRIAN (Mainly igneous and metamorphic rocks; no worldwide subdivisions.)				1,000	Green algae
				2,000	
				3,000	Bacteria; blue-green algae

Table 3 - A. The geologic column, major worldwide subdivisions, selected dates, and events in the evolution of life. (Dates are best estimates, after R. L. Armstrong, 1971, unpublished.)

proved. Newly discovered strata are treated similarly. In the late 1950s, when geologists began to study the rocks of Antarctica on a systematic basis rather than as explorers, they found in the almost unknown strata of that continent characteristic fossils by which they identified, in various areas, strata of many of the major subdivisions of the geologic column. This is one example of the improvement of the column that is always going on.

The chief reason why the column is idealized rather than real is that in no part of any continent are *all* the units, major and minor, present to form a complete sequence. This is because all strata are localized. The extent of any layer of sedimentary rock is no greater than the extent of the basin in which its sediments were deposited. True, the basins of some former shallow seas were very large. Some 100 million years ago, one of them extended widely from southern Mexico northward across North America to the Arctic Ocean, cutting the continent into two separate lands. Nevertheless, most basins have been much smaller than that one, and none ever occupied an entire continent at a single time. Every stratum is therefore limited in original extent. Many strata are limited further by erosion, which in many basins cut away part—even a very large part—of a layer before the next layer of sediment was spread over it. Usually evidence of the erosion is present at the interface between the top of the eroded layer and the base of the stratum above it (Fig. 3 - 3).

Here, then, is the geologic column, at once a diagram and a catalog. By the beginning of the 20th Century the major subdivisions shown in Table 3 - A were already known, as were the characteristics of their respective fossils. What was then still lacking was the element of time, and what were needed were dates.

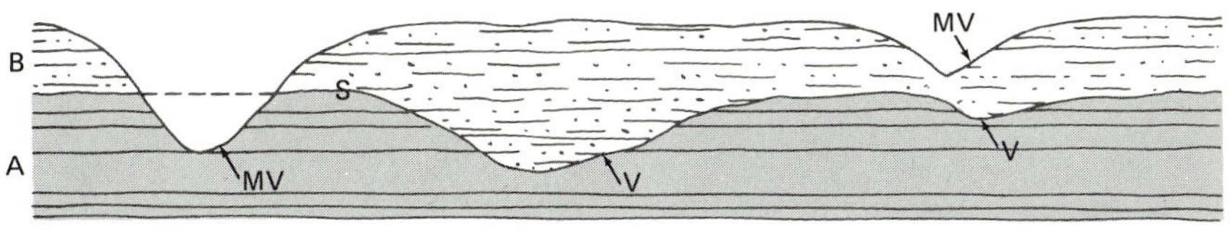

Figure 3 - 3. Two sequences of marine strata exposed along the side of a modern valley. After the older sequence A was deposited, the shallow sea withdrew, exposing the surface S to erosion. Valleys (V) were cut by streams. Later the sea invaded the region and submerged the eroded surface S, which gradually became buried beneath a new sequence of sediments (B). Again the sea withdrew, and erosion cut modern valleys (MV) into the exposed surface.

The relationship between sequence A and sequence B is one of unconformity (lack of continuity) and is visible at the interface S.

Geologic time

The need for a reliable means of measuring time was felt well before the geologic column had been completed in all its major subdivisions. Attempts were made to estimate the age of the oldest sedimentary rocks. One attempt took the form of dividing the aggregate thickness of the known marine sedimentary rocks in a continent by the aggregate yearly contribution of sediment by the continent's rivers today. The results were thought to represent the minimum time it had taken to accumulate the pile of sedimentary rocks. But this simple arithmetic was based on several factors that were unknown and therefore had to be assumed. The most misleading of these, that the rate of contribution of sediment was always the same, we now know is untrue. The numbers arrived at by this arithmetic, about 100 to 300 million years, were not very satisfying to most scientists, who thought it likely that the oldest sedimentary strata were a good deal older.

Radiometric dating. The next chapter in the dating of the Earth's history began in 1896 with the discovery of natural radioactivity. Research following that discovery put an end to estimates, because radioactivity provided a basis for directly measuring the actual ages of rocks. In Chapter 1 it was mentioned that about half the Earth's internal heat is probably generated by natural radioactivity. Chemical elements that in one form or another are naturally radioactive are present in the makeup of a considerable number of minerals. Among such elements are uranium, thorium, potassium, rubidium, strontium, and carbon. Radioactive forms of these elements possess this property, that their atoms spontaneously and continuously decay. They throw off energetic particles and become converted or degraded into *daughter* atoms, which have a configuration differing from that of the *parent* atoms.

For a radioactive form of any element, rate of decay is constant and can be measured with fair accuracy. In this fact lies the means of dating certain kinds of rocks. The explanation follows this reasoning: Rocks consist of minerals and minerals consist of elements, forms of some of which are radioactive. In a mineral that is part of an igneous rock, the radioactive decay began when the mineral crystallized from magma. Now for a particular radioactive form of a particular element, two values are known: (1) the amount present in the mineral at the start and (2) the rate of decay. A third value must be measured: the amount remaining today. With all three values known it becomes possible to calculate the time elapsed since the mineral crystallized from the magma and hence the approximate age of the igneous rock of which it is a part. By *approximate age* we mean an age determination having an uncertainty of less than 5 percent.

Similarly, by following the same procedure with a mineral from a metamorphic rock we can calculate the time elapsed since metamorphism forced the crystallization of new minerals from the elements present in it before metamorphism. Measurements of radioactivity therefore yield important information about the times when certain igneous and metamorphic rocks were made. But sedimentary rocks are different.

In what way are they different? Although a radioactive mineral can be part of a sedimentary rock, it was not *formed* in the rock. It was brought from elsewhere, a mere building block, a mineral foreign to and older than the rock of which it is now a part. Nor has it been recrystallized since it took up residence in the sedimentary rock. Therefore some part—perhaps even a large part—of its natural radioactive decay had already taken place before it became part of the sedimentary rock. Suppose we measure the extent of decay in the mineral and thus obtain the mineral's age. All that this age can tell us about the sedimentary rock *as a layer of rock* is that the rock is younger than our mineral. In short, a layer of sedimentary rock cannot be dated directly by radiometry—that is, by measuring radioactivity. This fact seems to place us in a dilemma, inasmuch as the geologic column is based on sedimentary rocks, and it is principally the geologic column for which we want dates.

But luckily there is a way out. The geologic column *is* dated, at least in part, as one can see in Table 3 - A, and the dates have been derived from igneous rocks. Each of the dates in that table is right at the boundary between two systems or two series. Most such dates are secondary, not measured directly but derived by interpolation between groups and clusters of dates that have been measured directly from samples of igneous rock. The bodies of igneous rock that are chosen for dating are in most cases bodies that occur in contact with known sedimentary strata. In such cases the character of the contact must show clearly whether the igneous body is older or younger than the adjacent sedimentary rock. Because this age relationship is known, the date of the igneous body obviously is an upper or lower limit—a roof or a floor—for the date of the sedimentary layer.

To see just how this works we can examine Figure 3 - 4, which shows a sequence of sedimentary strata. The systems to which they belong are known because of the fossils they contain; hence they are labeled in accordance with the geologic column (Table 3 - A). Among the upper layers are two ancient lava flows; in the lower layers are two dikes (Fig. 2 - 4) representing intrusions of magma. The ages of the two dikes, determined radiometrically, are 250 and 210 million years respectively. During its solidification, the hot magma that was parent to each dike "cooked" and chemically altered the adjacent strata throughout the narrow zones where magma and strata were in contact. The altered zone, however, is not present at the tops of the dikes. This means that when the magma of a dike arrived and solidified, the stratum that *now* overlies it had not yet been deposited; otherwise it too would have been altered by heat and chemical reactions. Therefore the two interfaces that include the tops of the two dikes must be the result of two long periods during which erosion removed whatever strata once overlay the positions of the present interfaces. These interfaces, then, are surfaces of unconformity like that in Figure 3 - 3. Referring to the radiometric dates of the two dikes, one can see that the strata of Triassic age, which are younger than one of the dikes but older than the other dike, must then be less than 250 but more than 210 million

Figure 3 - 4. Dating sedimentary strata by means of their relations to bodies of igneous rock whose ages are known.

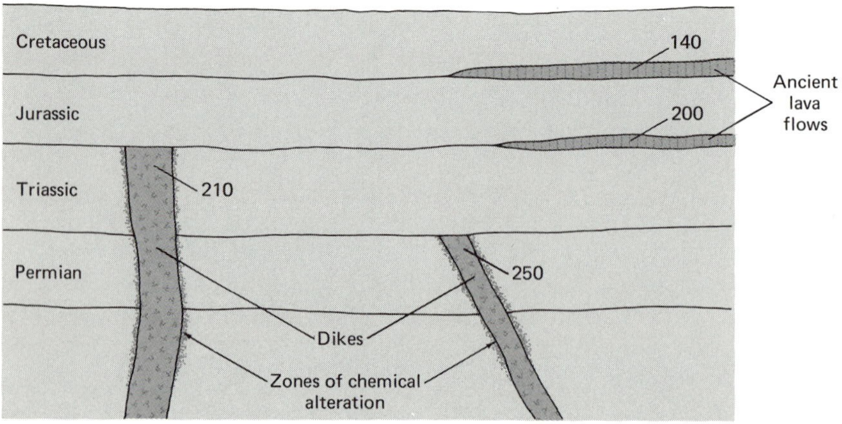

years old. Subtract one number from the other and we get 40 million years, the *maximum* duration of the Triassic Period that can be determined from this particular evidence.

Turning to the two ancient lava flows, we apply the same principles. The lava has altered the rock immediately beneath it, but the rock overlying the flows is not affected because it was not deposited as sediment until after each flow had solidified. The radiometric ages of the lava layers tell us that Jurassic strata here are no more than 200 million years and no less than 140 million years in age, and that the duration of Jurassic time (as determined here) was 60 million years. These numbers are continually being refined and made more precise by new radiometric dates of igneous rocks similarly related to strata in other parts of the world. The net result today is shown by the numbers in Table 3 - A, but in the future these numbers, too, will be improved. The dating of the geologic column, like the column itself, continually grows more accurate.

As Table 3 - A shows, the systematic dates on strata end downward at 580 million years, the base of the Cambrian System. The underlying Precambrian rocks include few strata and are made up mostly of an intricate pattern of igneous and metamorphic rocks. A great many of these have been dated; the oldest that have been measured up to now are at least 3.7 billion years old, but older ones will be found. As for the age of the Earth itself, it is not known with accuracy, but various data, including astronomic information, suggest that the Earth as a planet is some 4.5 billion (4,500,000,000) years old.

So great a number of years is hard to visualize, especially since Table 3 - A shows only a tiny proportion of Precambrian time and so tends to give a distorted impression of the earlier part of the Earth's history. A different sort of graphic representation (Fig. 3 - 5) gives a more realistic picture. The Empire State Building stands 1250 feet above one of New York's most famous streets. Let its height represent the 4.5 billion years since the birth of the Earth. Then all the building's 102 stories, plus 75 feet of the shaft on top, represent the time from the Earth's beginning to the start of the Cambrian Period. The rest of the shaft, 155 feet only, represents Paleozoic, Mesozoic, and almost all of Cenozoic time. A dictionary about 10 inches thick, laid on top of the shaft, would represent the time during which prehistoric man has inhabited the Earth. A dime placed on the dictionary would represent the time since the earliest

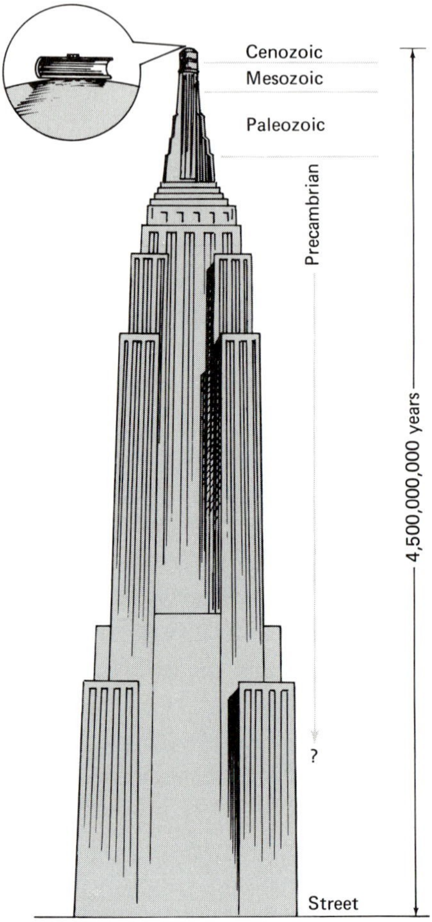

Figure 3 - 5. The Empire State Building used as a time scale for the Earth's history.

known civilizations in western Asia. The thickness of a scrap of typewriter paper laid on the dime would be the time elapsed since the birth of Christ.

The majestic length of geologic time makes possible the huge accomplishments that do not *seem* possible the first time one looks at them. Stand on the brink of the Grand Canyon (Fig. 3 - 1) in fair weather and look down into that valley a mile deep. The Colorado River seems but a narrow ribbon of brown muddy water. Yet during a rain it is easy to see where the mud is coming from, as rainwater trickles down the steep sideslopes carrying rock particles with it. With every rain the canyon becomes just a little bigger. The water of the river is sampled systematically. Year in and year out it is carrying 150 tons (dry weight) of those rock particles out of the canyon, *every minute.* One hundred fifty tons a minute adds up to nearly 80 million tons a year. How many years has the river been flowing there, while the rocky sides of the valley crumble and allow particles to creep and be washed downward into the stream? We don't yet know just when the river began, but that time was at least several million years ago. As one thinks of such big numbers the cutting of the Grand Canyon ceases to be mysterious. It takes on a different aspect and becomes understandable for what it is—a small segment of the rock cycle. The canyon and the rest of the river's drainage basin are being systematically cut away, and the waste from the cutting process is being carried downstream and deposited in a basin—that of the Pacific Ocean. The operation is like that seen in Figure 2 - 6. The forms are different but the principle is the same.

So geologic time and the rock cycle go hand in hand. Geologic time does not differ from the more familiar calendar time or clock time. It is measured by different means and not as accurately. But it is the same time, our familiar time extending back through the history of the Earth.

Radiocarbon. The dating of radioactive atoms in the minerals of rocks has one notable disadvantage. The rates of radioactive decay involved are so slow that with existing methods of measurement it is not possible to discriminate closely between very recent dates and dates of as many as a few hundred thousand years ago. Hence in general we lack reliable dates within the range of, say, the last 500,000 years—with one exception: dates obtained from measurement of radioactive carbon (radiocarbon or carbon-14). This form of carbon decays at a rather rapid rate, so that it is suitable for dating objects between as little as 100 and about 50,000 years old. And radiocarbon also enters importantly into the biosphere. This makes it possible to date organic substances instead of rocks.

Radiocarbon is continuously created in the atmosphere through the action of radiation coming in from outer space. It mixes with ordinary, non-radioactive carbon, forms carbon dioxide (CO_2), and diffuses quickly through the whole atmosphere, hydrosphere, and biosphere. Its proportion to ordinary carbon remains nearly the same throughout the entire system because rate of radioactive decay is in equilibrium with rate of production, as long as the latter remains constant. Hence the concentration, the "equilibrium proportion," of radiocarbon should be the same in samples of air, of fresh or salt water, of the wood or leaves of a tree, or of the body tissue of any animal—as long as the tree and the animal are alive. But with the death of an organism the proportion of radiocarbon in its substance diminishes, because the loss through radioactive decay is no longer replenished by new supplies brought in by life processes such as respiration and the taking in of food. By measurement and calculation one can determine, often with an uncertainty of only a few percentage points, the proportion of radiocarbon remaining in an organic sample and hence the time elapsed since death.

Radiocarbon dating is extraordinarily useful because it makes possible a calendar of the history of prehistoric man, other animals, and plants through about the last 50,000 years. It also dates many events in the latest of the great glacial ages, including a spectacular ice invasion of North America and Europe that reached its climax barely 20,000 years before our time. We shall make use of many radiocarbon dates, as well as the far older dates determined directly from rocks, as our story proceeds.

References

Faul, Henry, 1966, Ages of rocks, planets, and stars: McGraw-Hill Publishing Co., New York.

Kay, G. M., and Colbert, E. H., 1965, Stratigraphy and life history: John Wiley & Sons, New York.

Libby, W. F., 1961, Radiocarbon dating: Science, v. 133, p. 621–629.

Woodford, A. O., 1965, Historical geology: W. H. Freeman & Co., San Francisco, p. 191–220, discussion of radiometric ages.

4

The Lands Wash into the Sea

The Principle of Uniformity, on which our picture of Earth history is based, says that the same processes as those of today have operated in the geologic past. If we follow the principle, to perceive what has happened in the past we must understand what the external processes are doing today. If we know the characteristics of the sediment deposited by each process, we can recognize the same characteristics in sedimentary strata like those in the Grand Canyon. These, together with fossils in the strata, help to identify the environments in which the strata were formed, and thus make it possible to reconstruct sequences of events. This chapter and those that follow immediately show some of the ways in which geologic processes are working today and some of the characteristic features by which the various processes can be recognized.

Weathering

Chemical decay. What happens to a steel or iron nail or a "tin" (actually steel) can when it is taken from a protected shelf and left outdoors in a moist climate? It is common knowledge that within a few days it begins to show rust. A railroad track is kept bright by the car wheels that pass over it, but an abandoned track begins to rust at once. The yellowish-brown rust is the mineral* limonite (Table 2-A), created by the chemical combination of atmospheric oxygen and water with iron. Its creation involves two chemical reactions, oxidation and hydrolysis. The steel can and the steel rails are unstable in moist air, and when placed in such an environment begin to decompose. The decomposition forms limonite, a substance that *is* stable under the new conditions.

Similar reactions occur when an iron-bearing mineral such as biotite or hornblende is exposed to the atmosphere. The rock (perhaps granite) of which the mineral is a part soon begins to show brownish stains, signaling the appearance of limonite. Through oxidation and hydrolysis the iron from the biotite and hornblende is on its way toward assuming a more stable form, and the rock has begun to decompose. These reactions are among the many chemical changes involved in the general process called *weathering*, in which rocks (and sediments also) that are exposed to the atmosphere are broken down and decomposed. Weathering occurs on every part of the Earth's solid surface and below the surface as well, as deep as air and water can penetrate. The interface between rock and atmosphere is a boundary, and we have said before that boundaries provoke activity. In this case the hydrosphere also is present at the interface; rainwater acts upon rock as well as air does.

* Actually a group of several minerals.

The general process of weathering involves another reaction, *dissolution,* the results of which are frequently seen on statues, tombstones, and the walls of buildings made of limestone or marble. These two kinds of rock consist almost entirely of the mineral calcite, which is soluble in carbonic acid. The acid is created by the combination of rainwater with carbon dioxide, much of it generated at the surface and in the soil during the decay of vegetation by bacteria. The calcite simply dissolves and the products of the reaction are carried away by water that slowly percolates through the ground.

Another common reaction in chemical weathering likewise results from attack by carbonic acid. In this case the acid attacks feldspars, which are abundant in most igneous rocks and in many metamorphic rocks as well. For example, hydrogen ions derived from carbonic acid break into a molecule of orthoclase feldspar, and water combines with the remaining constituents to form hydrous aluminum silicate, one of the compounds found in the group of minerals that together constitute clay.

In fact, chemical weathering is a destroyer of igneous rocks, wherever they are exposed to the atmosphere, and a maker of clay on a huge scale. In its turn the clay is gradually washed away, appears as part of the mud in muddy rivers, and mostly ends up as sediment deposited on the floor of the sea. The destruction of feldspars and other silicate minerals in a rock weakens the bonds between those minerals and adjacent crystals of quartz. The quartz grains, which are stable in a surface environment and therefore resistant to chemical decay, are now loosened. Sooner or later they are washed downslope and are transported by streams to basins, in which they are deposited as layers of quartz sand. This grand sorting process, started by weathering, effectively separates quartz sand from the silicate minerals that are in the form of clay.

If chemical weathering is allowed to act on a body of bedrock through a long time, say one to several million years, in a region in which erosion is so slow that the products of weathering can accumulate in place, the result is a zone of very clayey regolith forming a blanket at the Earth's surface and grading imperceptibly downward, beneath the surface, into nonweathered bedrock (Fig. 4-1). This weathered zone may be from five or ten to more than 100 feet thick, and will have no sharp boundary at its base. The original sharp boundary between rock and atmosphere has become

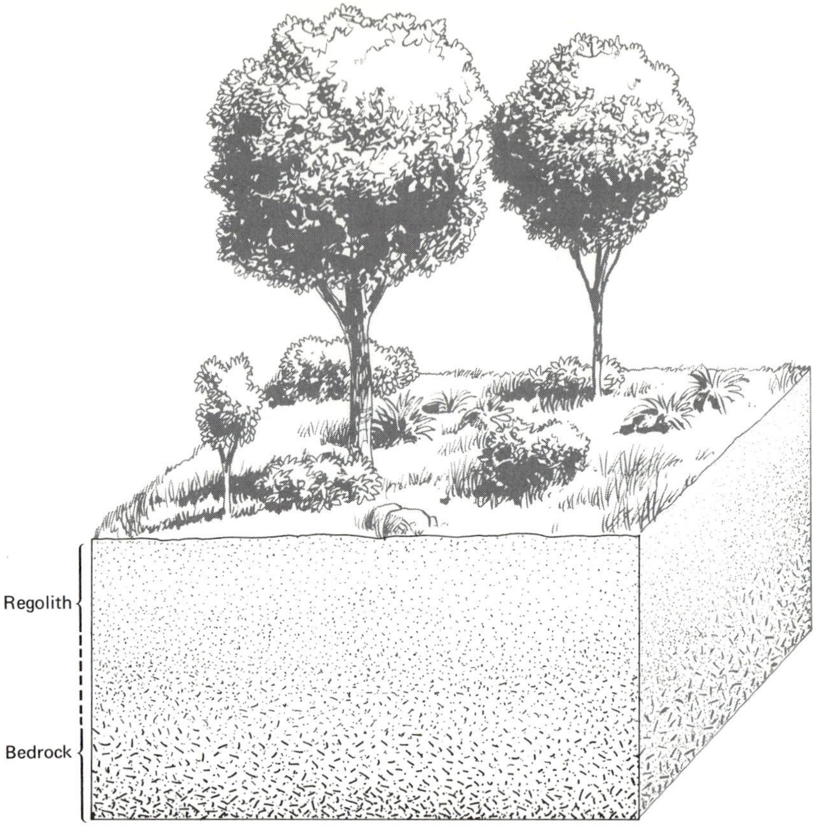

Figure 4 - 1. Thick zone of clayey regolith beneath the surface, created by long-continued chemical weathering of igneous rock.

At base of block, the rock (dark gray) is still fresh and free from the effects of weathering. In the middle the crystalline texture of the rock is still faintly visible, but upward the texture fades out into a mass of clay speckled with sand-size grains of quartz.

blurred by the creation of clay, a product of weathering that is stable in the existing environment. So clay remains at the surface while chemical weathering continues actively in the fresh rock beneath.

Only the top of the weathered zone in Figure 4 - 1 coincides with the surface; the rest of the zone is below ground. This reflects the fact that the atmosphere, and water from rainfall as well, penetrate the ground, in some places to great depth. Water and air cannot penetrate fresh solid rock, but they can move downward through the cracks that affect all rock, bringing about chemical change all along the faces of the cracks. Also they can slowly penetrate the tiny interfaces between adjacent crystals or grains. At every point along these routes, chemical activity weakens the rock (just as rusting weakens a steel nail) and creates more openings. The more surfaces there are along which water or air can invade the rock, the faster weathering occurs. If we think of a mass of rock as being divided into cubes of equal size (Fig. 4-2), we can understand that each time a cube is subdivided into eight smaller cubes, the aggregate area of cube surfaces is doubled. Because the chemical changes in the weathering process take place at surfaces (which are boundaries), opportunity for such changes keeps increasing. The subdivision of one cubic centimeter of rock into particles the size of the smallest clay mineral gives the former cube an aggregate area of almost one acre! So, as rock breaks up along cracks, it prepares the way for more and more chemical weathering.

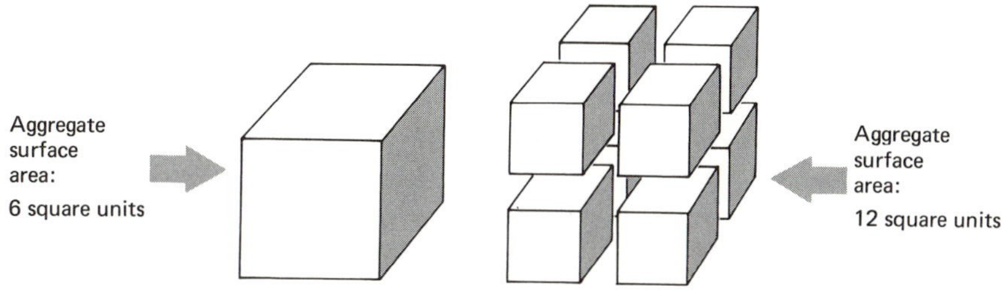

Figure 4-2. The cube (left) measures 1 unit on a side. Its surface area (6 sides) is 6 square units. It is sliced into 8 equal cubes (right), each ½ unit on a side and each, therefore, with a surface area of 1.5 square units. The aggregate surface area of the 8 small cubes is thus 12 square units—twice that of the original large cube, although the aggregate volume has not changed.

Especially on mountains and on steep slopes generally, erosion strips away clay and other weathered material. This exposes fresh rock to more rapid chemical attack. Chapter 2 showed how granite, solidified from magma at great depth, is thus gradually uncovered and exposed. The removal, century by century, of the thick covering of rock that originally overlaid the big body of granite has an important result. As the granite is unwrapped, its environment subtly changes. Temperature and pressure gradually decrease. At last the granite, with its protective roof destroyed by erosion, appears at the Earth's surface. It has emerged into a new environment characterized by air, water, and oxygen. With the removal of pressure the granite expands, cracks, and slowly bursts, not into its separate crystal grains but by the opening of cracks parallel with the surface (Fig. 4-3). Most of its minerals, formed deep in the crust and built to accommodate to high pressure and temperature, are poorly

Figure 4-3. High in the Sierra Nevada, California, a body of granite has cracked repeatedly, so as to form a series of "onion skins" parallel with the surface. For scale, note the man (center). (N. K. Huber, U.S. Geol. Survey.)

adapted to the "easier" environment at the surface and are vulnerable to attack by atmosphere and hydrosphere. Like the steel can taken off the shelf and left outdoors, these minerals are victims of a changed environment. They decay chemically and their components are dispersed and built into other substances that are adapted to existence at the Earth's surface. Similar things happen in the biosphere, as we shall see in a later chapter.

These chemical conversions of minerals at the Earth's surface leave traces by which they can be identified long after the changes occurred. Segments of the Earth's crust are continually being bent down in one region or another to form broad, shallow troughs or basins. Some of the basins become submerged beneath the sea. Where a bent-down segment has at its surface a zone of weathered rock like that in Figure 4-1, the zone becomes gradually covered with a layer of sea-floor sediment. Later, when that part of the crust is lifted up again, erosion eventually re-exposes the zone of weathering with its cover of marine sediment. At least two such zones are exposed today in the layers of marine sedimentary rock in the sidewalls of the Grand Canyon (Fig. 3-1). This shows that the region of the canyon has sagged down beneath the sea at least twice and has twice re-emerged, and that the periods of time before the downward movements were long enough to permit extensive chemical weathering of the rocks that were then at the surface.

Mechanical breakup. Although most weathering is chemical, some is mechanical. Bedrock and particles of regolith are split apart by the growth of plant roots large and small. Worms, ants, and termites burrowing in regolith bring huge quantities of small rock particles to the surface (one estimate: 10 tons per acre per year), literally turning regolith inside-out. Finally, at the tops of high mountains and in cold high latitudes ice is a formidable agent of mechanical rock-breaking. When water freezes to form ice its volume increases 9 percent. The thrust of water expanding and changing to ice as it freezes in a crack pries the sides of the crack apart. Where temperatures pass through the freezing point between day and night, great areas of land in high latitudes and in high mountains are littered with broken pieces of rock, completely concealing the bedrock underneath. The mechanically broken-up rock, unlike chemically weathered material, is nearly identical with the bedrock from which it was derived.

The cracking (Fig. 4-3) that occurs when, through long-continued erosion, coarse-grained igneous rock bodies are unroofed and exposed at the Earth's surface is not the result of exposure to the atmosphere and hydrosphere, and so we include it among weathering processes only by courtesy. Cracking occurs simply because the enormous load under which the rock formed has been reduced nearly to zero as the surface approaches the concealed body of rock.

Mass-wasting

In converting bedrock into loose particles by various methods, chemical and mechanical weathering prepare rock material for the next phase of the rock cycle: transport toward lower places. The regolith on a slope is not inert; it is moving. Gravity is pulling it downslope toward the nearest stream (Fig. 4-4, right). Occasionally a big mass moves suddenly as a landslide. Rushing downslope at a destructive 60 miles per hour, it gets into the newspapers, particularly if it has destroyed buildings, roads, and other man-made structures. Often regolith is washed downward, especially during heavy rainfalls, and at times it slowly flows as tongue-like masses of

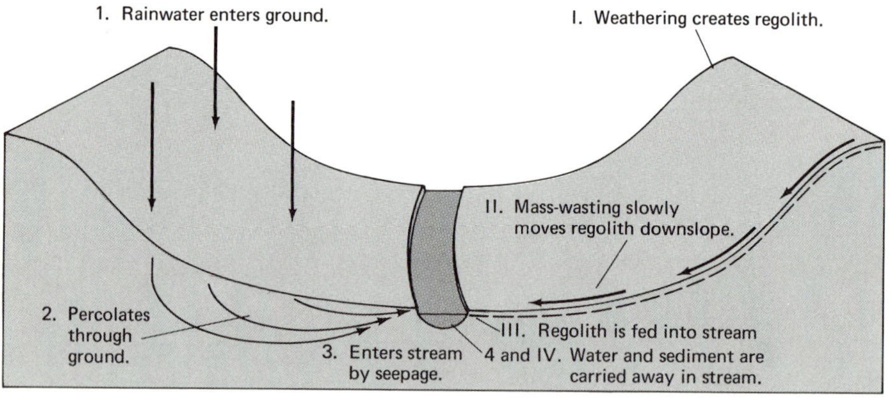

Figure 4-4. A stream is fed with water (as part of the water cycle) and with sediment (as part of the rock cycle), along its sides.

Left side: Paths followed by water only. Right side: Movement of regolith only. Of course, both water and sediment are entering the stream from *both* sides.

mud and stones. But most of the time regolith creeps downward imperceptibly, at rates as slow as one inch every five or ten years. All the activities that move regolith downslope—creep, landslides, and the rest—are together referred to as *mass-wasting.*

In human terms a rate of movement of one inch every ten years seems slow. But in the framework of geologic time it keeps the rock cycle going actively. At such a rate (one foot every 120 years) a given parcel of regolith would creep from top to bottom of a mountainside one mile long in about 630,000 years. But North America has existed as an area of predominant land through at least a thousand times that many years and probably far longer. Thus when we remember that virtually every land slope throughout the world's continents is a site of this creeping activity of mass-wasting, the total volume of sediment delivered from slopes to rivers in just one year is staggering.

Although thin and mostly slow moving, regolith covers some three-quarters of the world's land area; so the volume of loose rock matter moved by mass-wasting is huge. Although no one knows the exact figure, we might guess that the waste thus delivered to streams amounts to hundreds of cubic kilometers each year. Once it is in motion or, having moved, is deposited, rock waste becomes *sediment* by definition, no matter what process transports or has transported it. *Streams* are the means by which most sediment is transported across the land.

Streams

Variations of volume. Prepared by weathering and moved downslope by mass-wasting, rock waste sooner or later reaches a stream, where it begins to move in a very different way. In order to understand what happens, we must look more closely at what keeps a stream going. First look at an artificial stream flowing through a sewer. Sewers drain highways, city streets, and buildings. They carry off rainwater and a variety of wastes through concrete troughs and pipes. Between rains many sewers stop flowing because the concrete enclosures insulate them from the ground around them; the water in the ground cannot get into them. Streams are different. Except in deserts and other dry country streams keep on flowing, rain or shine.

Figure 4 - 5. One of several ways in which a stream is fed with sediment. This stream, in flood, undermines the sandy alluvium along one of its banks. (U.S. Department of Agriculture.)

Although their volumes vary continually, they keep flowing because they are continually fed with water that saturates the ground around them and that leaks into the streams. Most of the ground, whether bedrock or regolith, is permeable—that is, water can enter and move very slowly through the tiny, interconnected spaces within it. The ground absorbs water during every rainfall, and after a longer or shorter time of slow underground travel, this *ground water* seeps out into the nearest stream (Fig. 4 - 4, left) whether rain happens to be falling or not. In this way water in the ground regulates the flow of a stream by feeding it water all along its length between rains. It permits a stream to act as a transporting agency not just at rainy times, but at all times. And the sediment being transported gets into the stream in much the same way as the ground water does. It is fed into the stream all along its length by creep and other kinds of mass-wasting (Fig. 4 - 4, right; Fig. 4 - 5).

Transport of sediment. Once they are in a stream, rock particles move much faster than when they were creeping down a hillside.

But in the stream they move less steadily. Despite the effect of incoming ground water, stream flow does fluctuate with the seasons and even with changes in weather. Rains cause floods and drought makes a stream dwindle.

Running water, which is always a bundle of swirls and eddies, moves its sediment according to the sizes, shapes, and weights of the particles (Fig. 4-6). The "coarse" pieces (sand grains and larger sizes) travel in a different way from the "fine" ones (particles of silt and clay size). The coarse particles that can be moved at all roll, slide, or hop along the stream bed, while the tiny particles of silt and clay are carried above, suspended in the water, continually agitated by whirls and eddies and rarely touching bottom. Only swift streams have the force to move cobbles and boulders. Most such streams are found in mountains and flow down steep slopes; yet even then many may be capable of moving large boulders only during exceptional floods, when volume of water and velocity of flow are multiplied many times.

When any flood subsides, velocity diminishes and more and more sediment on the bed stops moving, particle by particle, the largest (and heaviest) first. Slightly smaller particles arrive from

Figure 4 - 6. Names and diameter ranges (in millimeters) of rock particles, as used by most geologists. Diameters are shown at true scale.

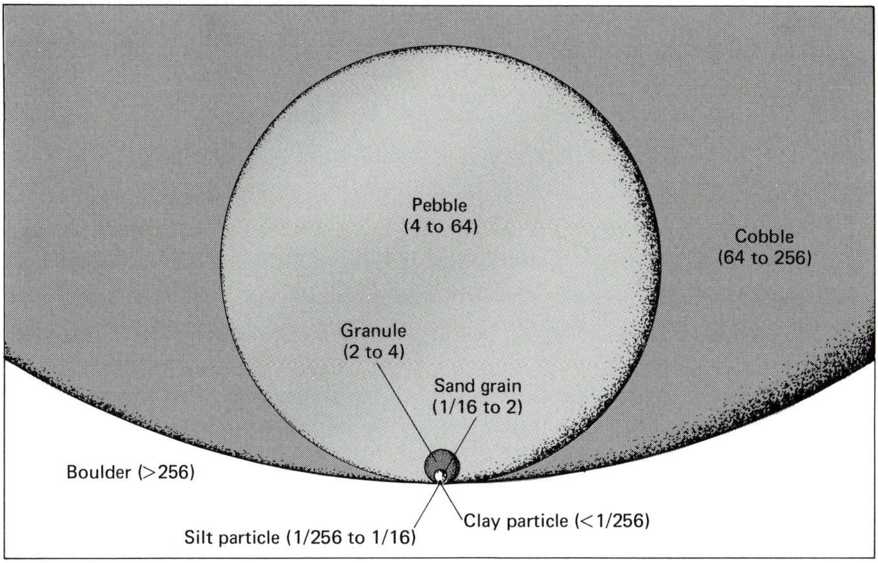

farther upstream and come to rest on top of those already motionless. The process continues, building up the bed as long as velocity decreases. What has happened is that the particles deposited by the stream have been *sorted* according to their size and weight by the sensitive response of the stream to the decreasing force of its flow. The coarser pieces have been stranded on the bed and the finer ones have been flushed onward, downstream. Sorting is a result of "washing" in water, a fluid—or in air, another fluid. Mass-wasting, which hardly involves fluids at all, does not sort rock waste. Sorting, then, is a very conspicuous part of the slow transformation of the regolith that lay on a hillside into an ideal sedimentary rock. Sorting is not the work of streams alone. Other kinds of water currents, and air currents as well, sort sediments effectively.

The sediment deposited by a stream is picked up again during the next flood, carried farther downstream, and sorted again as it is once more deposited. With each particle of sediment the process is repeated until the particle reaches its final resting place for that part of the rock cycle.

Regardless of grain size, the name for any sediment that is deposited on land by a stream is *alluvium.* Whether alluvium or not, any body of coarse sediment is called gravel or sand (Table 2-D) according to the size of its dominant particles. If it is later converted into sedimentary rock, the gravel becomes conglomerate and the sand becomes sandstone (Table 2-D).

A stream not only sorts its sediment; it rounds the sedimentary particles, as does also the surf along a sea beach. We see the result of rounding in Figure 2-6, in which *A* consists of fractured, angular, non-sorted particles of some older rock that might have been weathered mechanically and have slid or fallen down to accumulate at the base of a cliff. The rock represented is breccia; formerly it was angular gravel, a sediment. In contrast, *B* consists of smoothly rounded particles. Their corners have been knocked or worn off, and as they rolled over and over and were brushed by the innumerable sand grains being carried past them, they became evenly worn. All of them are of small-pebble size and therefore they must have been sorted. The sample *B* is now conglomerate, but was formerly rounded gravel. Its pebbles could have been deposited either by a stream or by waves along a beach; through mere inspection of the specimen we cannot tell which. Following along a stream from its headwaters toward its mouth and looking at the gravel exposed in banks and bed, one can sometimes see that the pebbles become, on the average, both smaller and better rounded in the downstream direction.

Looking at grain size and degree of rounding in some ancient layer of sedimentary rock deposited by a stream, and following the Principle of Uniformity, we can visualize something of the terrain through which the stream was flowing and get some idea of the distance through which the sediment was transported. These are significant elements of a former environment.

At places along the course of a stream where alluvium is well exposed and can be seen clearly, a close look will show that the deposition of stream sediment is more variable than we have suggested. Although it is true that as a general rule sediment becomes finer toward the stream's mouth, some or perhaps even all of the exposures we are looking at consist of layers of several different grain sizes cutting across one another and showing no consistent change in any direction. Generally this arrangement results from the repeated changes of volume and velocity that occur in all streams and that cause changes in the diameters of the sediment deposited at any given point along the stream's course. Such changes can be seasonal, daily, or irregular in time, but all of them influence the deposition of sediment. Furthermore the bed of any stream is unstable. We have watched sediment being deposited on it as a flood subsides. But when that flood began, volume and velocity increased and sediment previously at rest began to move along the bed or was picked up and carried downstream. In other words, the bed was being lowered by erosion. The wavering changes from erosion to deposition and back to erosion leave their record in the alluvium.

Most of the alluvium in stream valleys is only temporary. It stays for a time but is continually moved downstream. Eventually most of it reaches the sea. These relations will be seen in Figure 4-9.

The Mississippi River system. For a broader look at a whole stream system as an agency for sorting and transporting rock particles, there could hardly be a better example than the Mississippi River and its countless families of tributaries. The Mississippi River watershed, the land drained by the entire system of main river and

Figure 4-7. Sediment moves downslope and into the streams of the vast Mississippi River watershed. At the mouth of the main river the sediment builds a delta into the Gulf of Mexico. The part of the delta that has been built up above sea level extends far upstream, but the submerged part (see arrows) is bulkier.

tributaries, occupies 41 percent of the area of the conterminous United States, plus a small area in Canada (Fig. 4-7). It stretches from the Rocky Mountains to the Appalachians and from just south of the Great Lakes to the Gulf of Mexico. The water that falls as rain and snow on this vast area drains eventually into the Mississippi, passing through a series of narrow channels from headwater streams to the mouth of the main river.

This water always contains sediment, which it is sorting busily. The sediment, derived from the regolith and bedrock of the entire watershed, continually moves through the system at rates that vary from time to time and from place to place. It passes down the rivers to the sea, where fine sand and silt settle out as the river water diffuses into the Gulf. This sediment builds and daily adds to the huge Mississippi Delta that grows at the river's mouth. The finest sediment settles out on the floor of the Gulf beyond the delta itself.

Thus the rock material of even the greatest land area is removed, transported along well-defined lines, and from the mouth of the system is discharged into the sea and spread out on the sea floor, where it forms layers of sediment, much of which will later become rock. During their transport particles derived from the Rocky Mountains are shuffled and thoroughly mingled with particles brought from Minnesota and from the Appalachians. Some of the particles are distinctive enough to be recognized and traced back to the rocks from which they were eroded. By such means it might become possible in the distant future, when the configuration of mountains and rivers in North America will have become quite different from that of today, to re-create in broad outline the river system of the time in which we are living. Deltas similar to our present delta and marine strata similar to ours have been built again and again, on all continents, and later partly destroyed, throughout geologic history.

By measuring the amounts of sediment in samples of water from the mouth of the Mississippi, we can calculate the aggregate volume of the rock particles carried from watershed to ocean during a year's time. Spreading this volume over the area of the watershed, we find that on the average erosion is lowering the surface of the watershed by about one foot every 5000 years. If we assume that the average altitude of the watershed is 1200 feet and that this rate remained constant, the theoretical time needed to reduce the watershed to sea level would be 6,000,000 years.

But the actual time would be greater. As long as the level of the sea remains unchanged, lowering of a watershed by erosion reduces the slope between the headwaters of a stream system and the mouth of the main stream. Each stream flows more slowly on the gentler slope, and transports correspondingly less sediment in a unit of time than it did previously. For every foot by which a watershed is lowered, then, the rates of transport by its streams decrease. Hence the rates of erosion we measure today are not constant, as we assumed in the preceding paragraph; they diminish with time. So our projected 6 million years for eroding a great middle section of the United States down to sea level is too short.

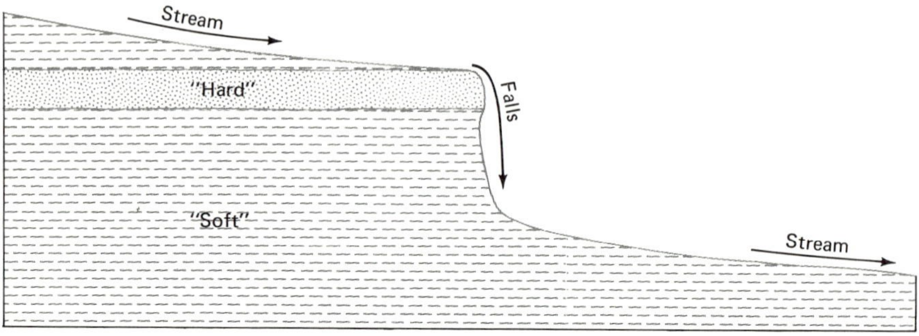

Figure 4 - 8. One of several kinds of local irregularity found in the profiles of streams. Here the presence of a hard, less erodible layer such as sandstone in a sequence of softer, more easily erodible strata such as claystone, delays downcutting and creates a cliff over which the stream falls. Niagara Falls is a pronounced irregularity of this kind.

Profile of a stream. If we plot the profile of a stream, regardless of size, from head to mouth, we find that it is a curve concave-upward. Of course there are irregularities, caused by the occurrence of "harder" and "softer" rocks that yield to erosion at different rates (Fig. 4-8) and by local movements of the Earth's crust, but the normal form is concave. The profiles of young streams are steep curves, whereas those of long-established streams (other factors equal) are gentle curves. This reinforces the truth of our statement that through the history of a stream or system of streams, profiles progressively become gentler.

Figure 4-9 illustrates this progressive change and shows that at any time during the process the amount of sediment carried through the system to the sea and deposited there matches that of the rock material eroded from the land. It shows also that the deposited strata become younger and younger toward the top, whereas the stream profiles become younger and younger toward the bottom, as indicated by the two series of numbers, which indicate relative time. The figure sketches the position of hilltops standing above the stream valley at the time of Profile 1. As the main stream cuts downward, so do its tributaries, and so also do mass-wasting processes lower the hills around them. The whole surface of the land is lowered and the hillslopes become gentler. If the process continues long enough, the surface can be worn down close to sea level, reaching a condition characterized by low, gently sloping hills and broad valleys with very gentle profiles. Such a surface is a *peneplain.* The

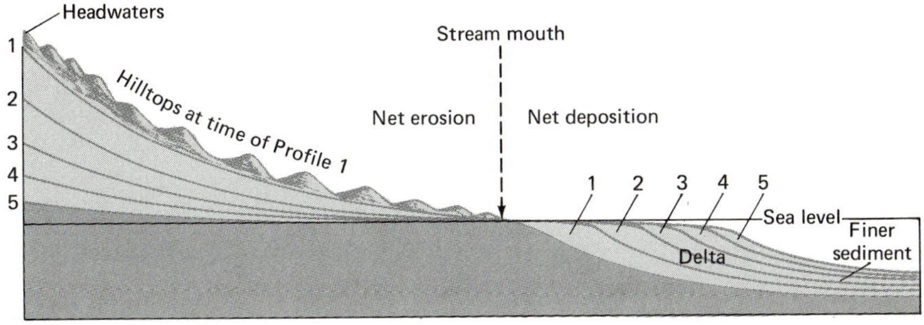

Figure 4 - 9. With the passage of time the profile of a stream becomes an ever-gentler curve, passing successively through positions 1, 2, 3, etc. The mouth of the stream, at sea level, remains almost unchanged as the profile upstream from it becomes gentler. At the time represented by position 5, the rock material shown in light gray has already been removed by erosion; that shown in dark gray remains to be eroded.

In the sea, the sediment deposited up through the time of profile 5 on the land is shown in light gray. Its volume should be about equal to that of the rock material eroded from the land. (Diagrammatic; not to scale.)

word means "almost a plain" but emphasizes smoothness too much. Peneplains are gently hilly like the example described below.

Peneplains. Although broad in area, a peneplain is a vulnerable thing. When uplift of the crust steepens slopes and starts a new cycle of erosion, the peneplain will be attacked; ultimately it will be destroyed. But if a segment of the crust with a peneplain forming its surface is bent down, at least part of the peneplain is likely to become buried beneath new sediment that will preserve it from erosion for as long as it does not become re-exposed. A peneplain buried in this way underlines strata of Cretaceous age (Table 3 - A) beneath much of the Atlantic Coast of the United States (Fig. 4 - 10).

Figure 4 - 10. The buried peneplain beneath the east coast of the United States. (Diagrammatic and not to scale.)

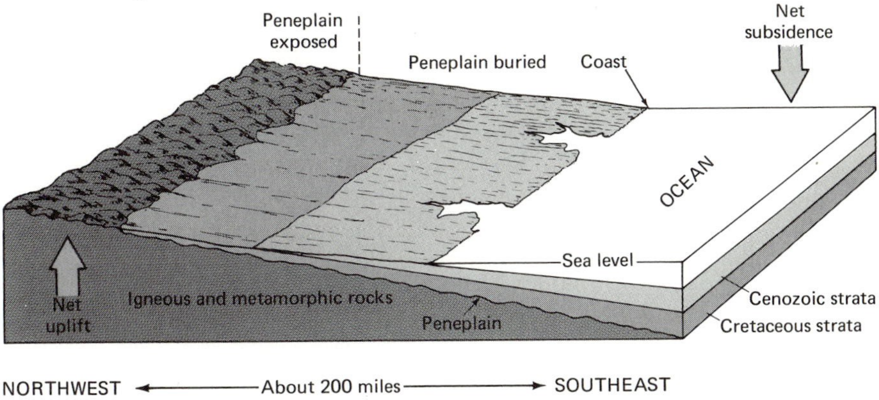

In the landward direction, where the strata that formerly buried it have been stripped away, the former peneplain is at the surface, still only moderately altered by renewed erosion. In the seaward direction it is still buried, but is encountered in deep well borings and by geophysical sensing beneath the surface. Its relief (the vertical distance between hilltops and valley bottoms) exceeds 400 feet in places, but is generally much less. As might be expected, hills are developed in the tougher, less erodible kinds of bedrock while lower areas are underlain by the weaker, more erodible materials.

The historical events implied by Figure 4-10 are these:

1. Creation of the peneplain, by streams and mass-wasting, through long-continued erosion possibly lasting as long as 100 million years.
2. Bending-down of the crust toward the southeast, causing submergence of part of the peneplain beneath the sea. Deposition of a thick body of marine strata of Cretaceous age.
3. Further bending, permitting the deposition of marine strata of Cenozoic age on top of the Cretaceous strata.
4. Uplift in the landward direction, causing the landward parts of the marine strata to emerge, steepening the slope, and stripping away the emerged strata by erosion, thus re-exposing the peneplain. Today the seaward inclination of the peneplain is a few tens of feet per mile. All these events can be read from a long, close look at Figure 4 - 10. Such interpretations constitute one of the chief ways by which the history of the Earth is read from the rocks.

A closely integrated system

Although the actions of rain, weathering, and streams are quite different, they are intimately related. An obvious relationship is that between annual rainfall and number of streams per square mile. In a desert, streams are widely spaced (and also dry most of the time); in rainy territory, streams are close together. If we neglect other factors, such as permeability of rock material, that influence spacing, we can say that in any area the spacing of streams is just that which will carry away the amount of rainwater that must run off over the surface—no more and no less. If there were more rainfall, more streams would be created; if less, some of the existing stream valleys would dry up and in time would probably be filled with mass-wasted material from the adjacent slopes.

Another relationship exists between the rate at which a stream lowers its valley by erosion (as in Fig. 4-9) and the rate of mass-wasting and weathering on the sideslopes of the valley. If because of uplift of the crust or some other activity the profile of a stream is steepened, the sideslopes must become steeper too. As a result, rates of creep and other processes of mass-wasting of weathered regolith must increase. But if regolith moves downslope at a faster rate, it must become thinner. The thinner regolith becomes, the more readily chemical weathering can attack the fresh rock beneath it, thus converting the rock into regolith at a faster rate.

All these activities are related in a sensitive way. Together they form a single integrated system. A change in one part of the system affects all the other parts as well.

Streams in the overall scheme

Probably most people think of the importance of rivers and smaller streams in terms of practical aspects of daily life: water supply, irrigation, trade routes, and recreation. Undoubtedly primitive man depended on streams and lakes for his water needs. Even today 75 percent of the water used in the United States comes from streams

Figure 4-11. Valleys and hills making a systematic pattern, San Gabriel Mountains, California. (Jerome Wyckoff.)

and lakes and only 25 percent from beneath the ground. Likewise rivers have provided water for irrigation and paths for trade for thousands of years. But in our appraisal of the Earth's dynamics and the Earth's history we think of streams in a broader way. Streams (along with mass-wasting) are the chief agencies that shape landscapes, by cutting valleys and leaving hills between them. Although some deserts with their heaped dunes of sand are shaped mainly by winds, most of the world's deserts are marked by telltale valleys. These indicate that even in deserts shaping of the surface is done mainly by streams. Some territories, particularly in North America and Europe, have been shaped by glaciers during a recent ice age. But by far the greatest area of the world's lands consists of systematic (and sometimes monotonous) patterns of valleys and hills (Fig. 4 - 11), the unmistakable work of streams.

Again, streams are the chief means by which rock material is transferred from lands to seas. They are an essential link in the rock cycle, a link that furnishes the material for the making of sedimentary rocks.

References

Gilbert, G. K., 1886, The inculcation of the scientific method by example: American Jour. Sci., v. 31, p. 284–299. (Reprinted, p. 24–32 in Cloud, Preston, ed., 1970, Adventures in Earth history: W. H. Freeman and Company, San Francisco.)

Hubbert, M. K., 1967, Critique of the Principle of Uniformity: p. 3–33 in C. C. Albritton, Jr., ed., Geol. Soc. America Special Paper 89. (Reprinted, p. 33–50 in Cloud, Preston, ed., 1970, Adventures in Earth history: W. H. Freeman and Company, San Francisco.)

Longwell, C. R., Flint, R. F., and Sanders, J. E., 1969, Physical geology: John Wiley & Sons, New York, p. 135–313.

Mather, K. F., 1964, The Earth beneath us: Random House, New York.

5

Seas, Coasts, and the Growth of Continents

River-borne sediment enters the sea

In Chapter 4 we followed the travel of sediment across a continent, down a river to the sea. Once in the sea, the river-borne sediment is mixed with other sediment that has been washed by waves from coastal rocks. The mixture is sorted, spread out, and deposited as strata. Once we understand how this is done we can compare the strata being formed today on the sea floor with ancient sea-floor sedimentary strata, and so visualize the environments in which the ancient strata were made. We can best start with the river sediments, some of which form deltas (Figs. 4-6, 4-7) at river mouths. As the illustrations show, the landward part of a delta, the part that is built up a little above sea level, is only a small part of the whole. As it diffuses into the ocean, the river water flows more and more slowly and the sediment it is carrying settles onto the sea floor in order of decreasing particle size. The finest sediment, which is suspended in the water, settles so slowly that some of it is carried far out from the land before it reaches the bottom. However, the rock particles do not necessarily drop straight to the bottom. In the shallow water near the coast the sea is far from passive. It is continually moving because it is agitated by waves and currents. These pick up particles that have reached the floor or are on their way down to the floor and shift them from one place to another. During the shifting and re-shifting the sediment continues to be sorted, beyond the sorting it was given by the rivers it passed through. As on the bed of a stream, the rock particles are picked up and dropped again and again. Sooner or later they are shifted to positions so deep that wave motion ceases to reach them, and there they stay. By this time they are thoroughly spread out and thoroughly sorted, forming layers that will eventually become rock strata.

The surf attacks coastal rocks

During this shifting and redepositing, the river-borne sediment is mixed and shuffled together with rock particles of different origin. This second lot of particles is eroded from rocks exposed along the coast (Fig. 5-1). The rocks are attacked by the turbulent surf that results when waves break against the land. Especially during storms surf is a hammer that can strike bedrock with a force measured in tons per square foot. The air in cracks is compressed by this force and the rock is split apart. Sediments, having much less resistance, do not have to be pried apart. They are simply washed away. By splitting and by washing away the surf cuts into the coast like a saw cutting horizontally into a standing tree. A tree eventually falls all at once. But the substance of the shore rocks falls piecemeal as the

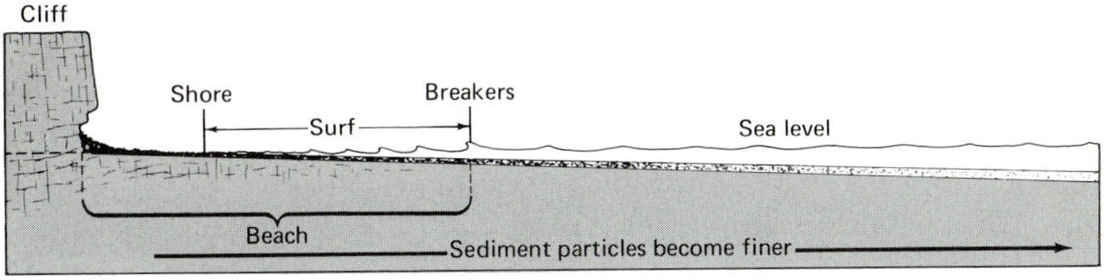

Figure 5 - 1. The surf at work on shore rocks as both a saw and a mill. As a net result, rock waste is gradually transferred seaward, in the direction of the arrow.

surf slices in at sea level and leaves overhanging cliffs unsupported. The fallen rock particles are the beginning of a beach. The waves, swashing up and sliding back again, move them back and forth, the smallest first, continually knocking and rubbing them together until they become smooth, round, and smaller in size. Sooner or later even the largest pieces are worn down by abrasive sand and pebbles to sizes small enough to be moved a little during the biggest storms. Then it is only a matter of time until they are worn or broken down to small, manageable size. Then they join the procession of other particles that move back and forth on the beach.

In what direction is the procession headed? Over the long pull the movement of sediment is seaward. The beach is a broad, thin carpet of sediment that begins at the base of the coastal cliffs and, passing under water, extends seaward to the line where waves break and become surf, stirring up the bottom all the way to the shore. The beach becomes finer-grained seaward. In many beaches cobbles and pebbles are near the cliff, with sand farther out. As the water deepens seaward, the energy in the surf decreases and so the maximum particle size that can be readily picked up by the surf decreases also. The fine particles, silt and clay, remain suspended in the water until they are dropped in deeper water seaward of the breaking waves. This is another grand example of sorting by natural forces. Like the sediment on the bed of a river, which gradually becomes finer in the downstream direction, the sediment of a beach becomes finer in the seaward direction. This sorting, preserved of course in sedimentary strata, gives us clues to the flow directions of ancient streams and to the positions of former coasts.

Sorting and rounding of the sand and pebbles on a beach and of bottom sediments beyond the beach is very thorough. Its thoroughness results mainly from the fact that before reaching their final resting places the particles are moved countless times, not merely back and forth in the surf, but alongshore as well. Waves, which depend on wind direction, are rarely exactly parallel with the shore. Usually the surf reaches the shore at an angle. Each sand grain and each pebble in the surf is therefore carried up the beach on a slanting path. But as the water washes back, the particle rolls seaward down the steepest path at right angles to the shoreline. In this way the particles of sediment in the surf move alongshore, sometimes for many miles, following zigzag paths as they are impelled by the swash and backwash of each wave (Fig. 5-2). If the wind changes and waves reach the shore at a different angle, the sediment may

Figure 5-2. Surf and beach at Point Reyes, north of San Francisco, California. The broken line with arrowheads suggests in a generalized way the zigzag path of a grain of sand as it is carried along the beach toward the camera. (M. W. Williams, U.S. National Park Service.)

travel in the opposite direction. The system is a great mill, grinding the particles to smaller sizes and rounding them more and more.

The sorting that goes on farther seaward is like the sorting of a delta. It depends mainly on depth of water and is part of the reason why sandstone, siltstone, and claystone are so well differentiated from each other in many successions of sedimentary strata.

Along coasts, then, both a saw and a mill are at work. The saw cuts into the land, the cliff retreats, and the waste from the cliff is given rough treatment in the mill. Whether or not it traveled down a river at an earlier time, the waste is sorted, carried seaward, and spread as strata on the sea floor.

Sediments on the deep-sea floor

Up to now we have said nothing about how far seaward the waste from the land, whether transported by a river or knocked about by surf, is carried. Yet the matter is important if we want to understand the history of sedimentary strata that accumulated on the sea floor. The way to find out is to sample sea-floor sediment at various distances from the land, and then to compare the samples as to origin.

Accurate sampling of such sediment involves many technical difficulties. Getting an undisturbed sample from the bottom, beneath an enormous thickness of seawater, is no easy matter. Only in recent decades have techniques for bottom sampling been developed so that we have a fairly good picture, not only of the bottom itself, but of what lies beneath it. Figure 5-3 shows the great difference between the continental shelves, over which depth of water averages less than 600 feet, and the deep-sea floor, which averages 2.5 miles and in some low places is more than 6.5 miles in depth. These two contrasting provinces have had different histories. We must examine them separately, beginning with their sediments.

The area of the deep-sea floor is five times as great as that of all the shelves combined. Much of what we know about the sediments at those great depths we have learned since about 1950. At that time there began a great international program of drilling holes in the ocean floor all over the world. The drills are not of a conventional kind. They are coring devices somewhat like the kitchen tool for removing the core from an apple. The device is lowered on

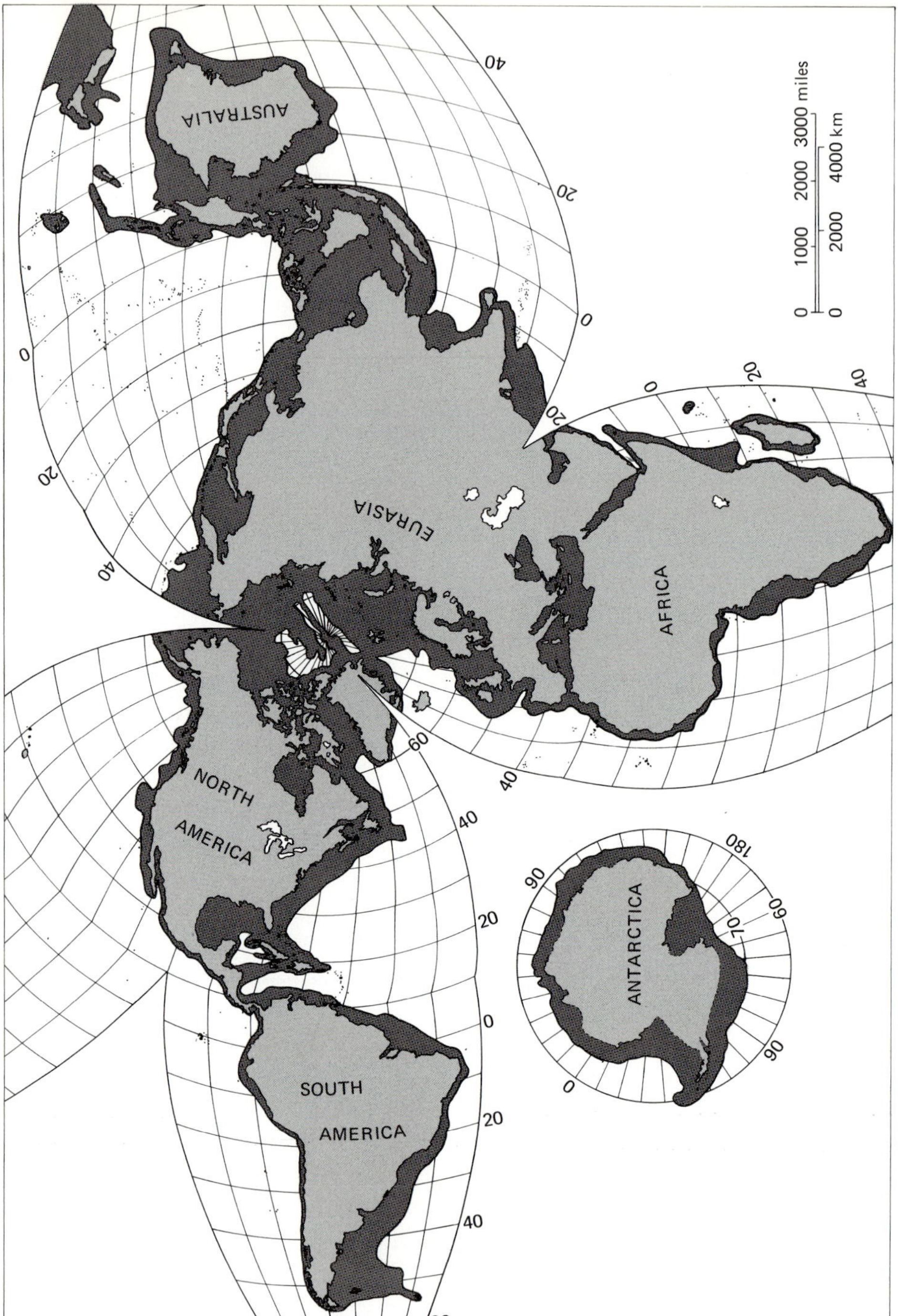

Figure 5 - 3. Continental shelves (black) together add about 10 percent to the areas of the continents. The map has been drawn on an unfamiliar projection in order to avoid distorting the areas of continents and shelves. All are shown on a nearly uniform scale. (Shelves sketched from Menard and Smith, 1966, *Journal of Geophysical Research*, vol. 71, p. 4306-4312. Base [except Antarctica] by permission. © University of Chicago, Dept. of Geography.)

the end of a cable to the bottom, driven into the sediment, and raised with a core of sediment inside it. Most cores are less than a few inches in diameter and some of them are more than 100 feet long. The sediment is layered, and some of the long cores contain layers that represent continuous deposition through tens of thousands of years of geologic time.

The sediment brought to the surface is so fine that 19th Century scientists (who obtained their samples by scooping sediment from the bottom with small buckets) called it *ooze,* an appropriate name for the sediment when it is saturated with water. The word has gone out of use, but it conveys a good idea of the weak, jelly-like consistency of the sediment.

After being dried, the cores are sliced into small samples for study under a microscope. When greatly magnified, the sediment is seen to consist of an array of varied and distinctive particles. After being sorted out and classified into three groups, the particles reveal where the sediment came from and how it was formed. A first large group, which is the greatest in quantity, consists of tiny shells and skeletons of marine animals and plants—a huge number of fossils of very small size (Fig. 5-4). Evidently this sediment is local, having been formed in the ocean itself. Many of the shells consist of the mineral calcite ($CaCO_3$). The calcium it contains has a long history. It is a product of the chemical weathering of rocks in the continents. Dissolved in ground water, it seeped into rivers, which carried it onward to the sea. There the small oceanic organisms extracted it and combined it with the carbon and oxygen dissolved in sea water as carbon dioxide. With this combination the organisms built their shells of calcite.

A second group of fine particles is foreign to the ocean, having been derived from the lands with little or no chemical change. But these particles originated in different ways. Some of them are clay minerals that formed during the weathering (Chapter 2) of feldspars and other minerals at the surfaces of continents and were brought to the sea by rivers. They are so extremely small that they were carried far from the lands of their origin before they came to rest on the bottom. Still other clay particles, likewise the product of weathering, were picked up by winds from dry places and were blown seaward, only to fall into the water and make a slow descent to the sea floor. Other fine particles were shot into the air from volcanoes and eventually reached the bottom of the sea. These, of

Figure 5 - 4. Shells of sea animals so small that they make a dime look as big as a saucer. The shells, built of calcium carbonate, were washed out of very fine sediment ("ooze") that had been scooped up from a depth of 3000 feet on the floor of the Caribbean Sea. (Yale Peabody Museum.)

course, are not products of weathering. They are "new" minerals derived originally from magma.

This second group of sea-floor sediment particles, then, originated on the continents and was delivered to the deep sea by rivers or by transport in the atmosphere. A third group, insignificant in quantity, consists of minute particles that entered the Earth's atmosphere as meteorites and eventually dropped into the ocean. These originated neither in the ocean nor in the continents; they are strangers to the Earth, travelers from distant outer space.

Sediments in the continental shelves

Having sampled the sediments in the ocean depths and having sorted out their ingredients, we are ready to look at what the continental shelves have to show. Our samples from the shelves are more varied. We have not only cores and other samples from the shelf surfaces but also the records of drillings deep into the shelf bodies. Both the surface samples and the deep samples tell essentially the same story: The shelves are not at all like the deep-ocean floors. They consist mostly of layers of sand, silt, and clay, and of their cemented equivalents sandstone, siltstone, and claystone. Hence some of the sediment is in the form of rock and some of it is not yet cemented. Almost all this sediment was derived from the continents by the various kinds of weathering and was transported, mostly by rivers, to the shallow water over the shelves. There the sediment particles, being mostly comparatively coarse and therefore rather heavy, soon settled to the bottom. These things are happening today and they have been happening throughout past ages, as the strata down within the bodies of the shelves tell us. We see that now and in the past sediment poured into the sea by rivers settles out on continental shelves, leaving only a small proportion of the finest sizes to reach the floor of the deep sea. Hence, although rivers are by far the greatest movers of sediment from continents to oceans, they lack the ability to move sediment far beyond the continental coasts. So as a continent is eroded its rock waste is carried outward by rivers and settles out as an apron or fringe close around the land mass. The waste goes into the building of shelves and gradually adds to the size of the continent.

The way shelves grow

We can get a clearer idea of how shelves grow if we look more closely at a particular example, that part of the Atlantic shelf of the North American continent that lies off North Carolina. Thanks to geophysical sensing beneath its surface and to deep holes drilled in it in the search for oil and natural gas, we have an unusual amount of information about it.

In Figure 5 - 5 the fringe of marine sediment beneath and just off the coast is visible. The sediment is thick—at least two miles in places—yet when examined in detail it is seen to have physical characteristics and fossils that indicate a shallow-water environment. The strata simply could not have been deposited in water two miles deep. How then could they have accumulated to such thicknesses as two miles? We can solve this puzzle if we suppose that along the continental margin the crust slowly subsided while deposition proceeded, one layer after another. If that happened, the water need never have been deep at any time. A glance at Table 3 - A tells us that if subsidence occurred it must have happened very slowly. We can say this because the Cretaceous Period, during which much of the sediment was deposited, lasted more than 80 million

Figure 5 - 5. The coastal plain and continental shelf in North Carolina are underlain by Cenozoic, Cretaceous, and older strata deposited in a shallow (shelf) sea. Beneath these strata, which have been bent slowly downward and are very thick, are metamorphic rocks of Paleozoic and even greater age. Note similarity to Figure 4 - 10. Deposition of the strata that form the shelf has increased the width of the continent. (Data from publications of W. F. Prouty, K. O. Emery, and other authors.)

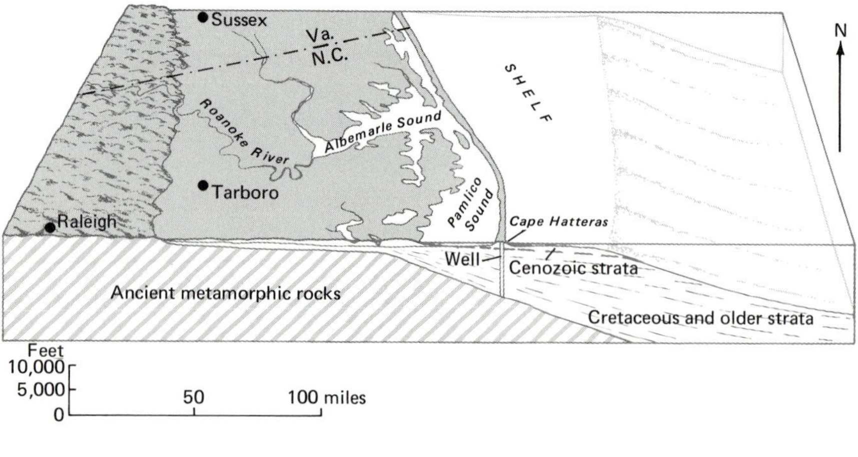

years. The overlying Cenozoic strata are comparatively so thin that it looks as though a great deal of the subsidence and deposition had taken place before the later Cenozoic sediments began to be deposited. If (looking again at Figure 5 - 5) we suppose that a thickness of 10,000 feet of sediment was deposited in a period 80,000,000 years long, the average rate of deposition would have been 1 foot in 8000 years or 1 inch in 666 years. Such a rate, 1/10 inch in the adult lifetime of one person, is almost negligible in human terms. But through a time span of geologic proportions the accumulating sediment becomes very thick and very heavy, and this has important consequences. Especially important is the weight of these tremendous loads of rock matter. It causes the shelves to sag downward, leaving room for still more waste to be deposited on top.

Also, at some places along coasts (although not in the part shown in Figure 5 - 5) the strata beneath the shelves have been bent so as to create long narrow basins parallel with the shore. Such basins are pockets in which sediment arriving later is trapped. Their presence furnishes another reason why most of the rock waste shed from the continent through erosion is deposited close to the continental edge rather than being carried farther seaward. Indeed it has been calculated that about nine-tenths of all the sediment poured from the lands into the sea comes to rest on the continental shelves and their seaward slopes. Only one-tenth is fine enough to be carried onward and deposited on the deep-sea floor.

Evidently, then, sediment cast off from a continent comes to rest around the continental margin. The sediment forms shelf-like masses that gradually extend the continent outward. Its volume is huge. The combined areas of the world's shelves at the present time equal about one-third the area of all the world's lands. But not all the material that forms the shelves is still in the condition of sediment. A good deal of it has been converted into sedimentary rock. The change is mainly the work of slowly circulating ground water, which has gradually deposited cementing substances in the spaces between the grains of sediment. In time other processes will lift up these "new" sedimentary strata, converting them into an integral part of the continent itself.

This does not by any means end the history of continental growth. The newly added rock has now become part of the land and is subject to the destructive forces of erosion. Inevitably its waste will be carried seaward again to help build new shelves.

We can predict with some confidence that this will happen because the record shows that it has happened again and again in the past. Most of the marine sedimentary strata that we find in the continents today (strata like those in Figure 3-1) represent the kinds of sediment that are deposited in shallow water—that is, water less than about 600 feet deep—not the kinds we find on the floors of the deep oceans. Their grain sizes, the features of their stratification, and the fossils they contain all tell us the same story. We are dealing mainly with shallow-water marine strata, whether deposited in a shelf around a continental margin or in some trough or basin in a continental interior.

This means that much continental rock matter is being recycled without having much to do with the deep ocean. Indeed the continents seem to have been added to, through periods billions of years long, by accretion; that is, by incorporation of new rock into the continent, as sediment deposited piece by piece. But the pieces of new rock are mostly shelf materials consisting of compacted waste derived from erosion of the continent itself. In a sense, therefore, the continent has been rebuilding and enlarging itself partly by feeding upon its own waste.

But this is not the whole story. The fine sediment from the floors of the deep oceans does get back into the continents, although in a much changed form, and in a remarkable way that we shall describe in Chapter 6.

Successive additions to the North American continent

If this has actually been the history of a continent such as North America, and if the principal pieces added to the land mass were former continental shelves, then the oldest rocks of the continent should be found in its central part and the youngest ones near its coasts. The radiometric dating of rocks described in Chapter 3 confirms this deduction. The dates of more than 100 samples of granite and metamorphic rocks exposed in North America, when plotted on a map, form a distinct pattern (Fig. 5-6). Near the center of the continent all such rocks are more than 2.5 billion years old—more than half the entire life span of the Earth. Encircling these central

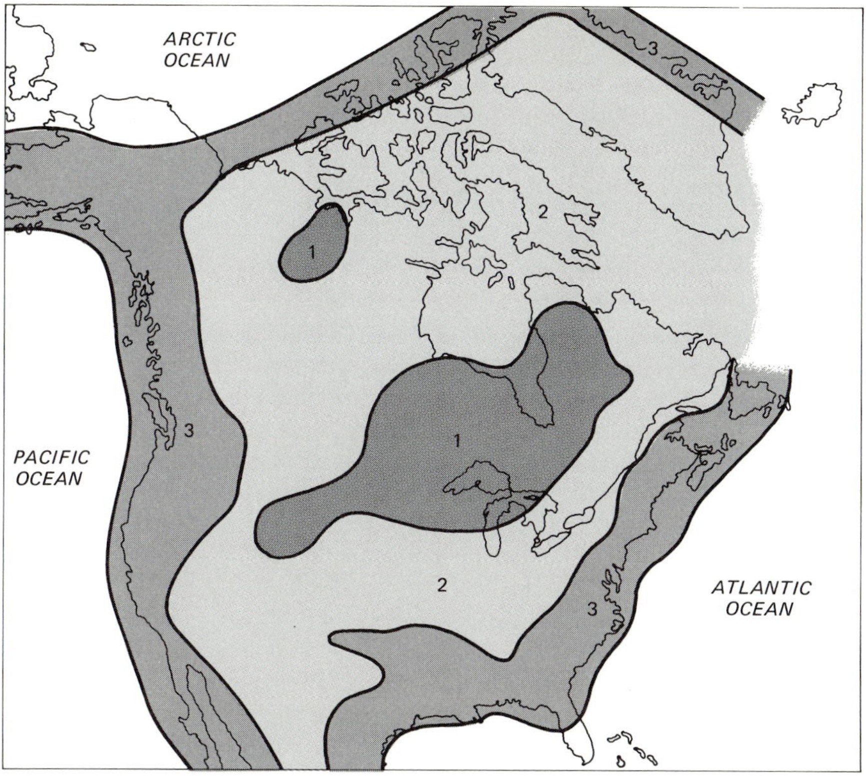

Figure 5 - 6. Three geologic provinces in North America were built successively, as shown by the ages of granitic and metamorphic rocks determined by radiometric dating. Together they form a concentric pattern. Explanation:

3. Less than 600 million years old.
2. Between 1 billion and 2.5 billion years old.
1. More than 2.5 billion years old.

(After Engel, 1963, figs. 2, 4.)

rocks is a wide belt of younger ones, all of which are more than 1 billion years old. Finally a third belt at the margins of the continent consists of rocks less than 600 million years old.

Six hundred million years ago is a critical date, because it approximately marks the beginning of Phanerozoic time (Table 3 - A). That is the time span represented by the fossil-bearing strata that in much of North America thinly cover older metamorphic rocks. We can read the main events in the history of Phanerozoic time directly from the strata and their fossils. But beneath those strata the older metamorphic and igneous rocks, with their dates, tell us that North America was built by the addition of ring-shaped belts to its outer part. (However, continents have lost pieces of themselves also.) Furthermore, each of the belts of rock shown in Figure 5 - 6 has been more thoroughly sorted than the belt next inside it, because its materials have passed more times through the rock cycle. Thoroughness of sorting is indicated by the minerals the rocks contain (each mineral having known amounts of known chemical elements).

Chemical sorting

The sorting has been of two kinds. First and more obvious is the mechanical sorting performed by streams as they carry sediment from the continental interior toward the shelves, and by waves and currents after the sediments have reached the sea. Second, less obvious but just as important, is chemical sorting. This is done, not by streams, but by chemical weathering in which chemical elements are dissolved and are carried to the ocean first in ground water and then in streams. Some of the elements in solution are precipitated chemically or biochemically in shallow basins or on shelves, while others continue to the deep sea. The residual material left behind is different from the rocks that existed there before weathering because of the loss of some of its elements.

This chemical kind of sorting is rather like what happens to the body of a wild animal after death. The body tissues are eaten by various scavengers or converted by bacterial decay. The tissues are, in effect, carried away. What is left behind at last consists of durable parts: bones, teeth, and perhaps horns, hooves, or claws. The chemical compositions of those durable parts are different from the average composition of the body of the living animal. What has happened is biochemical sorting.

But although they have been sorted, the carbon, hydrogen, oxygen, and other chemical elements that constituted the dead body still exist as constituents of the bodies of other organisms. Although one animal has ceased to exist, life goes on as before. In a similar way a continent continues to exist. Its rocks and its minerals are being destroyed, but the elements of which those minerals were built are being sorted and re-sorted into different combinations, creating new minerals and new rocks. And so the continent survives.

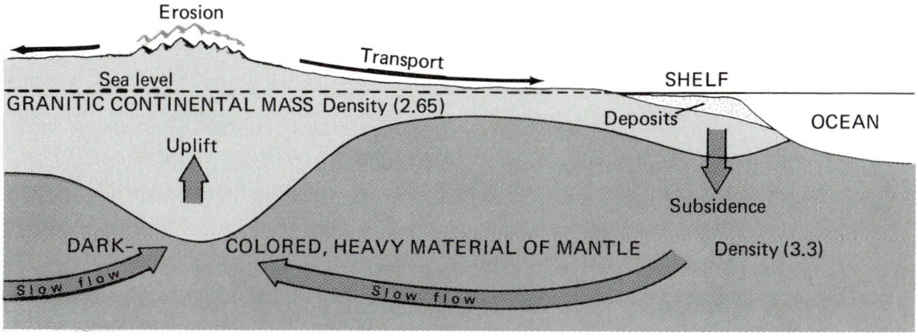

Figure 5 - 7. Granitic continent floating in the heavy material of the mantle. Its thickness is greatest where its upper surface is highest. Erosion of the high mountain mass removes sediment and reduces pressure on the heavier material beneath. Waste from mountains is transported and deposited in the continental shelf, creating an extra load. This causes subsidence, pushing mantle material aside. That material flows slowly toward area of reduced pressure beneath the eroded mountains. The result is uplift that will renew the mountains.

Floating continents

We said in Chapter 1 that a continent is a kind of granitic raft floating in basalt and in material even heavier than basalt. This is shown in a much idealized way in Figure 5 - 7. Now we can understand that the raft is self-perpetuating. It would not be accurate to compare a continental raft with a rigid wooden raft, because the floating continent is not rigid. It is flexible, and because it is flexible it more nearly resembles a thick floating slick of heavy oil or asphalt. There is, however, an exception to continental flexibility. The uppermost part of the continent, which bears very little weight of overlying rock, is not flexible but brittle. As anyone can see in a stone quarry or other large excavation, its surface is marked by many cracks.

The fact that continents float explains something that otherwise would be a puzzle. The well-known team consisting of weathering, mass-wasting, and streams is believed capable of stripping a continent down nearly to sea level in no more than 20 million years. Since the Earth is far older than that, why haven't continents been reduced to great peneplains long since?

What happens is illustrated by a canoe containing three people. It glides up to a dock and one person steps out. With the removal of his weight the canoe rises a couple of inches, as water flows in from the area around the canoe to replace the water that had been displaced by the weight of the third person. Similarly, with the removal of rock material from part of the surface of a continent, that part of the continent rises. The rock material removed is deposited on the continental shelf and areas of the shelf subside. Underlying matter flows from beneath the subsided areas to new positions beneath the region from which rock was eroded. That region rises correspondingly.

But whereas the canoe is rigid and rises as a unit, the continent is flexible. The part of it that has lost most by erosion can rise, while other parts rise less or not at all. This explains why the continents have not all been reduced nearly to sea level, forming vast peneplains. As they are worn down, they are repeatedly renewed. They have continually kept themselves from being wholly submerged, so that living things, once established on the lands, have been able to survive and evolve instead of being wiped out. Indeed the continuity of life on the lands is pretty good proof that floating continents have been in existence during at least the last half billion years.

In this chapter we have established that continents are lighter in weight than the basaltic stuff in which they float. Heft a chunk of granite in one hand and a chunk of basalt of equal size in the other and feel the difference. We have seen that the rock cycle, led off by chemical weathering, has recycled the material of the continents without changing overall composition appreciably.

How did the continents come to be lighter than the rest of the crust in the first place? The answer to this question is related to their physical history, which forms the subject of the next chapter.

References

Engel, A. E. J., 1963, Geologic evolution of North America: Science, v. 140, p. 143–152. (A similar paper is reprinted in Cloud, Preston, ed., 1971, Adventures in Earth history: W. H. Freeman and Company, San Francisco, p. 293–302.)

Longwell, C. R., Flint, R. F., and Sanders, J. E. 1969, Physical geology: John Wiley & Sons, New York, p. 315–393, 534–539.

6

Origin and Development of Planet Earth

The Earth's beginning

Lithosphere. Although there is reason to believe that Planet Earth is about 4.5 billion years old, we are not yet sure of the way it and the Solar System of which it is a part came into being. All we have is a theory, and as it is brought into line with newly learned facts about the Solar System, the theory keeps changing.

In its broadest essentials our theory says that the entire Solar System—the Earth, the other planets, and the Sun—formed as a family at the same time. It says also that the formative process was not unique or even unusual, but is apparently common within the physical universe and indeed is happening today. The stuff from which the solar system was built was a nebula, a huge, whirling, disk-shaped cloud of cold gases and "dust." This cold, diffuse material gradually condensed into a group of individual clusters, which ultimately became the Sun and the planets. The cluster that later became the Earth, now 8000 miles in diameter, then had a diameter of several million miles, so far apart were its particles at that time. But the particles came closer together through gravitative attraction and lost energy in countless collisions. The heavier particles "fell in" like hailstones toward the central part, and this gravitational compression created so much heat that the mass finally collapsed into itself, forming a solid body.

But the solid body of the Earth was very hot, much hotter in all its parts than it is today. We suppose that the particles of iron in this young, hot, compacting Earth melted and formed droplets. Being especially heavy, the iron droplets sank toward the Earth's center. There, with other minerals, they accumulated and formed what is now the Earth's core.

The less heavy particles, having "fallen in" less far, formed a thick layer surrounding the heavier core. The layer was rocky rather than metallic and consisted of silicate minerals. It was what today we call the mantle, described, together with the core, in Chapter 1.

The escape of heat from the hot interior was impeded (and is still being impeded today) by the thick rocky layer of silicates, which are poor conductors of heat. So the mantle heated up. Although the temperature became high it was unable to melt anything in the deeper part of the mantle because the weight of the great thickness of rock overlying it created too much pressure to permit melting. But in the upper or outer part, where the overlying rock was thinner and the pressure therefore less, the rock began to melt, but only in part. The minerals that melted were only those whose melting points were lowest. Melting, in other words, was *selective.* It created magma. But because the magma was derived from only *some* of the minerals in the surrounding rock, it differed from that rock. Its composition was that of basalt, and it was also a little lighter in weight than the unmelted rock around it.

Being lighter, the bodies of new basaltic magma tended to rise.

There were many bodies and they rose very slowly. But little by little they solidified at or near the Earth's surface, where temperatures were low. As they solidified they gradually formed, over the entire Earth, a crust consisting of basalt.

In summary, the basaltic crust is the product of selective melting. Its composition differs slightly from that of the constituents of the mantle. This selective melting of some constituents of the primitive rock, creating basalt, probably represents the first of the complex processes by which primitive rock has been differentiated little by little into the rock that forms the Earth's crust. A second process consists of chemical weathering. A third process consists of selective melting of basalt to form granitic rock, lighter still than basalt. All three processes, apparently, are still going on. But we shall have to wait until the end of this chapter to explain the matter more fully.

Atmosphere and hydrosphere. So much for our theory of creation of the outer part of the lithosphere. But the atmosphere and the hydrosphere—how were they created? Although at first thought it seems odd, it is likely that these two fluid envelopes originated in much the same way as did the solid crust: through rise of magma to or toward the Earth's surface. Such an origin is possible because all magma contains gases. We said a moment ago that magma was formed in places as a result of melting of deep-lying material by heat emanating from the hot interior. In the process of melting, some minerals readily dissociated; that is, chemical compounds within them parted company. This is a different thing from the sorting of mineral crystals by their different weights. It is the breakdown of substances under the influence of temperature, pressure, and chemical environment in the magma. Among the substances released in this way were hydrogen (H_2), nitrogen (N_2), carbon monoxide (CO), carbon dioxide (CO_2), and water (H_2O). Rising magma brought these substances up to the surface and discharged them there. Thus began the building of the Earth's atmosphere, an atmosphere that at that early time was entirely volcanic. It consisted of methane, water vapor, ammonia, hydrogen, and nitrogen, and perhaps some carbon monoxide and carbon dioxide. These same elements and substances are found in gases emitted by volcanoes today. The most obvious difference between such an atmosphere and the atmosphere that exists at present is that all the oxygen the early atmosphere contained was locked up in chemical combinations. None of it existed in the free form. This is significant because an oxygen-poor atmosphere has an important bearing on the origin of living matter, as we shall see in Chapter 7.

As for the hydrosphere, it is not difficult to explain its creation from an atmosphere that contained abundant water vapor, contributed as volcanic steam. Water vapor would begin to condense and fall as rain. Runoff from the rain would form streams, which in turn would end up as lakes in the lowest parts of the Earth's surface. The lakes would grow and merge, creating an ocean. During the process, the water cycle would have begun.

An atmosphere containing carbon dioxide and water, coming into contact with the new basaltic rocks exposed at the surface, would have caused weathering. The newly created streams would have carried the products of weathering, rock particles and solutes (substances in solution) to the new lakes. The solutes in the water of the growing and merging lakes were the salt (actually a mixture of many salts) present in the new ocean water, the beginning of the salt content in the seas of today.

Although this chain of events seems likely because it conforms well with our knowledge of the Earth, it is still only theory because it is far from fully established. Also, very little is known about the timetable involved, about how long it took for atmosphere and ocean to evolve to their present size and composition. There is reason to believe that no important change in these fluid envelopes has occurred during the last half billion to one billion years at least.

In summary, it seems likely that the constituents of the Earth's crust have been created by a continuous succession of volcanic events accompanied by weathering of rocks, subsidence, and remelting at great depth—in other words, accompanied by the rock cycle. It seems likely also that the constituents of the atmosphere and of the hydrosphere as well have been "sweated out" little by little from the Earth's hot depths. They have moved up as gases dissolved in rising magma and been exhaled through volcanoes. This upward transfer has been going on ever since early in the Earth's history, and "new" water is still arriving at the surface today. By now most of the water exhaled by volcanoes is "old" water, which has passed many times through the water cycle and which, lying in the ground, has been warmed up by volcanic heat. But if only half of one percent of all volcanic water is "new," it has been estimated that at its present rate of arrival at the surface it could have created the entire hydrosphere, ocean and all, within the span of 4 billion years.

This theory of the origin of the Earth's crust, its water sphere, and its gaseous sphere accords very well with the Principle of Uniformity that governs all geologic activity. It leads us to the far-reaching conclusion that even in very early times the same activities were going on in much the same ways as those we see about us at present.

The Moon. The character and history of the Moon have a bearing on our theory of the Earth's origin. Data on the Moon, obtained mostly from the Apollo program beginning in 1969, so far have shown nothing incompatible with the Earth's history as read from geology. Apparently the Moon, a very small member of the Solar System, was formed at the same time as the Earth, and as a body independent of the Earth. Its outer part, at least, seems to have been molten for a time. The bedrock so far found is dark-colored igneous rock, some of it basalt and some of it gabbro-like (Table 2 - B). It shows evidence that as the molten matter solidified, the lighter-weight constituents persistently rose toward the top of the melt while the heavier ones settled toward the bottom. This is what we think happened, to some extent, at least, during the making of the crust of the Earth as well.

But here resemblance to the Earth ends. The Moon is covered with a layer 5 to 10 feet thick of what is popularly called "Moon dust;" we call it Moon regolith. It consists of loose debris caused by mechanical disintegration of the solid rock underneath it. Some of it is in large and small chunks but most of it is powder-like. This disintegration was not caused by mechanical weathering like that on the Earth. It is the result of explosive pounding, resembling the effect of bombardment by enormous bombs. But the "bombs" were large and small meteorites crashing onto the Moon's surface, some of them of enormous size. Impacts of meteorites are the cause of the Moon's craters as well. Of course meteorites strike the atmosphere of the Earth also, twenty million or more each day. But most of them are so very small that they "burn up"—disintegrate—as they are made hot by friction generated by their passage through the atmosphere. From time to time a large meteorite makes it to the Earth's solid surface. When that happens, a crater results.

"Moon dust," the Moon's regolith, shows not the slightest sign of chemical weathering, the commonest feature of the regolith of Planet Earth. This is because the Moon lacks both an atmosphere and a hydrosphere. The lack is not hard to explain. The Earth has enough mass so that its gravitative attraction is strong enough to hold the gases that form its atmosphere. But the Moon's mass is equal to only .012 that of the Earth, not enough to retain gases of any kind. So the surface of the Moon is devoid of chemical weathering, and for that same reason it is devoid of life.

The continents and crust plates

Chapter 1 noted that the continental crust is much thicker than the oceanic crust as well as being different in composition, and Chapter 5 explained how individual continents increase in size and why they are afloat. But up to now nothing has been said about the shapes and the positions of continents and ocean basins. The reason is that continents have moved about and oceans have widened and narrowed so much that our map of the world today is very different from maps of the world of, say, a hundred million years ago. Such big changes would have been very hard to understand if we had not first developed a general picture of the Earth's constitution.

So continents are not fixed but are moving. Although, surprisingly, such movement was first suggested 350 years ago and has been proposed several times since then, the idea was not generally accepted by scientists until well after 1960. Most people thought the crust was so rigid that continents couldn't move. Now we all know that they do. To understand what has been happening let us begin with some of the evidence that movement does occur.

Evidence that continents move. Kinds of evidence that suggest that at least some continents now separated were formerly connected, or that they have changed position, include these:

1. The shapes of the margins of some continents that face each other across an ocean suggest a former fit, like two adjacent pieces of a jigsaw puzzle. The South America/Africa pair is the most obvious example. Because the true outer edge of a continent is not today's coast but rather the outer slope of the continental shelf, the fit should be checked not on an ordinary map but on a map that shows the shelves (Fig. 5-3).
2. Some pairs of continents show fits of other kinds: special kinds of rocks, ancient mountain belts, identical fossil animals and plants. Such fits imply that the two continents were once joined.
3. Fossil animals and plants that must have lived in warm climates

occur in the rocks of some continents at latitudes where today's climates are colder.

4. Strong evidence of widespread former glaciers occurs in strata some 250 to 300 million years old, in eastern South America, southern Africa, India, and Australia, at latitudes where such glaciers could not exist today because of warm climates. Similar evidence occurs in Antarctica, where glaciers do exist today. It looks as though all these continents were formerly parts of a single continent, which broke into pieces. The pieces drifted apart. Some of them drifted into warmer climatic belts, whereas the Antarctica piece edged into its unique position at the South Pole.
5. Evidence of ancient compass-direction indicators locked into many strata shows that continents have moved in relation to the magnetic North Pole. This statement needs explanation:

The magnetic record. When we want to know direction we use a compass. The needle, being magnetized, aligns itself with the Earth's magnetic field; it points toward the magnetic North Pole. Some rocks, likewise, are natural compasses. Crystals of magnetic iron minerals the size of sand or silt particles, carried as sediment in a stream, behave in a similar way. As they settle to the stream bed they align themselves in the north-south direction, and in many cases the alignment is preserved even after the stream sediment has become sedimentary rock.

The magnetic iron minerals in an igneous rock such as basalt have a comparable history. As lava or magma cools and solidifies, it reaches a temperature below which such minerals can become magnetized parallel to the Earth's field. This magnetic orientation (which has nothing to do with the orientation of the crystal itself within the rock) preserves the position of the magnetic field at the time when the rock cooled. The iron minerals, like those in the sediment deposited by the stream, are compass needles fixed in place, pointing toward the position of the pole.

Scientists have found the former compass-direction pointers preserved in strata of many different ages, and have had the patience to measure them at hundreds of localities and at a whole series of stratigraphic positions (which, of course, represent a whole series of times in the past). The resulting directions are different from those a compass would show today at these same localities. The directions indicate that during at least the last 200 million years or so the continents have moved through long distances, not only with respect to the pole but with respect to each other. The path followed by a continent can be traced, not yet accurately but in a general way.

During the course of these studies of ancient magnetic directions a surprising fact appeared. It was found that in many strata deposited during the last few million years (the time span studied so far) polarity was reversed. The magnetic particles in the rock were pointing toward the South Pole rather than toward the North. Radiometric dates of the same strata showed that the times when the reversed patterns were made were the same everywhere. Evidently, then, the Earth's magnetic field, which controls the patterns, had reversed itself, with the two poles exchanging places. There have been many such reversals.

What causes the field to reverse? We do not yet know, but it has been suggested that the changes may be caused by movements within the liquid part of the iron-rich core of the Earth. Whatever the cause, the reversals affect the whole Earth simultaneously.

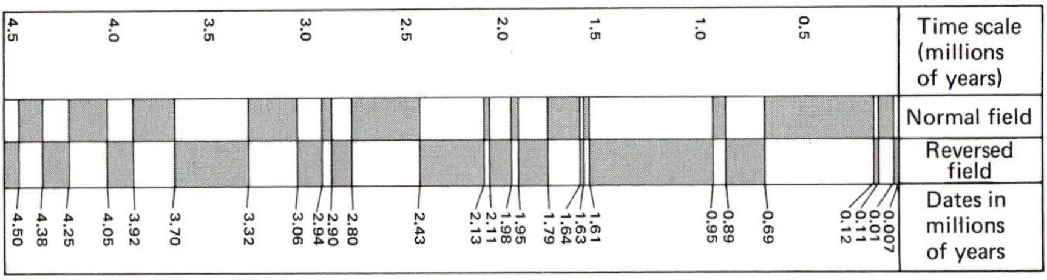

Figure 6 - 1. The time scale of alternating changes in the Earth's magnetic field. It was constructed from the radiometric ages of many rock samples whose polarization had been measured. (After Cox and Doell.)

For this reason it has been possible to construct a worldwide calendar from the reversals. The calendar (Fig. 6 - 1) shows the spans of time during which the magnetic field has been normal (like that of today) and the complementary times when it has had a reversed pattern. The discovery of the reversals, which made the calendar possible, is entirely distinct from the measurement of magnetic directions from which the movement of continents is inferred.

Comparing the magnetic calendar with the geologic column (Table 3 - A) one sees that the present 4.5-million-year span of the

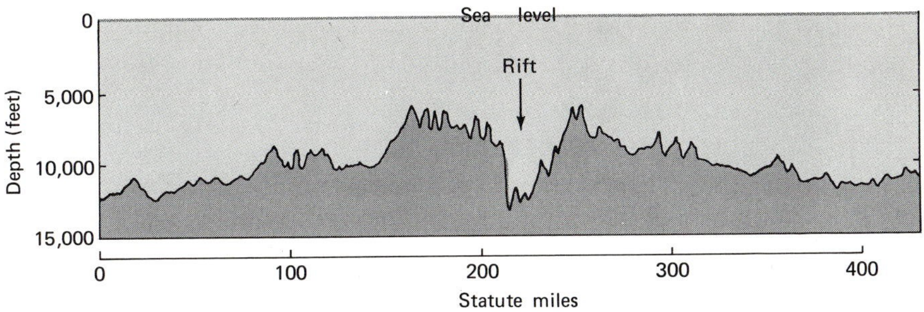

Figure 6 - 2. Profile along an east-west line through the middle of the Atlantic Ocean floor. It shows a rift, here 6000 feet deep, along which basalt lava is welling up from below.

calendar includes all of Recent time, all of Pleistocene time, and part of Pliocene time. Armed with the knowledge that magnetic reversals occur and are imprinted in the magnetic pattern of solidified basalt on the continents, we can now profitably look at the magnetic patterns shown by basalt on the ocean floor.

Mid-ocean rifts. The most remarkable of the many discoveries made about the ocean floor since about 1950 is that it is not a monotonous flat plain; it is as varied in form as are the continents. Among its many features is a deep rift, a nearly continuous valley 40,000 miles long and in places thousands of feet deep (Fig. 6 - 2), which neatly bisects the Atlantic Ocean (Fig. 6 - 3) and traverses the Indian and Pacific Oceans as well. Instead of following the very low parts of the ocean floor, the rift follows the central, highest part of a long, high ridge. This odd relationship is seen clearly in Figure 6 - 2. The rift is much deeper than the deep valley seen in that figure. It extends far below the surface, 50 to 100 kilometers down into the Earth's body, and an unusual amount of heat from the interior reaches the surface along its length.

Beneath the thin carpet of sediment that covers most of the deep-sea floor, the floor on both sides of the rift consists of basalt. When research ships carrying magnetometers began to sail east and then west on courses hundreds of miles long across the region of the rift, they were measuring the magnetic polarity of the basalt on the floor thousands of feet beneath them. An astonishing pattern began to show up in the magnetometer data. The pattern consisted of alter-

Figure 6 - 3. The rift in the Atlantic Ocean floor bisects the ocean. It is offset and displaced to right or left along transverse lines. These are the traces of breaks that extend down through the Earth's crust. For continuation of the rift beyond the Atlantic, see Figure 6 - 6.

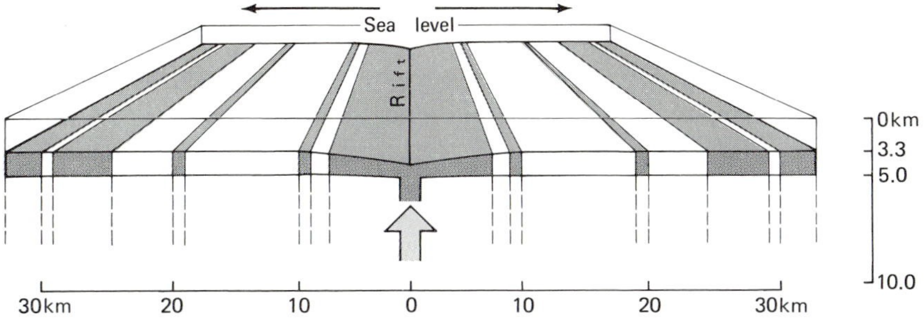

Figure 6 - 4. Diagram showing stripe-like bodies of basalt built in a mid-ocean rift by solidification of up-moving lava (wide arrow). The basalt is continually moved away to right and left (slender arrows) as ocean floor spreads outward from the rift. Each stripe preserves the magnetic polarization it acquired at the time it solidified. Dark stripes indicate normal polarity; light stripes, reverse polarity. (After F. J. Vine.)

nate stripes or bands of normal (north) and reversed (south) polarization. The stripes parallel the rift, and those on the left side form a mirror image of those on the right side of the rift (Fig. 6 - 4). The sequence of wide and narrow stripes duplicates the top part of the calendar of reversals shown in Figure 6 - 1. Because of this duplication the date of each stripe must be the same as the date of the corresponding unit in the calendar. The youngest stripes of basalt lie immediately next to the rift, and are flanked on their outer sides by older stripes. Older and older stripes appear as we move either right or left away from the rift.

This amazing arrangement can mean only one thing. Evidently new crust is continually being created by solidification of the lava that wells up into the rift from far below. Also the crust must be moving away from the rift in opposite directions, one part toward the east and one toward the west, like two conveyor belts moving outward from a single line (Fig. 6 - 5). The newly solidified lava retains the normal or reversed polarity that was imprinted on it by

Figure 6 - 5. A series of four conveyor belts that are moving in directions opposite to each other. Compare Figure 6 - 4.

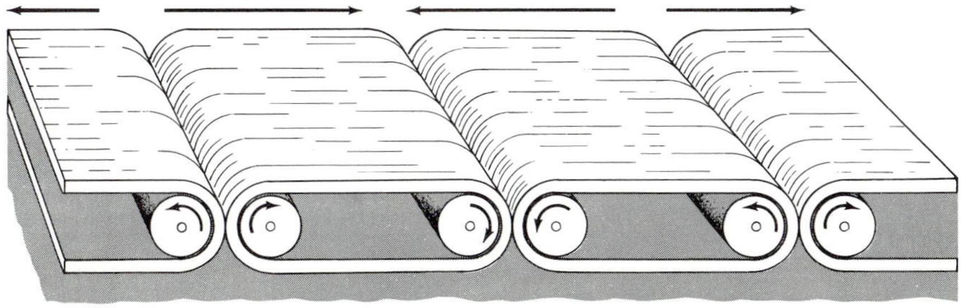

the Earth's magnetic field at the time when it solidified. Whenever the Earth's field reverses, a new stripe begins. This is because the stripes are defined, not by the lava (which is welling up and solidifying continuously and always has the same composition) but solely by the polarity of the magnetic mineral grains within it. In other words, without magnetometer measurement one could not recognize the stripes. Picture two conveyor belts, moving in opposite directions, placed end to end (Fig. 6-5). If each were loaded at their common point of origin with equal amounts of sand, first black, then white, and if the color were changed at irregular intervals of a few seconds, the pattern of the sand on the moving belts would resemble that of the basalt stripes beneath the sea floor.

Or think of a rift as a long crack in a wooden board. Someone standing below fills the crack with wood filler or putty, but the wood keeps pulling apart and the crack reopens and must be refilled. If this went on long enough the filling, multiplied many times, would become extremely wide. In the real case of the Atlantic Ocean floor the combined fillings have become as wide as the ocean itself.

What is the speed of this extraordinary process? Its present speed is calculated from the time scale of the magnetic reversals (Fig. 6-1). It turns out that the crust beneath the western half of the Atlantic is now moving northwest at one to two centimeters per year. If we assume the rate is 1.5 cm/year and has never changed, this movement alone has widened the Atlantic Ocean by 1500 kilometers during the last 100 million years. Eastward movement on the eastern side of the rift would have done still more to widen the Atlantic. Indeed, it has been argued that between 150 and 200 million years ago (that is, late in Triassic or early in Jurassic time) the Atlantic didn't yet exist; instead, the east coast of North America was firmly joined to northwestern Africa.

However, the Atlantic crust is a comparatively slow mover. Calculation from the magnetic-reversal time scale tells us that in the Indian Ocean movement reaches 3 cm/year, while in parts of the Pacific Ocean it is 6 cm/year or more. By the widths of the black areas Figure 6-6 shows the areas of new crust that have been created along the mid-ocean rifts within the last 10 million years. Thus it shows relative speed of movement in various oceans.

Another look at Figure 6-3 shows the many east-west lines along which the Atlantic rift has been offset. These are clean, deep-going breaks caused by tearing apart of the crust by the pull of the two

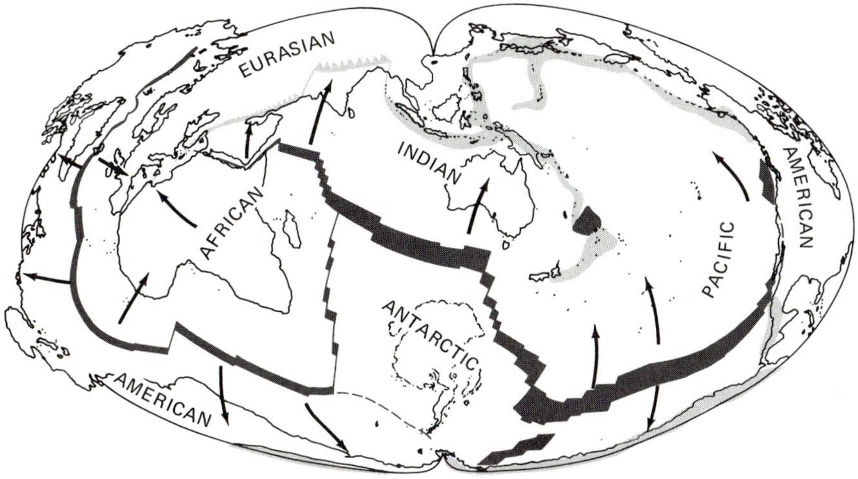

Figure 6 - 6. Major crust plates move as indicated by arrows, *away* from rifts (black bands), where new lava is welling up, and *toward* zones (gray bands) in which crust rock is slowly plunging into the Earth's body.

The widths of the black zones show the width of new crust added during the last 10 million years. The combined area of these zones equals 5 percent of the area of the Earth. During that time equal amounts of crust have been destroyed as they plunge downward into the Earth in the gray areas.

The odd-looking map projection was made in order to keep the scale as nearly uniform as possible. Compare Figure 6 - 7. (After H. W. Menard.)

opposed movements away from the rift. They are one of the many kinds of geologic features called *faults*. The friction involved causes many earthquakes.

The system of trenches and mountain chains. Clearly, then, parts of the crust are moving. Not just the continents, but whole plate-like parts of the crust, 50 to 100 kilometers thick, with the continents floating in them piggyback style. We might compare a *crust plate* (also called *crustal plate*) with a continent partly sunk into its body, to a broad cake of ice having a plank or a log frozen into it, floating with many other ice cakes in a lake.

But toward what are the plates moving? Because crust material is continually being created stripe by stripe at the mid-ocean rifts, something must be happening to the crust to keep it the right size, to prevent it from becoming too extensive to fit the Earth's body. Something, somewhere, must be destroying it as fast as it is being made.

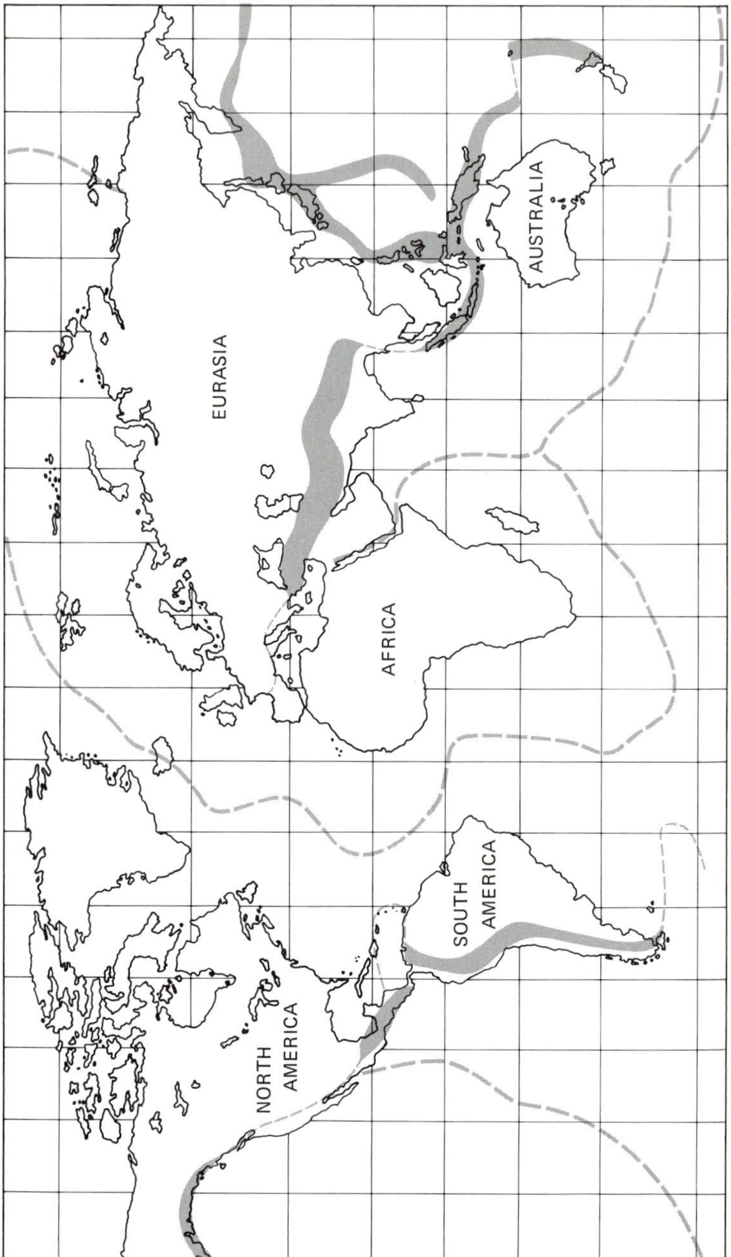

Figure 6-7. System of trenches, chains of volcanic islands, and high mountains shown (in gray) on the more familiar Mercator map projection (exclusive of polar regions). Dashed line indicates general position of the oceanic rift, but with many of its offsets omitted. (Compiled from various sources.)

The sites of its destruction are not far to seek. Draped around the globe is the long, narrow chain shown in gray in Figures 6 - 6 and 6 - 7. It stretches from Antarctica to Cape Horn, along the western flanks of both Americas, through the Aleutian Islands, Japan, the Philippines, New Guinea, and New Zealand. A branch leads westward through Indonesia, the Himalaya, Iran, Turkey, and the Mediterranean region.

This tremendous chain is made up of three elements. These are (1) deep trenches in the ocean floor, one of them 6.75 miles deep, far deeper than the height of the highest mountains; (2) arc-like rows of volcanic islands like looped strings of beads; and (3) the long series of the world's highest mountains. The chain formed by these three elements is a zone of great instability. Clear evidence of instability is the fact that nearly all the world's deepest and biggest earthquakes occur right along this zone. Among earthquakes that make news today, those in Chile, California, and Alaska and those in Iran and Turkey figure prominently.

As a whole, the great chain is complementary to the enormously long oceanic rift. Plates of crust are moving toward it along both its sides. They are being bent down and fed into the Earth, and are being re-melted there (Fig. 6 - 8) much as a wax candle is melted if it is held horizontally and pushed slowly toward a hot flame. The

Figure 6 - 8. Diagram illustrating the creation and destruction of crust, and collisions between crust plates. Parts of four plates (A, B, C, D) are seen; arrows indicate directions of their movement. An oceanic rift (R) is accompanied by faults. A hole has been cut through the model near the margin of Plate A to show how the near part of the plate is overriding the plunging edge of Plate B, whereas the far part of A is plunging beneath B as the two plates collide. In another collision, Plate D is overriding C.

Earthquakes and the creation of chains of mountains accompany the collisions of plates. Diagram is schematic and not to scale. Its length might represent 3000 miles. The piggyback continents are not shown separately from the rest of the crust. (After Oliver and Sykes.)

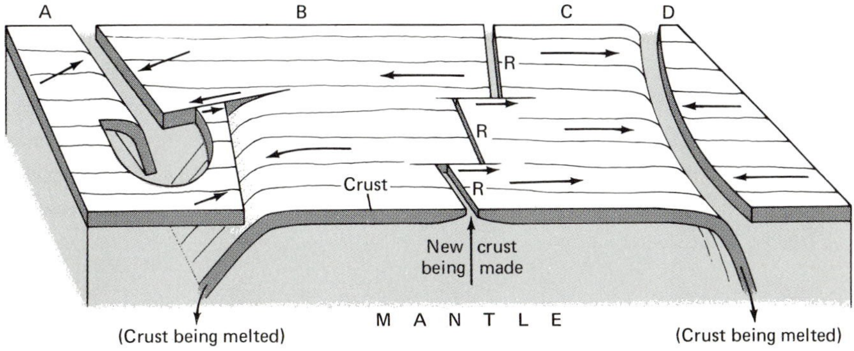

conveyor belts are being destroyed by melting at the far end of the line as fast as they are being added to at the near end. So the total area of crust remains unchanged. The creative activities are confined to the oceanic rifts and the destructive activities to the trench-and-mountain chains respectively. In the broad areas between, in the main parts of the conveyor belts, activity is much milder, consisting mostly of the erosion of rock, the deposition of sediment, and gentle movements of uplift or subsidence.

Geometry of crust plates. The crust plates, some of which are named in Figure 6 - 6, have been compared with the pieces of a jigsaw puzzle. The comparison is not very close, because the pieces of a puzzle keep their characteristic outlines, whereas crust plates are being added to along their trailing edges and are being sliced off by melting along their leading edges. So their shapes do not remain constant. Some of them may be no more than fragments of what they were formerly, and others (such as the Eurasia plate) may consist of several older plates welded together. Likewise the analogy between a plate and a conveyor belt, although useful, is not very close either, because plates rotate. For example, the varying width of black areas in Figure 6 - 5 shows that during the last 10 million years the Pacific plate has rotated clockwise. In this respect crust plates more closely resemble great cakes of ice moving with the current in a lake, always touching, colliding, bumping together, slowly turning, and even melting at their edges and later refreezing. It is very likely that the world's great chains of mountains are the result of collisions between crust plates. For example, the Alps system (Fig. 6 - 9), which shows the effects of colossal buckling, crumpling,

Figure 6 - 9. Section cut through part of the Swiss Alps, showing the intense deformation undergone by sedimentary strata at some time after the Eocene Epoch. Whole slices of crumpled, mashed, and squeezed rock were pushed for miles across rock underneath, and then the slices themselves were deformed. The force that accomplished such deformation was enormous. (Modified by Longwell and Flint, after Arnold Heim.)

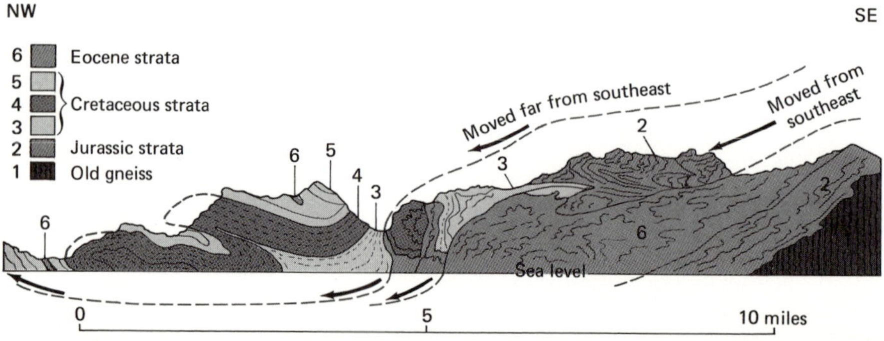

and squeezing of once-horizontal strata, may have been created by collisions of the northward-moving African plate with the Eurasian plate. Likewise the Himalaya may be the result of a collision of the Indian plate with Eurasia. The Himalaya is still active, rising at the rate of a foot or so per century. Layers of weak sedimentary strata of the kind that characterize many continental shelves (Fig. 5 - 3) would not stand much chance if crushed between two great masses, each some 30 to 60 miles thick, coming together slowly but with enormous inertia. A small rowboat caught between two heavily loaded harbor scows in a head-on collision would probably be mashed beyond recognition, even if the scows were moving at less than four or five miles per hour.

When we speak of scows floating in water we must remind ourselves that what we are describing is *not* continents that float freely in and move through basalt. Instead we are describing thick plates of basalt that move past something heavier underneath them, and that carry continents embedded in their backs. Hence not just continents, but entire thick plates of crust with their carpets of deep-ocean sediment described in Chapter 5, must meet along the line of the trench-and-mountain chains. Some of the mountains (such as the Alps) contain great masses of metamorphic rock that because of their chemical composition are believed to be the squeezed remains of deep-sea sediment. Too light in weight to sink, such sediment probably piles up along the trenches as the basalt plate beneath it plunges downward into the Earth. Like the scum on dirty water, the sediment stays afloat and is eventually so thoroughly squeezed that it is transformed into metamorphic rock.

By working backward and by using the magnetic data we have described, which show apparent former positions of the pole, we can trace the probable positions of continents as far back as 100 million years or so—that is, into Cretaceous time. The continents were closer together then than they are today, and the Atlantic Ocean was much narrower. Attempts to trace the movement still farther back to 200 million years ago (to early Jurassic or still earlier time) give much less certain results. One reconstruction of the world map of 200 million years in the past (Fig. 6 - 10) shows today's continents joined to form a single "world continent," which broke into pieces like the ice on a frozen lake in early spring. The illustration shows the suggested paths by which the broken pieces reached

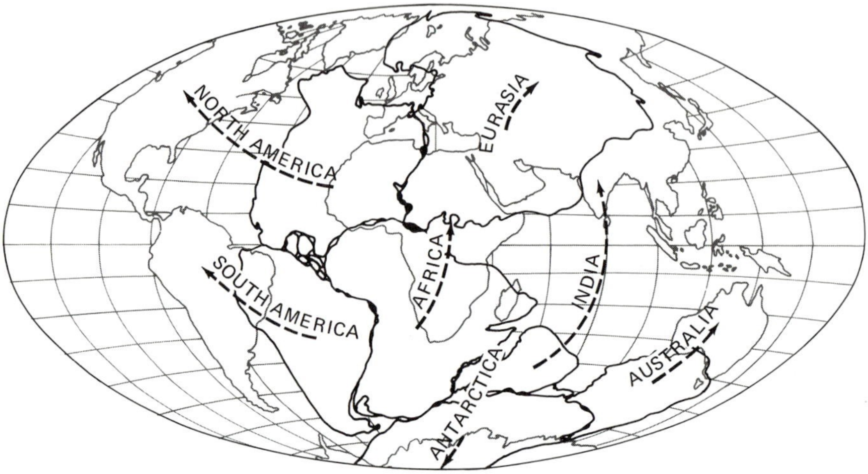

Figure 6 - 10. Map illustrating the idea not yet proved that around 200 million years ago (early in the Jurassic Period) the world possessed a single huge land mass, which then broke up into plates. These moved in directions indicated by arrows and became the existing continents. Fitting together of the pieces was done by computer. (Adapted from Dietz and Holden.)

their present positions. Although the map is certainly interesting and provocative, we must realize that it is still in the realm of theory. The plates themselves, however, are probably far more than 200 million years old. Indeed, plates may have first formed as much as 2.5 billion years ago. If they did, they have existed throughout more than half the Earth's entire history.

What makes it all go? No reader can have visualized this dynamic array of rifts, trench systems, moving crust, plates, and mountain chains without wondering what makes it go. What is the mechanism and what supplies the huge amount of power that must be required to operate it? Although we know at least what the system is, we do not yet know what makes it work. The power is certainly derived from internal heat, which rises to the surface in so much greater quantity through the oceanic rift than it does elsewhere. It is thought likely that most of this heat has been generated by radioactivity in the outer part of the Earth's body.

But if internal heat does supply the power, how is the power applied so as to drive moving crust plates? Several mechanisms have been suggested. One idea that has had wide appeal is the mechanism of convection, a process that can be watched in a pan of very thick soup just hot enough to simmer. Heated at the base of the pan, soup rises slowly in irregularly spaced columns or cells. Reaching the surface, it spreads out, cools, and sinks again along other paths to replace rising hotter soup.

It is suggested that convection is occurring in the Earth's mantle. In its simplest form the idea runs like this: Rising slowly beneath the oceanic rifts, fluid-like heavy matter reaches the crust and spreads away beneath it. As it moves away from the rift it drags overlying crust plates along with it. The areas beneath the line of trenches and mountain chains represent the descending part of the convection system. The fluid-like mantle material, now somewhat cooled, moves downward, balancing the rise of hotter material beneath the rift. According to this idea the conveyor belts (Fig. 6-5) move because friction along their undersides drags them forward.

The idea of such a mechanism is pure hypothesis, and there are some valid objections to it. Other, less simple explanations have been proposed. But we will have to wait for an improved idea that will be generally acceptable to scientists.

Now that we have visualized the movement of crust plates, we can see more clearly how the granitic continental rock is being differentiated from heavier basaltic rock. Crust plates consisting of basalt with lighter-weight continental rocks embedded in it are plunging downward into the trench systems. As they get deep enough down, they are being partially melted to form magma of lighter weight than pure basalt. The magma, of course, tends to move upward toward the surface, and eventually the resulting rock becomes chemically sorted by weathering. The effect of the moving crust plates is therefore to help make the continental crust lighter and more granitic in character with the passage of time.

Conclusion

This ends our survey of the way in which we think Planet Earth was formed, of the origin of the continents, and of the extraordinary, active system of rifts, moving plates, and mountain chains. We have traced the Earth, however sketchily, to its present interesting condition as a physical body around whose surface much varied activity is going on. Our next step is to try to understand how the surface came to be peopled with a huge variety of living things. That step must begin with the nature of organic matter.

References

EARTH ORIGIN:

Abelson, P. H., 1966, Chemical events on the primitive earth: Nat. Acad. Sci. Proc., v. 55, p. 1365–1372.

McLaughlin, D. B., 1965, p. 669–698 in Kay, Marshall, and Colbert, E. H., Stratigraphy and life history: John Wiley & Sons, New York.

Urey, H. C., 1952, The origin of the earth: Sci. American, v. 187, p. 53–60.

MOON:

Mutch, T. A., 1970, Geology of the Moon: Princeton University Press, Princeton, N.J.

MAGNETIC REVERSALS:

Cox, Allan, and others, 1967, Reversals of the Earth's magnetic field: Sci. American, v. 216, p. 44–54.

MOVEMENT OF CRUST PLATES AND CONTINENTS:

Bott, M. H. P., 1971, The interior of the Earth: St. Martin's Press, New York.

Ernst, W. G., 1969, Earth materials: Prentice-Hall, Inc., Englewood Cliffs, N.J.

Hammond, A. L., 1971, Plate tectonics . . . : Science, v. 173, p. 40–41, 133–134.

Heirtzler, J. R., 1968, Sea-floor spreading: Sci. American, v. 219, p. 60–70.

Menard, H. W., 1971, p. 1–14 in Turekian, K. K., ed., Late Cenozoic glacial ages: Yale University Press, New Haven.

Morgan, W. J., 1968, Rises, trenches, great faults, and crustal blocks: Jour. Geophys. Research, v. 73, p. 1959–1982.

Wilson, J. T., and others, 1972, continents adrift: W. H. Freeman and Co., San Francisco. (Paperback) (Reprints of 15 articles from Scientific American about the movement of crust plates.)

7

The Early Biosphere

Organic matter

We have seen how the atmosphere is believed to have been made by the sweating-out of some of its primitive constituents from deep within the solid body of the Earth, and how the hydrosphere seems to have resulted from condensation of atmospheric moisture. The atmosphere is a simple mixture of elements, CO_2, and H_2O vapor; the hydrosphere, mainly H_2O, is simpler still. Compared with them the biosphere is complex. Despite the fact that it all consists of the same basic substances, the biosphere is complex because of its chemical and structural variations. In the days when knowledge of organic matter was very limited, before modern biochemistry discovered a multiplicity of variations and specializations, the basic organic substance was called *protoplasm.* The chemical elements that constitute this substance (really a large group of substances) are mainly hydrogen, oxygen, nitrogen, and carbon. Indeed, about 99 percent of all organic matter consists of these four elements. Together with them in protoplasm are sulfur, phosphorus, and small amounts of some two dozen other elements. These constituents occur in an extremely large number of combinations. Qualitatively, the overall composition is not unlike that of seawater. It includes no unique element peculiar to life, no single element that does not occur in inorganic matter as well. Hence, in its most basic chemistry the substance of the biosphere does not differ from the makeup of the other two fluid spheres. It differs only in the way in which its basic components are selected and organized. The character of the difference fits well with the idea generally held by scientists that living matter has been built up from inorganic materials.

Cells

Living matter consists of structural and functional units called cells that might be compared with the bricks or blocks that are assembled to form a building. They are made in various sizes and shapes, and so are able to form many different functional structures. Each cell is surrounded by a cell membrane, a semi-permeable molecular "wall" through which water and nutrients can pass. Some simple organisms consist of only one cell; other organisms are enormously complex systems of cells that have many diverse functions. Regardless of its degree of complexity, an organism preserves its distinctive structure and form while through it flows a stream of matter (derived from sources on the Earth) and energy (derived from the Sun) in a series of chemical reactions. These reactions do many things. They nourish, they build new cell material, they eliminate wastes, and they reproduce individuals. In other words, the organism *lives.* The life span of a particular cell may be less than a few minutes or more than a few decades; eventually it is destroyed. But as a type the cell survives, because it possesses the ability to reproduce itself. So it enjoys a continuity, an immortality, that lies beyond the reach of any single organism.

Three main groups of living things

Living things are classified in three large groups, two of which, the plant kingdom and the animal kingdom, are familiar to everyone. As a broad generalization we can say that an animal can move around under its own power, whereas a plant cannot. A more fundamental difference between these two groups, however, is biochemical. Plants build their own food by chemical *reduction* in the process called *photosynthesis*. That is, they act on carbon dioxide (CO_2) so as to separate oxygen from carbon, and in the process they use water and solar energy. They build organic compounds, in which the solar energy is stored as potential energy, while the leftover oxygen is released as free oxygen (O_2). Animals are different. Unlike plants, they do not build their food. They nourish themselves with food already built up by other organisms. They do this by eating plants or else other animals who themselves eat plants. Broadly speaking, therefore, they are parasites. By *respiration* they extract free oxygen from the atmosphere or hydrosphere and with it *oxidize* the organic compounds in their food. (Plants respire too, but to a lesser extent.) Oxidation is a process of slow combustion that releases energy from the carbon-hydrogen bonds in organic molecules. This is the energy that had been stored by plants in their substance. Oxidation makes this energy available for use in activity of many kinds. In the respiration process CO_2 is exhaled, ready to be recycled once more by plants. Thus the relationship between plants and animals is largely reciprocal and consists basically of reduction and oxidation, with oxygen a currency that continually changes hands. In the reduction reaction, plants emit O_2; in the oxidation reaction, animals emit CO_2.

The third group of organisms is far less familiar. It consists of the kingdom of protists, which are neither plants nor animals, and includes such organisms as green algae, bacteria, and the single-celled protozoans that were once thought to be animals. Most protists are either single celled or have rather little-differentiated bodies. Green algae are believed to be the probable ancestors of plants. Indeed protists as a group are believed to be ancestors common to both plants and animals, and hence to be closely related to the question of the origin of life.

Let us go back for a moment to the recycling mentioned earlier. The recycling of elements in the biosphere is similar to what happens in the inorganic spheres. In Chapter 2 we described the rock cycle as a series of transformations repeated over and over again, with the same chemical elements changing hands and entering into different combinations, from which they are split off later to form still other combinations in unending cycles. The recycling of carbon and oxygen, as these elements become temporarily parts of the bodies of plants and then of animals, only to enter into plants again, is a biologic cycle. This cycle takes its place with the rock cycle as an activity of the dynamic Earth.

Living matter defined

How is living matter to be defined? Perhaps the closest approach to a definition at present is the statement, "Living matter consists of things that can reproduce themselves, mutate, and reproduce the mutations." These capabilities represent transfers of energy and transfers of information. We will deal with mutations in Chapter 8; here we need only note that the ability to reproduce itself is the outstanding characteristic of an organism. No inorganic substance can do it. We might add, not as part of a definition but as an interesting fact, that in general organic molecules are much larger and more complex than inorganic ones, and therefore they embody greater capacity for change. And change, continual change, is what the whole of the long record of life shows from its beginning to its present state, and change is going on in the moment in which we are living.

The earliest fossils

As we are concerned with the early history of the biosphere, a sensible way to begin is to review what the earliest fossils tell us about life before Phanerozoic time began. Although such fossils are very few and very far apart in time, nevertheless they are firm evi-

	Development of living matter	Time (billions of years before the present)	Development of atmosphere and hydrosphere
Biological evolution	Today (Phanerozoic)	0	Today
	*Earliest known fossil marine animals		Free oxygen sufficient for animal respiration
	*Fossil green algae (multi-celled; sexual reproduction)	1	↑
	*Fossil bacteria and blue-green algae	2	Free oxygen invades atmosphere; starts to screen out Sun's ultraviolet rays. Weathering of rock begins
			Free oxygen in hydrosphere increases; oxidizes iron
		3	
	*Earliest known fossils (single-cell protists (bacteria) and simple plants (blue-green algae)		Algae liberate free oxygen to hydrosphere
Chemical evolution			Fermentation(?) adds carbon dioxide to hydrosphere
	More complex hydrocarbons		
	Protein molecules	4	Primitive atmosphere and hydrosphere containing no free oxygen
	Amino acids		
	Formation of		Planet Earth

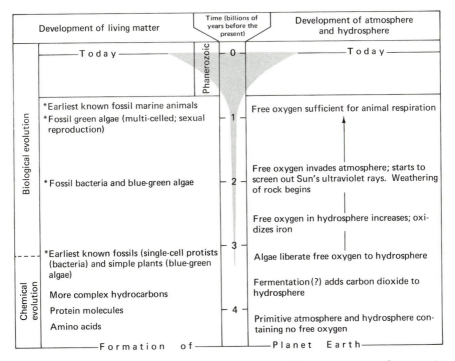

Table 7 - A. Critical events in the early history of living matter and events in the development of atmosphere and hydrosphere. Width of shaded area represents relative number of kinds of organisms. Stars denote fossils, which constitute real data, fixed in time by potassium/argon dates. (Compiled from various sources.)

dence; and since all of them have been dated, at least crudely, by potassium/argon measurements, they provide a start for a calendar of the early history of living things.

The oldest fossils identified in the rocks thus far (Table 7-A) consist of marine organisms: protists, represented by bacteria (Fig. 7-1), and plants, represented by single-celled, blue-green algae. These fossils occur in South Africa in rock at least 3.2 billion years old. Their date means that at a point when less than one-third of geologic time had elapsed, the sea was already populated with living things, however simple. This was no mean feat; bear in mind that when the atmosphere and hydrosphere were first formed they apparently lacked carbon dioxide, which is required by plants for photosynthesis.

The next-oldest fossils, found in Ontario in rock about 2 billion years old, likewise consist of bacteria and blue-green algae, as

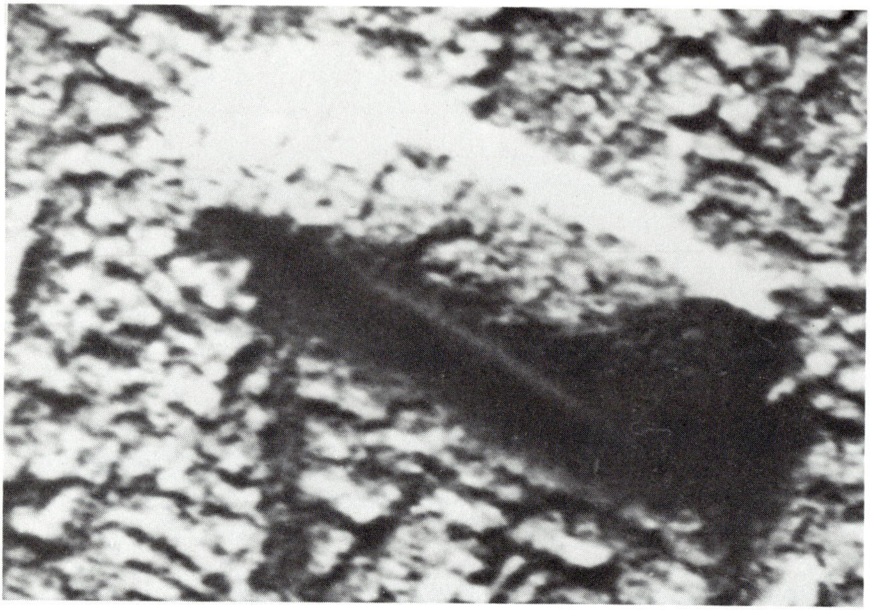

Figure 7 - 1. The oldest known fossil, dated at 3.2 billion years before the present. A bacterium (named *Eobacterium isolatum*) from the eastern part of South Africa. Less than 1/1000 of a millimeter in length, it was photographed with the aid of an electron microscope. (Courtesy E. S. Barghoorn.)

Table 7 - A shows. But in rock in central Australia, probably about 1 billion years old, are fossil plants of several kinds. These include the earliest known organisms having many cells rather than merely single cells. They also include green algae, a type of plant that possesses complex cells of a sort that imply capability for sexual reproduction rather than reproduction of simpler kinds. This characteristic makes possible genetic diversity and so leads to the development of an endless variety of organisms. Such variety apparently did not exist much earlier than a billion years ago. But by Cambrian time, at the beginning of the Phanerozoic some 400 million years later, living things were already embarked on an explosive increase in numbers and variety, toward the condition of life in the world today. The explosive increase may have resulted from the "invention" of the sexual type of reproduction.

Up to now we have mentioned only protists and plants. What about the appearance of animals? In South Australia hundreds of fossil specimens of primitive sea animals have been discovered in

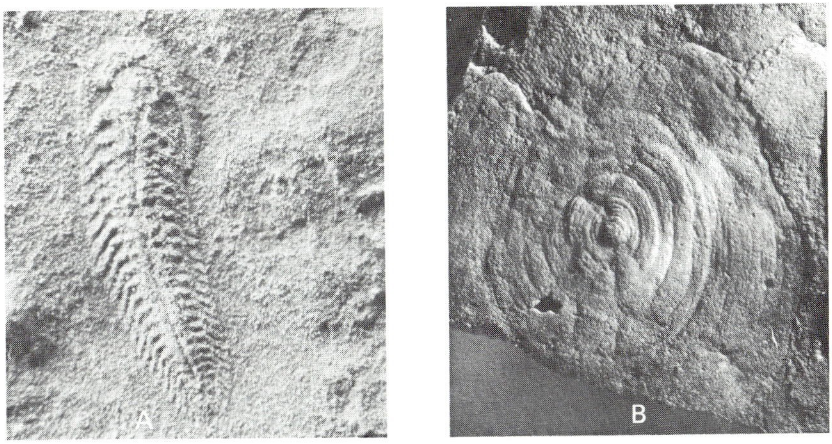

Figure 7 - 2. Two of the oldest fossil animals in the world, from strata in South Australia. (Courtesy M. F. Glaessner.)

A. A segmented worm, *Spriggina floundersi,* about 2 inches in diameter.

B. A jellyfish-like animal, *Cyclomedusa davidi,* about 1 inch long.

strata older than Cambrian and perhaps more than 800 million years old. All of them are impressions made by animals stranded at low tide on the soft sediments of ancient mud flats and beaches along a shore (Figs. 7 - 2, 7 - 3). The animals had soft bodies, lacking shells or other hard supports. The impressions preserved in sandstone are those of a sort of jellyfish, of a branching form resembling a kind of coral, of other forms unlike any living animal, and finally of the bodies of worms and tracks made by worms on a shallow

Figure 7 - 3. A group of the oldest known animals, restored as they might have looked when living in shallow water along an Australian shore.

bottom. In this varied collection the worms are the most significant element because they are a highly developed sort of segmented worm. Their presence implies that simpler animals must have existed in the sea still earlier—earlier enough so that they could have had time to evolve from extremely simple kinds to the more advanced kinds represented by segmented worms.

It is hardly surprising that the simpler animals implied by these fossils have not yet been found in still older rocks. Small, soft-bodied animals can be preserved only under a very rare combination of conditions: extremely fine sediments, quiet protected water, later covering up with more sediments without erosion, and finally, much later, just enough erosion to expose the fossils to view without destroying them all. So probably most of the still older fossil animals have been destroyed, and only a lucky chance would reveal any of the remainder to us.

The existence of the several kinds of animals found in Australia means that respiration was already common practice in the biosphere. And respiration implies, in turn, that the hydrosphere, and therefore of course the atmosphere, then contained free oxygen, a constituent conspicuously lacking at the time of the primitive Earth. As we shall see presently, the oxygen could have been supplied by plants, which, as the record of fossils tells us, had already existed for 2 billion years or more.

Meager though it is, this record of very old fossils provides us with firm information, not only about kinds of organisms but also, by implication, about changes in the composition of the sea. With this information in mind it becomes possible to speculate about the chemistry of the early ocean, about the reactions that might have been involved in the beginning and earliest history of life.

Theory of origin of the biosphere

At the beginning of this chapter we said that living matter contains no element not present in the inorganic realm of the Earth. Let us add to this statement that the four chief elements (hydrogen, oxygen, nitrogen, carbon) in living matter are among the most abundant the Earth possesses. So the basic stuff of living matter is common; it is readily available in quantity.

In what way were these elements combined to form the biosphere? We do not yet know. But we already have enough information to narrow down the possibilities and to make ever-better-educated guesses. The bewildering array of different kinds of organisms living today includes every gradation from highly complex animals down to simple forms consisting of single cells. The gradation does not stop there. It continues downward to include single large protein molecules, represented for instance by certain viruses, which exist at times in an inert, crystalline state and at others in a fluid state mobilized for life activity.

But a protein molecule itself consists entirely of even simpler things, the compounds of carbon, hydrogen, nitrogen, and oxygen, bonded together in definite geometric patterns, that are called amino acids. These acids form part of every living thing, but are they themselves "living"? Perhaps this question belongs rather to philosophy than to science. At any rate, scientists have succeeded in inducing the creation of amino acids from what are believed to have been the constituents of the Earth's primitive atmosphere, a gaseous mixture of hydrogen (H_2), methane (CH_4), ammonia (NH_3), and steam (H_2O). Such a mixture was made and was subjected to strong electric discharges. Then it was condensed to a liquid. Amino acids and other hydrocarbon acids as well were found in the liquid. Other experiments have gone farther and synthesized from inorganic materials many of the compounds that constitute living matter.

The results of such experiments, coupled with what we know about the structure of living things and about their biochemical activities, have led to the belief that organisms evolved by chemical evolution over a long period of time from inorganic materials consisting of combinations of chemical elements and inorganic compounds. The combinations are likely to have occurred in seawater, which biologists have called a "warm organic broth," although it must have been a very thin sort of soup. Soup or not, seawater is a fluid medium in which chemical reactions could occur with the least difficulty. Water is a solvent and a splendid medium for reactions among organic compounds, which themselves are liquid systems. Furthermore, atoms of carbon have an extraordinary ability to bond to each other in a wide variety of patterns. So altogether the situation provided for a very great degree of flexibility.

As we noted in Chapter 6, the primitive atmosphere and hydrosphere (including the broth) lacked free oxygen; so all the oxygen present existed only in combination with other elements. Not only is oxygen necessary for the life processes of animals; the oxygen in today's upper atmosphere creates a protective screen against the Sun's ultraviolet rays, which would otherwise destroy living things. We live today under an effective umbrella or, better, parasol provided for us by free oxygen, a protection that did not exist during the Earth's early history. So in the early lifeless world, even if the living things of today could somehow have been created, they could not have survived because the parasol did not then exist.

Before going further we had better review the timetable we have to work with. As we have said (Table 7-A), fossils in the rocks show that at a time about 3.2 billion years ago both protists and very primitive plants already existed in the sea. Because the lithosphere was formed about 4.5 billion years ago, this gives us a span of about 1.3 billion years within which living matter must have formed and evolved. This span was twice as long as the entire Phanerozoic, with its long and complicated record of development of the biosphere, and in itself is more than adequate. Within that 1.3 billion years living matter must have been created in an environment that lacked free oxygen, and then must have evolved to the level of protists and primitive plants. In considering these events we must also look ahead and realize that later on O_2 was added in large quantity to the atmosphere, for without that addition there could have been no animals.

The events in what has been called *chemical evolution* can be explained logically in theory, though it remains for the future to establish how much of the theory may be right.*

The materials necessary for creating living matter were present in the primitive ocean, the organic broth. In the absence of free oxygen the Sun's ultraviolet radiation struck the lands and the surface waters of the sea. Either such bombardment or discharges of lightning could have provided the high energy needed for creating amino acids like those made in laboratory experiments. The amino acids could have combined to form protein molecules which, as they react and combine readily, would have promoted further changes, the building up of more complex hydrocarbons.

* For a suggested explanation in more detail see Wald, 1954, the explanation we follow here.

Somewhere in this sequence we could probably say that "life" began. But at just what point it began depends on how life is defined. Probably the first living thing consisted only of a single cell. Possibly it was a very simple bacterium or something that resembled it. Whatever it was, it had to have nourishment. Its nourishment could have consisted of hydrocarbon molecules from the mud of shallow sea floors. From this diet it was but a short step to absorbing, or "eating," first the waste products and then the dead bodies of fellow organisms. The next step would logically have been the consumption of organic matter that was still living.

In this supposed sequence of events we see the beginning of the splitting up of living things into two groups. One group lived by absorbing the waste products and the dead substance of other living things. The other group, in a way more sophisticated, preferred a diet of live organisms. This basic splitting up, the beginning of diversification, might have led later on to the distinction between plants and animals, although the idea is certainly much oversimplified.

The changes involved in this and other early phases of evolution are best stated in terms of biochemical reactions. Among the early changes is likely to have been the one involved in the reaction called *fermentation.* As we observe it today, fermentation is carried on by primitive organisms, including some bacteria, that live in environments devoid of free oxygen. Because they live in such environments these organisms give us a clue to the kind of life that was led by organisms in the primitive ocean. During fermentation hydrocarbons are broken apart and rebuilt, a small amount of energy is released as heat, and one of the resulting products is carbon dioxide (Fig. 7-4).

Figure 7-4. Theoretical chain of processes by which major groups of living things might have developed. Output of energy in respiration exceeds that in fermentation, as indicated by size of lettering.

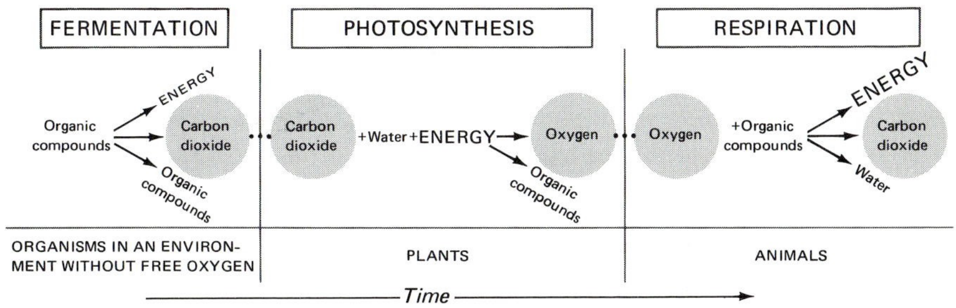

That product, whether a result of fermentation or of some other reaction, was of great importance to the theory of the way in which early life developed because carbon dioxide in quantity, added to what little there may already have been in the sea, would have given the environment a new capability. It would have made possible the photosynthetic process mentioned earlier in the present chapter. That process, a characteristic of most plants, builds various organic compounds from water, carbon dioxide, and energy trapped from sunlight (Fig. 7 - 4). If carbon dioxide were steadily added to the sea, plants, which could not have existed in the chemically more primitive ocean, could develop, building their tissues with it.

Here there is a further important point. One of the products of photosynthesis is free oxygen, O_2. So with plenty of plants to produce it, O_2 in turn could begin to accumulate in the sea. In the new presence of free oxygen the process of respiration (Fig. 7 - 4) would now have been possible. Respiration, the reverse of the photosynthesis process, not only produces carbon dioxide usable by plants, but also releases a large amount of energy—about 35 times more than fermentation does. This energy in usable form becomes available for growth and movement. Respiration, with its evident advantages, would have made it possible for animals to exist. Animals made good use of the abundant energy that came from respiration. They learned to move about, some of them very rapidly, in pursuit of food. Now, to be successful, movement demands coordination of body parts, very close control, and quick, complex decisions. For that one must have a brain, another possession, and a very precious one, that distinguishes animals from plants.

The theory of the origin of the biosphere, then, begins with chemistry that becomes biochemistry through this ideal sequence of reactions:

amino acids → protein molecules → more complex hydrocarbons → fermentation → photosynthesis → respiration

It is an ingenious, chemically reasonable chain of events. The chain is logical, but we must not forget that it is still nothing more than hypothesis. The "inventions" * of these processes may have

* In writing informally about the development through evolution of some new structure or process in an animal or plant, scientists often use the word "invention," even though "inventing" seems to imply forethought and intention. But the word is convenient, and we will use it frequently in the same sense. In doing so we imply nothing more than the operation of natural selection.

taken place in ways somewhat different from the ones we have outlined. But they did, somehow, come into operation. Most scientists believe that the cause was a long series of chance chemical combinations, purely random events, such as collisions of molecules moving within the "broth."

The word *chance* must not lead us to suppose that such a chain of events is "chancy," by which we mean "extremely unlikely," or even "almost impossible." Chance really means only *probability*. When we talk of chance occurrences (such as molecules bumping together) we are dealing with random events. Any random event, however unlikely it may be that it will occur in a single trial, becomes more likely to happen as the number of trials is increased. As J. B. S. Haldane once pointed out, by repeated shuffling of 7 cards each bearing one of the letters A, C, E, H, I, M, N, the word MACHINE would be formed, on the average, once in 5,040 trials. An event that has only 1 chance in 1,000 to occur in a single trial will, in 10,000 trials, have 19,999 chances in 20,000 to occur at least once, a score that in popular language would no longer be considered "chancy." It amounts to near-certainty.

Now, how many times 10,000 trials were possible within the span of time available? Well, there was the span of about 1.3 billion years from the formation of the Earth to the appearance of plants, plus another 2 billion years or so to the time when animals are known to have existed (Table 7-A). How many times 10,000 trials of each of many such random events could have occurred within a period of 3.3 billion years? One's imagination boggles at the thought of trying to calculate so great a number. No one familiar with statistics rejects the idea of chance chemical combinations simply because there wasn't enough time. There was a huge abundance of it.

At some point in the series of changes, whether they were the ones we have outlined or some others, the chemical evolution of the earliest world became biological evolution when organisms began to reproduce themselves. The boundary between the two sorts of evolution is shown arbitrarily in Table 7-A at the level of the oldest known fossils, although its proper place must be farther down.

At a time in its development, probably between 1 and 2 billion years ago, biological evolution became greatly enriched by the "invention" of sexual reproduction. With that invention the possibilities for adjustment of organisms to their environments became much greater, as we shall see in Chapter 8.

Review

It is time now to look again at Table 7 - A, which contains in capsule form much that we have been discussing. Once the primitive atmosphere was in existence, we see in the table the aggregation in the sea, the "organic broth," of amino acids, protein molecules, and more complex hydrocarbons. At some point late in this sequence a process such as fermentation began, adding CO_2 to the ocean. Among simple, single-celled plants (the blue-green algae that today are fossils in South Africa) photosynthesis was already in operation, liberating free oxygen to the sea. In addition to the plant fossils there is other evidence of this. Sedimentary strata between 2 and 3 billion years old include, on a worldwide scale, unique layers of iron oxide precipitated on shallow sea floors by the combination of iron and oxygen contained in the seawater. Younger strata lack this type of sedimentary iron but do contain red iron oxides of kinds that are created today when oxygen in the atmosphere combines with iron during chemical weathering. Therefore it seems likely that as plants became more complex their photosynthesis created more free oxygen than the sea could absorb. Excess oxygen began to leak from the sea into the atmosphere and oxidation of rocks in chemical weathering began to occur.

At the same time the oxygen newly accumulating in the atmosphere began to form the parasol that protects the Earth's surface against ultraviolet radiation. When the quantity of free oxygen became great enough, the screen could become fully effective. Plants, which hitherto had had to live in the deeper and darker parts of the ocean, could spread and populate the whole of the surface water as they do today. Besides this, the abundance of free oxygen made respiration possible on a large scale, and so presumably led to the evolution of marine animals. With guaranteed protection from lethal radiation and with free oxygen available for breathing, the way was prepared for two great steps. The first was the creation and spread of *plankton,* the myriad microscopic plant and animal life that today still crowds the surface waters of the open ocean and that provides food for larger marine organisms. The second step that could now begin was the movement of living things from the seas to the lands. Probably the movement was accomplished in this way: Some of the plants and animals that were already crowding the waters just offshore moved into the intertidal zone, the zone that is submerged at high tide but is exposed to the air at low tide. When they had become adjusted to life in that zone, their logical next move would have been the final move ashore. Sea plants and sea animals thus made it onto the land for permanent residence.

As we have said, the fossils and their dates shown in Table 7-A are facts. The other entries in the table represent varying degrees of probability (strengthened, however, by the occurrence of the two kinds of iron-oxide strata, which are also facts). The table as a whole is what scientists call *reasonable*. The mixture of fact and theory is logical. Of course more fossils are sure to be found and the story will be modified correspondingly. But even in its present state it gives us a consistent outline of the way living matter may have developed before the beginning of the Phanerozoic. One thing we must remember, although it may be difficult to realize. What has been sketched is not just the dawn of the Earth's history. It is *nearly all* of the Earth's history, embracing about 87 percent of geologic time. So the whole of the remainder of our story deals only with the final 13 percent of the span of the Earth's existence. The reason for such disproportion lies in the abundance of the record of life in the Phanerozoic part of the geologic column. Because the Phanerozoic is rich in fossils, the variety of Phanerozoic life is comparatively easy to read and offers much fascinating detail. But the older rocks have been widely and thoroughly metamorphosed, so that most of the fossils they once contained have been destroyed. Other parts of the older rocks, which may very well contain some fossils, have been covered up by sedimentary strata deposited later, and so are hidden. For those reasons a large part of the oldest 87 percent of the Earth's history still lies buried in obscurity. So we are obliged to fill the great gaps between our bits of firm knowledge with thoughtful guesses. We hope the guesses, one by one, will be replaced by firm knowledge.

One other thing about the events we have sketched is important. Not only do those events illustrate that environments (such as an atmosphere without free oxygen) strongly influence the biosphere, they also indicate that the biosphere strongly influences the environment. The grand example here is the strong probability that the free oxygen we enjoy in today's atmosphere was put there by organisms, literally "breathed out" by plants in their industrious practice of photosynthesis. So we, as animals, owe the plant kingdom not only the food we eat but also apparently our very origin. Without the atmospheric oxygen we could not have come into existence.

References

Barghoorn, E. S., 1971, The oldest fossils: Sci. American, v. 224, p. 30–42.

Cloud, P. E., 1968, Atmospheric and hydrospheric evolution on the primitive Earth: Science, v. 160, p. 729–736.

Glaessner, M. F., 1961, Pre-Cambrian animals: Sci. American, v. 204, p. 72–78.

Henderson, L. J., 1958, Fitness of the environment: Beacon Press, Boston. (Reprint; originally publ. 1913.)

Oparin, A. I., 1968, Genesis and early development of life: Academic Press, New York.

Rush, J. H., 1957, The dawn of life: Hanover House, New York.

Wald, George, 1954, The origin of life: Sci. American, v. 191, p. 45–54.

8

Evolution of Living Things

The complex chain

The building and melting down of crust, accompanied by the movement of continental plates, is an astounding concept because of the sheer size and weight of the moving pieces and the colossal budget of energy that is used in making the system run. But the process of change in living things throughout the Earth's history is no less astounding because of the incalculable numbers of individual creatures it has involved and because of the intricate complexity of the biochemical activities that together constitute the stream of life. Throughout Earth history individual animals and plants have been buried and their shapes and structures have been preserved as fossils fixed in strata. In our time many fossils have been recovered and reassembled. Placed in sequence, they are seen to constitute continuous progressions. Some of these fit with and continue into the changes we observe in the generations of organisms that are living today. Together these progressions make up a complex chain with many strands that has been lengthening continuously, link by link, since self-reproduction gradually replaced chemical evolution a good deal more than 3 billion years ago. The theory of this chain, the theory of the evolution of organisms, makes the history of life on Planet Earth understandable rather than a bewildering jumble of fossil forms. Also, as we noted in Chapter 3, it has made possible the correlation of strata between regions, and so has led to the construction of the geologic column.

So when we attempt here to outline the history of living things, we begin logically with the process of evolution that has guided the history we read from fossils. We begin with the evidence on which our concept of the process is based, and then continue with the mechanisms by which the process works.

The evidence of evolution

The belief that evolution has always occurred and is occurring today has been held almost universally among scientists during the last hundred years. Research in several fields of science has contributed to the tremendous body of evidence that supports the belief, and we will review some of this evidence. The belief is based on the relationships among fossils and among embryos, on features of anatomy, and on experience with the controlled breeding of plants and animals.

Fossils. Perhaps the most convincing support for the theory of evolution comes from the assembled record of fossils. When the groups of fossils collected from different strata are compared, it is apparent that progressively more complex organisms are contained in progressively younger strata, from base to top of the geologic column. This is true, for example, of the fossils in the strata of the Grand Canyon. We noted in Chapter 7 that the oldest known fossils are protists, the next oldest include simple plants, and the next oldest simple animals. This increasing complexity continues upward through the strata (and onward in time), with man appearing last of all in strata of late Cenozoic age. We can hardly fail to accept this relationship as evidence that ever since it began, biological evolution has never stopped creating new forms and new structures.

Now, if the various fossils are compared not with their "home" strata but with each other and with the organisms that are alive today, another striking relationship appears. When they are arranged with the most similar kinds grouped together, the arrangement resembles the trunk, branches, and twigs of a tree, with kinds that are living today forming the tips of the twigs (Fig. 8 - 1). As we follow these organisms downward from the twigs, we find the change from one to the next to be very small, but the cumulative change from a twig to a large branch is usually great. The fossils change downward, through larger and larger branches, into ever-simpler forms, with the simplest at the base of the tree trunk. By one path or another, all these forms of life are related to each other like the names of one's ancestors in a genealogical chart. Even the tiny protist is an ancestor of man, as can be seen if we follow the line far enough downward.

Of course this *"tree of life"* is still very incomplete. Some of the connections between branches and twigs are not yet continuous, although such gaps are not shown in Figure 8 - 1. These gaps in the series of fossils are links in the chain that are still missing from our record. But gradually, one after another, the missing links are being discovered and fitted into the chain. One of these is *Ichthyostega,* earliest and most primitive of known fossil amphibians, found in Devonian strata in Greenland in 1948. It is so similar to one of the Devonian fishes that only its limbs show it to be a land animal (Fig. 12 - 11). It shows a connection between fishes and amphibians.

Another link that is no longer missing is *Archeopteryx,* the most primitive bird we know of, first found in Jurassic strata in Germany in 1861. A "feathered reptile," it is a link between reptiles and birds, and is classified as a bird only because it is feathered. Finally, there is the very early man, *Australopithecus africanus* (Fig. 19 - 5),

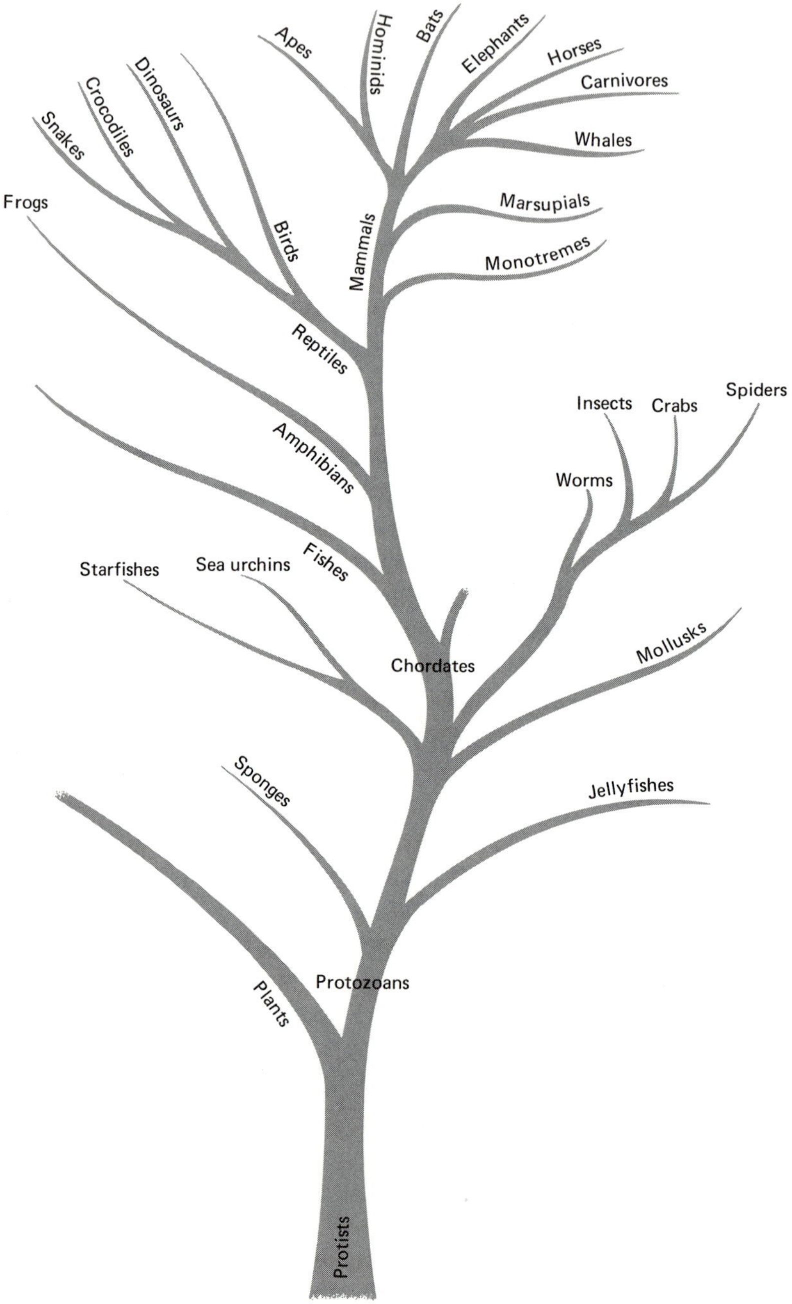

Figure 8 - 1. The "tree of life," showing the branching relationship among forms of life, both living and fossil. The "tree" shown here is very incomplete, representing only a few familiar kinds of animals and no details of plants.

Figure 8 - 2. A coelacanth, a "living fossil." This specimen was caught in March 1966. (Yale Peabody Museum.)

found in South Africa in 1924, that is far more primitive than the human fossils known before that time. It became a new link with much more primitive ancestors of man.

Some of the strands of the evolutionary chain have turned out to be much longer than had been supposed. In 1938, fishermen in the Indian Ocean caught a strange-looking, very large fish, which proved to be one of the coelacanths (pronounced *seé-la-kanths)* (Fig. 8 - 2), a primitive kind of fish having curiously lobed fins. This group is well known among fossils from Devonian strata up through Cretaceous strata, and was supposed to have become extinct before Cenozoic time began. But the discovery of a living coelacanth has added some 70 million years to their long history. These fishes are close relatives of the lobe-finned Devonian fishes that evolved into amphibians, as we implied above, and that form part of the chain of man's ancestors.

Another "living fossil" is a redwood tree, *Metasequoia* (Fig. 8 - 3), which prior to 1941 was thought to have died out in mid-Cenozoic time. In that year, however, it was discovered alive and well in the remote interior of China. This tree, a relative of the familiar *Sequoia,* has now been grown from Chinese seeds in other parts of the world.

Similarities among embryos. A different kind of evidence of evolution is found in the study of embryos. The developing embryos of the various kinds of vertebrate animals are strikingly similar to each other during the earlier period of their growth, and become less and

Figure 8-3. *Metasequoia* growing in the United States from a seed sent from China. (P. E. Olsen.)

less similar as their growth proceeds (Fig. 8-4). The closer the relationship between the kinds of animals (as that between man and apes), the longer the similarity between their developing embryos persists.

An especially noteworthy feature is the similar gill slits that are visible in early phases of animal embryos. The slits, however, change differently in different animals as they develop. In fishes the slits and their related organs become a device for breathing with gills. But in birds and mammals, which breathe in a different way, the

Figure 8 - 4. Developing embryos of five different vertebrate animals, showing how similarity decreases during development. Gill slits are very apparent. (Not to scale. Drawn from various sources.)

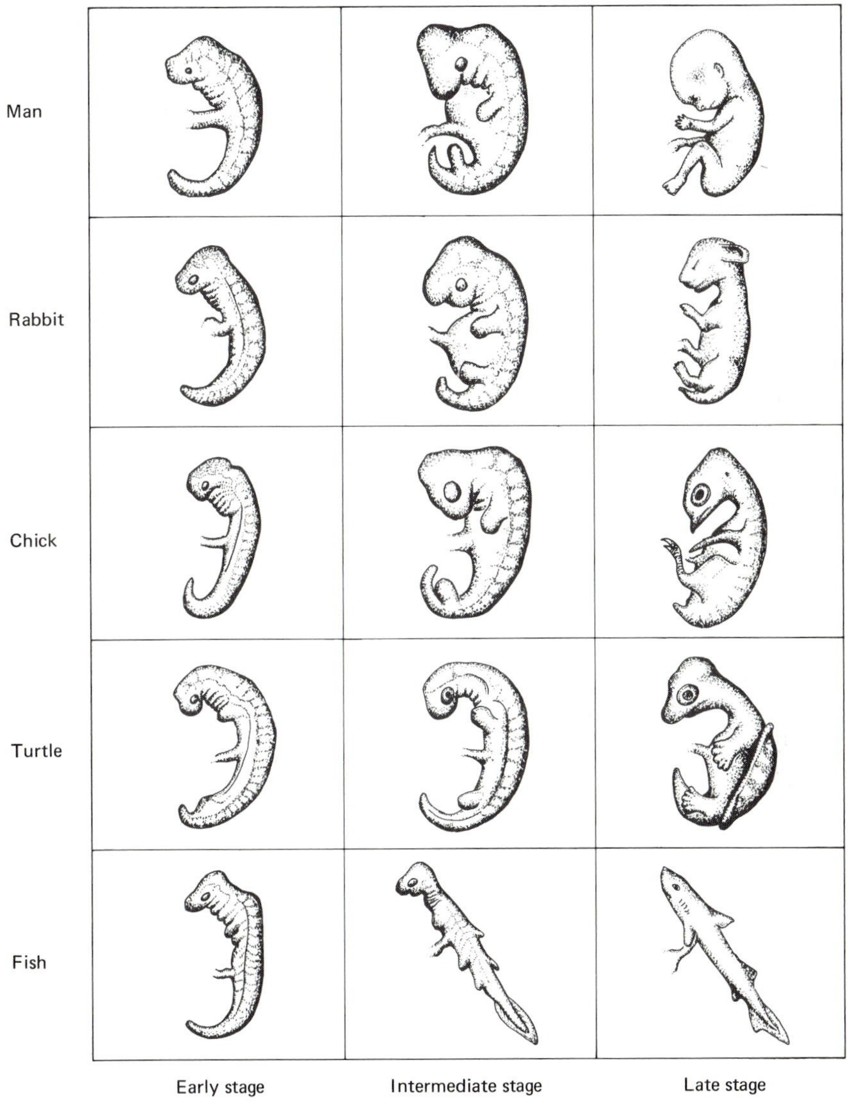

slits are useless for breathing. Instead, they or their remnants are remade into organs for creating and for hearing sounds. In man, for example, parts of the ear and throat are developed from embryonic gill slits. This is strong evidence that man and other air-breathing

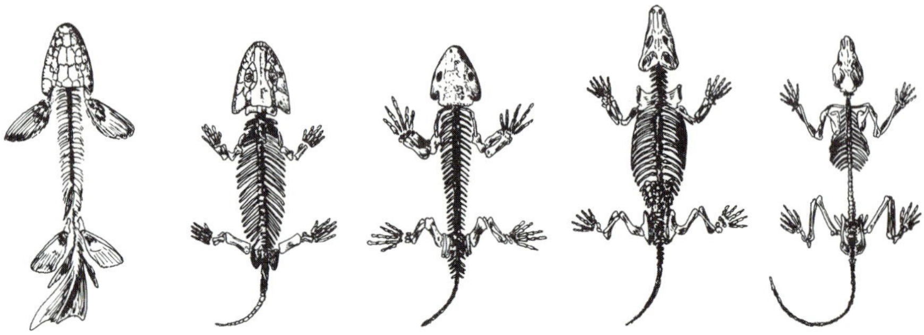

Figure 8 - 5. Skeletons of a fish and seven land animals, drawn to scale and arranged in similar positions to show similarity of parts and organization. (After W. K. Gregory.)

vertebrates have retained embryonic features they inherited from distant gill-breathing ancestors that lived at least 360 million years ago.

Similarities among skeletons. When the skeletons of ancient and modern vertebrate land animals are compared (Fig. 8 - 5), they tell a story like that told by the comparison of embryos. The skeletons consist of similar arrays of bones put together in the same basic pattern. Only the sizes of the individual bones and their relative proportions differ from one kind of animal to another. The common pattern, so remarkably consistent, can mean only one thing: the skeleton of a living land vertebrate is basically a very old-fashioned structure. Although through the years it has been stretched here and compressed there, at one time enlarged and at another reduced, it has never been scrapped and completely replaced with something new. It has only been remodeled again and again, and that very slowly, by numerous very small changes. Here once more is compelling evidence that the higher animals of today have evolved slowly from remote ancestors, inheriting much of what those ancestors possessed.

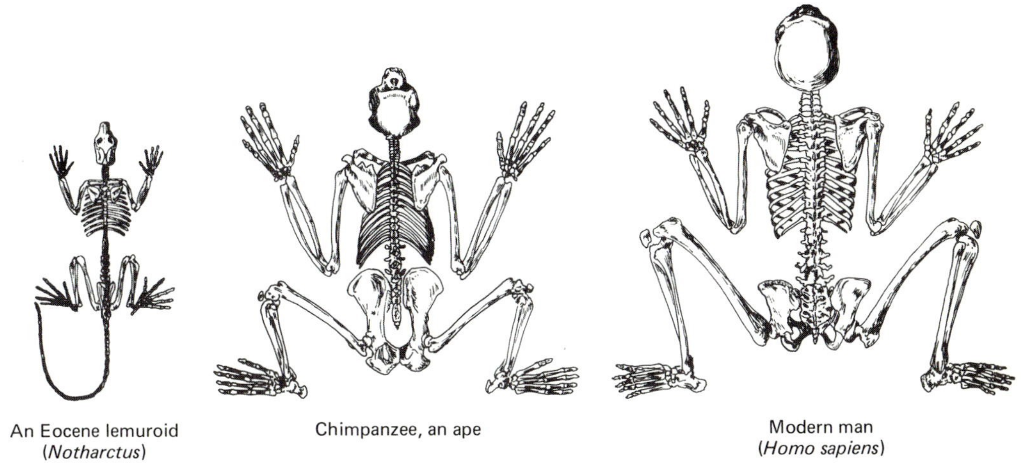

This can be seen even more clearly if we look in greater detail at some skeletal part, a forelimb for instance (Fig. 8-6). The basic pattern of the forelimb is present in all vertebrates, but in each kind it has been altered in different ways that make it useful for the types of locomotion necessary for life in different environments, such

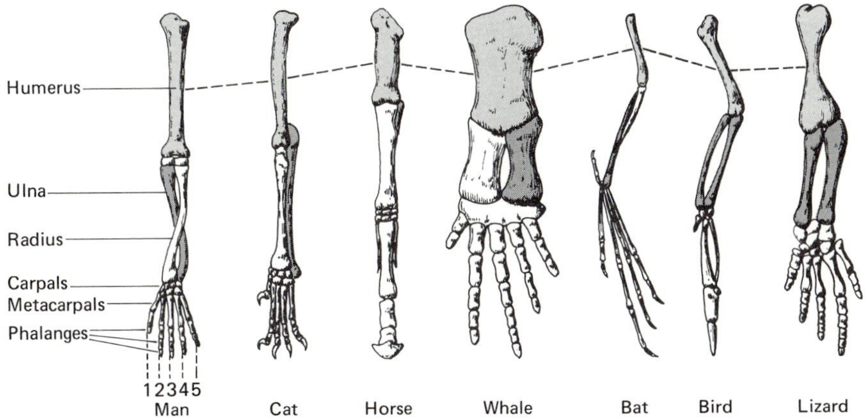

Figure 8-6. Forelimb bones of various vertebrate animals. All have the same basic pattern, with varying degrees of modification for locomotion: for flying (bat, bird), swimming (whale), rapid running (horse), and sprinting (cat). In lizard and man the basic pattern is little modified.

as the air (birds), the sea (whales), wide grassy plains (horses), and woodlands (cats). So environments have left their marks on the skeletons, as we shall soon see.

Vestigial structures. Both bones and soft parts of many living animals include structures that have no apparent function, but that are similar to structures in other living animals or in the fossils of extinct animals, in which their function is obvious. In those animals in which such structures are useless, they are called *vestigial structures* because they are vestiges of something once useful to an ancestor. In the human body alone there are more than 150 such structures, of which perhaps the most widely known is the appendix, a blind alley that can become infected and in many cases may have to be removed surgically. In some lower animals, however, the corresponding structure functions as a useful part of the digestive system. Less well known is the coccyx. This consists of seven vertebrae, complete with muscles and nerves, joined together at the base of the spine. It is a vestigial structure that in elongated form constitutes a tail in most other animals. The small, nonfunctional splint bones on the legs of a living horse are the vestiges of functional toes in the horse's three-toed fossil ancestors (Chapter 16). Whales and some snakes have vestigial bones that correspond to bones in the hind limbs of four-legged animals. Some flightless birds such as the ostrich have vestigial wings, remnants of the functional wings of other birds. These examples are just a few of the very large number of vestigial structures that have no rational explanation except that they became vestigial through evolution.

Domestication of animals and plants. By controlled breeding, through periods ranging from a few years to a few thousand years, man has succeeded in developing many varieties in single species of animals and plants. A familiar example consists of the very large number of "breeds" of dogs developed from what was apparently a single species resembling the wolf. The variety of "breeds" in terms of size, form, and adaptation for various purposes is great, although all still belong to a single species. These artificially controlled changes demonstrate the heritability of the characteristics selected by the breeders. The changes imply that changes of similar kinds have been taking place throughout the long history of living things.

Taken together, the lines of evidence we have mentioned make an extremely strong case for the theory of evolution. Indeed, evolution is hardly a theory any more. It is a fact.

The process of evolution

After this review of the very strong evidence that evolution is a reality, we can turn to the process itself, looking at three aspects of it. The first consists of molecules that can construct copies of themselves and transmit instructions in code. The second consists of variations that occur in individual organisms and that can be inherited. The third consists of environments and their influence on the statistics of populations. If these three aspects are understood, it is not difficult to understand the process by which evolution, so clearly demonstrated by the record of fossils, has occurred and is occurring.

DNA molecules. In the body of an organism, every individual cell that has a nucleus contains a quantity of an important substance called deoxyribonucleic acid (DNA for short). A molecule of DNA is unusually big for a molecule; yet to be visible in an ordinary-size photograph it has to be enlarged several hundred thousand times. Photographs made with an electron microscope show its shape to be that of two intertwined spiral ladders (Fig. 8-7). This big, spiral DNA molecule has a remarkable capability for self-duplication. It can split, apparently down the middle of the ladder, into two exactly equal halves. Then each half can construct a replacement half, and so complete itself as a whole ladder. The result, then, is duplication —two complete ladders where only one existed before. Duplication is possible, it is believed, because each half of the original ladder consists of four chemical components that can be arranged in different sequences. Like the dashes and dots of the Morse Code long used in telegraph messages, the order in which the chemical components are arranged constitutes a genetic code, which spells out as "information" or "instructions" the form that the new half of the ladder is to take. The coded "messages," then, are a sort of blueprint. They specify the kinds of chemical building blocks that are to be used. Also, they specify the patterns in which the blocks are to be put together. The patterns form the particular chemical compounds needed by a cell or group of cells in order to develop and to

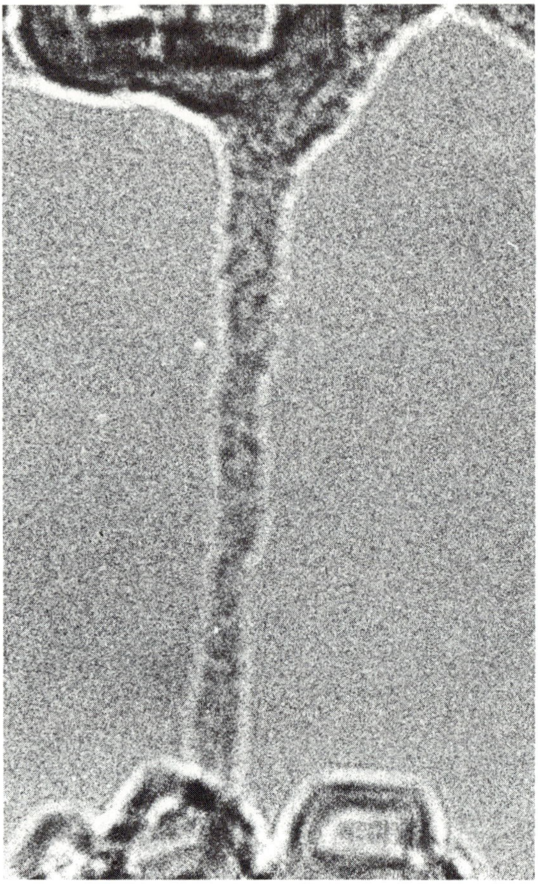

Figure 8 - 7. A DNA molecule magnified 1,750,000 times, seen with the aid of an electron microscope.

Taken from the body of a fish, this molecule forms the long, narrow, vertical image that crosses a gap between two groups of carbon crystals, part of the preparation for making the photograph. The double spiral is just visible but the rungs of the ladders are not. If a large automobile were photographed at the same magnification (1.75 million times) the photo would reach from New York to San Francisco. (Courtesy L. L. Ban, Cities Service Company. Page 93 *in* Thomas, J. M., and Roberts, M. W., eds., 1972, Surface and defect properties of solids—Volume 1: The Chemical Society, London.)

function. The building blocks, of course, are derived from the food eaten, digested, and ready for use by the organism. The amount of food must therefore be large. The difference in weight between a fertilized human egg cell and a full-grown 150-pound man represents a weight increase of nearly 50 billion times.

This is the way in which an individual organism (a collection of specialized cells) grows throughout its lifetime, by the splitting and reconstruction of cells under the guidance of coded instructions contained within the molecules in the cells themselves. As cells subdivide, the instructions are duplicated and reduplicated, so that every nucleated cell in the body has its own copy for immediate use. Thus in each individual organism, not only its growth but every one of its traits and characteristics is specified. If such great detail is hard to visualize, we need only remember that a DNA molecule, although quite a big molecule, is very small in terms of ordinary human experience. With a diameter of about .0000003 inch, it can exist in the human body in almost inconceivable numbers. The complexity of the myriad activities that are going on at any moment in a big city like New York, where there are 8 million people doing perhaps 4 million different things, is not much when compared with the creation, movements, and destruction of molecules of DNA and other compounds in the body of a single human inhabitant of that city. Because the number of such molecules is certainly much greater than 14 trillion times 4 million, the human activities in New York are few and simple by comparison.

In summary, DNA molecules, with their built-in instructions and their ability to duplicate themselves, specify the master plan for the growth and functioning of the cells that constitute an organism, and hence all the traits and characteristics of the organism. We must next see how the coded instructions, and therefore the traits, are transmitted from one generation to the next. These instructions lie at the root of heredity.

Heredity with variation. As specified by the genetic code, characteristics of form and behavior are transmitted from parents to offspring, but in the process they are changed, at least somewhat, because of the cell geometry that is involved in sexual reproduction. In an individual of any kind of organism that reproduces sexually (and that includes almost all animals and plants), the sex cells contain long thread-like *chromosomes*, each of which contains a large number of genes, the units that control heredity. A gene is a part of the DNA molecule, that part whose pattern contains all the instructions for duplicating that particular individual—its form, its growth, its physiology, and its behavior.

Now when a sperm cell from a male parent unites with an egg cell from a female parent, the result is a complicated joining of the cells to produce a new, individual composite cell having a combination of the chromosomes from each parent. The fertilized cell multiplies by subdivision. In all these multiplications the new combination of chromosomes is duplicated exactly. The coded pattern possessed by this combination is of course different from that of either parent. Accordingly, the new individual will have traits that can combine those of each parent in almost any proportion. The chromosomes in a single human individual are capable of producing an estimated 2^{1000} possible combinations. This is an almost incredibly large number, a number much greater than the total number of protons and electrons in the entire universe. Because of such almost unlimited potential combinations, it is not possible for two individuals to be exactly alike. Each will have received genes from each of its parents, and these genes will have formed a combination that is unique. The combination provides a unique set of coded instructions, according to which the traits of the individual possessing them will develop and will be passed on in some proportion to the succeeding generation.

The essential result of this process is *variation.* Because each individual in each generation differs somewhat from its parents, variations are occurring continually in a pattern of numerous small, inconspicuous changes, all of which can be inherited.

Natural selection. At this point environment enters the scene. Each individual in each generation, with his own particular packet of variations, is born into some environment. He responds to the environment by adjusting to it, but the extent of his response will depend partly on the variations he has inherited. Some individuals in a large population will fit more successfully into the environment than others will, simply because one or more of their particular variations gives them a slight advantage within their surroundings. So, statistically, one group of individuals will have a very slightly better chance to survive and reproduce than will the rest of the population. Because the variations can be inherited, the next generation will have a slightly larger proportion of individuals that possess the favorable variations. If the environment remains unchanged it will be only a matter of time (although perhaps a long time) until the whole population will consist of "favorable" individuals.

This is so important a matter and is so frequently misunderstood that we must repeat it here. In any population, variations are appearing in each generation. From this continual offering of variations the demands of the environment tend to select the best (that is, those most nearly adapted to the environment). The results are quite likely to give the impression that the individuals within the population are consciously trying to conform. Nothing, however, could be farther from the truth. The process of conforming is no more than a series of blind reactions. Evolution has no program. But the fineness of detail in which the conforming organisms respond to the imperatives of the environment is little short of marvelous.

A simple but clear example of this process is the recent history of the moth population in Britain. When observations and counts made today are compared with those made about 1850, it appears that about 60 species of British moths have changed in appearance. As a result of the Industrial Revolution, buildings and even trees became grimy with soot. In 1850 less than 10 percent of the moth population was dark colored, but now more than 90 percent is dark colored. The original range of colors was the result of variation. However, as light colors show up more distinctly against a dark background, birds were able to find and eat more of the light-colored moths, leaving fewer of them to survive and reproduce. For moths, the critical factor in their environment was the presence of predatory birds. The moths did not consciously adapt *themselves* to the darkening of their environment; they *became* adapted to it because of the normal activity of birds, and the adaptation shows up in the statistics of color in moths.

This unequal survival, which is caused by pressures in the environment and which through reproduction influences the makeup of succeeding generations, is called *natural selection.* It is a process of sorting and shuffling. Although confined to organisms, natural selection is similar in its effects to the inorganic selection, the sorting and shuffling by other natural processes (Chapter 6), that operates to create a continental crust. In environments far below the Earth's surface, certain minerals are "selected" because of their relatively low melting points and the melt moves upward. At the surface, minerals are "selected" because of their susceptibility to chemical destruction and their components are carried away. Indeed, natural selection becomes more understandable when we view it as no more than one of the many complex natural processes that characterize the Earth.

In natural selection the change from one generation to the next is extremely small, but like the cutting of the Grand Canyon or the sorting of the particles derived from the weathering of granite, the cumulative result can be enormous. As we have noted in earlier chapters, geologic changes are occurring continually at the Earth's surface and are causing changes in environments. These in turn generate pressures that cause plants and animals to adapt to them.

We must remember that such adaptation is unconscious and is not the result of any purpose. It is automatic because it consists of little more than alteration in the probability of survival for each individual in a population. Each, with his own unique combination of characteristics, faces the demands of his environment. Some individuals survive, others are weeded out, and each new generation is therefore a little better adapted to its surroundings than its parents were.

All over the world we see specific kinds of animals and plants that, through adaptation, possess structures and traits that are specialized to particular environments. A polar bear, with its thick fur coat, the all-over layer of blubber beneath its skin, and its skill as a swimmer, is well equipped for life on Arctic sea ice. A camel, with its long legs and built-in techniques for conserving water, is well equipped to live in broad continental regions having little water and little vegetation. Other organisms are adapted to life in tropical forests, to life in the air over wide oceans, to life in a pool within a cavern, and so on through a long list of environments. Each of these specializations is the product of evolution guided by natural selection.

If we study fossils carefully, we can see the effects of adaptation to the same range of environments through much of the past. The environments were distributed differently because of geologic changes in both continents and seas. These ancient adaptations are a secondary effect of the Principle of Uniformity.

Mutations. If the variations we have described were the only heritable changes that occurred in organisms, evolution would occur. However, another sort of heritable change does take place. It is called *mutation* and it adds to the variety of the living material that is exposed to the selective influence of environments. Mutation that occurs in an egg cell or sperm cell in an individual can be inherited by offspring. Such mutation consists of a change in the coding machinery within the DNA molecule, so that some gene in a new individual differs from the corresponding one in either parent. It occurs suddenly and unpredictably, and because it is heritable it results in changes in structures, traits, or characteristics that show up in subsequent generations. Once exposed to environment, these changes, if favorable, are perpetuated or, if unfavorable, are "selected out." The basic chemistry of mutation is not understood in every detail, but one can see that mutational changes could have an important influence on the direction taken by an evolutionary chain of organisms. They almost certainly do.

Discussion

Evolution consists basically of *adaptation* to environment in response to the directing force of natural selection. Changes, whether by mutation or by the new combinations that are formed with every generation, occur continually. These provide the raw material upon which natural selection works. All sorts of new designs are tried out. Those that are successful in the environment are retained; the unsuccessful ones are gradually lost.

When Charles Darwin first described natural selection in 1859, he discussed the process in terms of the fitness of organisms for life in their environments, of competition between individual organisms and between groups of organisms, of the achievement of success by some groups, and of the elimination of the less fit. Not unnaturally, this concept was later taken over by other people and applied to human behavior and human relations. In such an application evolutionary changes become, at least in part, a matter involving morality. So we must distinguish clearly between the meaning of terms such as competition, fitness, and success in the context of evolution and their meaning in the context of human relations.

In terms of geologic history, *success* in a group of organisms depends on just two factors: (1) ability to reproduce effectively, and (2) ability to occupy and hold territory against competition. In the history of dinosaurs, flowering plants, or glacial-age mammoths these matters are non-moral. Morality, a human concept, plays no part in the business until well on in the history of man.

We need to add here that study of the record of fossils has not led to the recognition of any overall purpose in evolution. It has not shown that evolution has been following any particular direction towards any specific goal. In other words, apparently evolution has no program. Very evidently it has not followed straight lines. Its course branches repeatedly (as we saw in Fig. 8-1) and the branches have many bends, but it has never repeated itself exactly. We explain this pattern as opportunistic, the result of a continuous series of responses by living matter to environmental opportunity.

How long has it taken the process of evolution to create the huge variety—more than a million species—of animals and plants now living? Near the beginning of the process, the first protists, as we noted in Chapter 7, gradually appeared. This was more than 3 billion years ago. Nearer to our time the world of living things had reached essentially its present state by the end of the Pliocene Epoch, perhaps two or three million years ago. Between those two times most of the evolution of life must have occurred. That the progress of the Earth's organisms to their existing state of adaptation and variety could have been accomplished within a span of somewhat more than 3 billion years surprises no one who has studied the processes involved. Scientists believe that the time has been ample. Indeed, had the geologic processes that create environmental changes worked more rapidly than they do, probably the same degree of evolutionary development we see today could have been achieved within a shorter time.

The rate of evolutionary change, of course, has been extremely variable from time to time, from place to place, and from lineage to lineage, depending as it does on the rates and places of change in the Earth's environments. For example, our collection of fossils shows that throughout the last 63 million years of geologic time—that is, throughout the Cenozoic Era—invertebrate sea animals evolved slowly, changing little, whereas most land animals evolved much more rapidly, changing greatly. The difference reflects differences in rates of change of environment. Land environments responded promptly to uplifts, the making of mountains, and changes in temperature and rainfall, while environments in the sea remained comparatively stable. Yet in terms of the human calendar all rates of evolution *seem* slow. In the period, perhaps 10,000 years long, during which dogs and cattle have been domesticated and bred by artificial selection, not a single new species has been recognized. The artificial changes amount to no more than the creation of varieties within single species.

We must repeat this fundamental principle. As long as an environment is stable, its populations of organisms remain stable also, with little evolutionary change. But when the environment begins to change (that is, becomes unstable) its populations become unstable also. They begin to adapt rapidly; in other words, evolution speeds up. Environments, then, afford opportunities for the testing out of the biological "inventions" that are always appearing. As we trace the further history of living things as learned from fossils, we shall see many examples of such "inventions," some of which have been successful while others, having failed the test of adaptation, have disappeared from the face of the Earth.

The close dependence of organisms on environment, the sensitive way in which living things keep readjusting to environmental changes, so clearly demonstrated by the fossils collected from ancient strata, carries with it a lesson for modern man. Areas of the world in which man has become industrialized to a great degree are today undergoing exceptional environmental changes. Some of those changes are more obvious, and more disadvantageous for at least some organisms including man, than the changes brought about by change of climate and other natural processes. Change of both kinds can be expected to provoke corresponding responses (both adaptations and extinctions) in the local biosphere, including man himself. The possible consequences should not be neglected by the one species that is causing the changes and is capable of controlling them, with results beneficent or disastrous for the plants and creatures of the Earth.

References

Blum, H. F., 1951, Time's arrow and evolution: Princeton University Press.

DeBeer, G. R., 1964, Atlas of evolution: Thomas Nelson & Sons Ltd., London.

Dobzhansky, Theodosius, 1950, The genetic basis of evolution: Scientific American, January 1950, p. 2–11.

Moody, P. A., 1962, Introduction to evolution: 3d ed., Harper & Row, Inc., New York.

Smith, H. W., 1961, From fish to philosopher: The Natural History Library. Anchor Books, Doubleday & Co., Garden City, New York. (Paperback.)

Stebbins, G. L., 1966, Processes of organic evolution: Prentice-Hall, Englewood Cliffs, N.J.

Volpe, E. P., 1967, Understanding evolution: W. C. Brown Co., Dubuque, Ia. (Paperback.)

Yanofsky, Charles, 1967, Gene structure and protein structure: Scientific American, v. 216, p. 80–94.

9

The Nature of Fossils

The fossil record

Our discussions of the geologic column and of organic evolution have made frequent mention of the fossil record. This term includes the whole series of collected, identified, catalogued, and described fossils of all kinds maintained in museums, universities, and other research institutions around the world. It comprises hundreds of collections and millions of fossils large and small, from huge dinosaurs and whales down to microscopic animals and plants. The fossil record includes not only the fossils themselves but everything written about them, whole libraries of scientific books and articles in which fossils are carefully described and illustrated and their positions in the geologic column discussed. Already 100,000 species of fossil marine invertebrate animals have been fully described, and more are being added to this number continually.

So the fossil record is a very big and very real affair involving thousands of scientists and technicians, as is apparent to anyone who makes a thoughtful visit to one of the institutions devoted to the study of fossils. The inquiring visitor wants to know what fossils are like, how animals and plants can be preserved for millions of years, and how fossils are reconstructed if reconstruction is necessary. The answers to these questions should make our history of living things much more comprehensible.

What could have been learned about the history of the Earth if no samples of the world's former populations had been preserved as fossils? Of course we would have known about the geologic cycles, the Principle of Uniformity, and the movement of plates of the Earth's crust. But our geologic column would be very primitive and far less useful than it actually is, our reconstruction of former environments from the study of strata would be far less complete, and our tracing of the paths of evolution through time would be largely a matter of guesswork.

Preservation

The existence of fossils (and in what vast numbers!) we owe to preservation—the conservation of things that were once alive. Although most of the individual organisms that formerly lived on the Earth have been destroyed, here and there bodies or parts of bodies were protected against oxygen, that universal destroyer. They were buried quickly beneath sediment and then were converted to rock by deposition of mineral matter from the slowly percolating water that saturates the ground. This, in broad outline, is the history of most fossils. Let us look more closely at preservation and see just what happens.

When an animal or plant dies, most of the substances of its body are vulnerable to rather quick destruction. Other organisms (among which bacteria are conspicuous) consume them and chemical decay decomposes them. In organic matter as in inorganic minerals (Chapter 4), a common reaction in chemical decay is oxidation. We see it in the wood of a fallen tree trunk that has lain a few years on the ground and has become brown and punky. The fossils in the strata represent the tiny proportion of former living things that by chance escaped this common fate. They are the exceptional bodies that were protected against scavengers and against oxidation at or soon after the moment of death, usually by being submerged in water, as is explained below.

The great majority of fossils are those of marine invertebrate animals—clams, snails, and the like. The soft tissues were eaten alive by predatory creatures or consumed by scavengers soon after death. But the hard shells remained and were covered up by sediment drifting down onto the sea floor. On land the chances of preservation were poorer. But even there some animals and plants, alive or dead, fell or were swept into water, which protected their bodies against oxidation. Others were suddenly buried in accumulating sediment, which soon afterward became saturated by ground water. Stream floods are often responsible for burial of both animals and plants (Fig. 9-1). The rapid deposition of volcanic ash, and sometimes even of flowing lava, is another cause.

In short, either quick burial or the presence of surrounding water, or both, are the chief factors that start the preservation of a fossil. A third factor is almost equally important—the presence in the dead organism of hard substances such as shells, bones, teeth, or horns that resist destruction by scavengers or by chemical decay. Nearly all fossil mammals consist of skeletons only; the soft tissues are rarely preserved, and then only in very special environments. However, if a whole body should be buried it can be preserved without alteration of the minutest detail (Fig. 3-2). Even the play of colors on the wings of fossil insects has been preserved intact through hundreds of millions of years.

Means of preservation. The making of a fossil does not necessarily end here. Other things can happen to it, including the effects of processes that completely alter its composition and others that destroy its substance but preserve its outer form.

Figure 9 - 1. This sandy creek bed, dry or nearly dry most of the time, becomes a muddy torrent during heavy rains. Animals are drowned, swept downstream, and quickly buried in alluvium. (Courtesy of Eli Lilly and Company.)

The simplest example is a body that became encased in ice, asphalt, or some other natural substance that preserved the material of the body itself. Dramatic examples are the 36 mammoths and several woolly rhinoceroses found in northern Siberia at various times during the last 200 years, preserved in frozen ground dating from the latest of the glacial ages. In at least one find the meat was still fresh enough to be eaten by sledge dogs. A popular but incorrect story has it that some of these mammoths were found in the ice of glaciers. This is not true; as far as we know, all were encased in frozen mud. Probably the animals became mired just before the autumn freeze-up. The mud froze to a great depth and then, early in the following spring before the mud had time to thaw, river floods covered it with a thick blanket of alluvium. The blanket prevented the thawing of what lay beneath. Radiocarbon dates of the skin and hair of these fossils show a spread of more than 30,000 years between the times of death of the earliest and the latest of the dated mammoths. Thus if we consider only the 36 mammoths found frozen, the miring and preservation of an animal need have occurred, on an average, only once every 800 years. In comparison, the number of mammoths that died more conventionally must have been very great. It has been estimated that the tusks of 20,000 to 30,000 glacial-age mammoths were found on the surface and were exported from Siberia as ivory before the year 1900—and perhaps a greater number of tusks and whole bodies still lie buried in frozen mud.

Other examples are the bodies of the glacial-age woolly rhinoceros, preserved even to their characteristic long hair, in asphalt that seeped out of the ground and eventually hardened. The preservation of mammal bones by the million in tar seeps in California is described in Chapter 18. The chemistry of such preservation in tar is much the same as that involved in the mummification of human bodies by the ancient Egyptians. The embalmers used petroleum, which excludes air and so prevents oxidation.

A far more common case—in fact, nearly the universal case—is one in which the soft parts are destroyed but the hard parts become petrified. The word means "made into rock," and conversion to rock is exactly what takes place. The processes are chemical, occur down within the ground, and are part of the whole slow conversion of sediment into sedimentary rock. Slowly moving ground water deposits mineral matter in the spongy inner parts of bones or in the cavities of the cells of plants (Fig. 9-2). Or ground water dissolves the original substance of shell or bone and replaces it with some other substance. Usually the replacing substance is silica, occasionally in the form of opal; but rarely, where the chemical environment is favorable, it may consist of some exotic substance such as pure

Figure 9 - 2. Petrified wood. Section of a small log, the cellular structure of which has been filled with silica and thereby preserved. (National Museum of Natural History.)

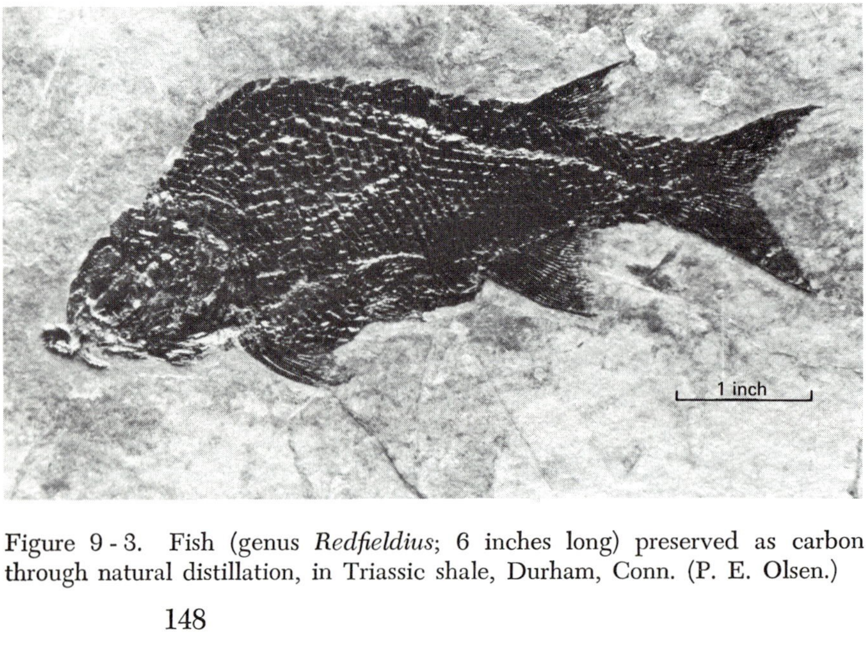

Figure 9 - 3. Fish (genus *Redfieldius*; 6 inches long) preserved as carbon through natural distillation, in Triassic shale, Durham, Conn. (P. E. Olsen.)

silver. Finally, petrifaction can be accomplished by natural distillation deep underground. In this process the hydrogen and oxygen in the original organic matter are slowly driven off as various gases, leaving only the carbon behind. In this way the form of the organism (frequently a fish) is preserved in nearly pure black carbon (Fig. 9-3).

Although most of the fossils in our record are shells and bones that have been petrified, there are other kinds of preservation. In layers of silt and clay deposited on the floors of deep lakes the impressions of leaves are common (Fig. 9-4). The leaves sink to the bottom and eventually are covered with sediment. The leaf substance decays but the impression of the leaf form remains. Sometimes the forms of insects are preserved in a similar way (Fig. 12-9).

An unusual sort of impression was discovered in a layer of basalt, a lava flow of Oligocene or Miocene age, near Coulee City, Washington. The flow, advancing into the shallow water of a small lake 20 or 25 million years ago, encountered and enveloped the body

Figure 9-4. Impression of the leaf of a sycamore tree in siltstone of Eocene age in Utah. The leaf itself has vanished; the dark color is mineral matter deposited by ground water while the original sediment was being converted into rock. (Bill Ratcliffe.)

Figure 9 - 5. The bloated death pose of the fossil rhinoceros from Washington, restored from a plaster mold. (W. M. Chappell and others, 1951.)

of a dead and bloated rhinoceros more than six feet long that was lying there on its side. The lava molded itself around the carcass and cooled so rapidly in the water that it did not destroy the organic tissue. In a similar way the lava enveloped a number of nearby trees. When discovered in 1935, the fossil consisted of a small cavern having the shape of a bloated rhinoceros (Fig. 9 - 5), and still contained part of the jaw and teeth as well as other bones of the animal. A plaster mold of the cavern was made, thus restoring the bloated carcass for scientific study.

Footprints and tracks (Fig. 9 - 6) are another kind of fossil even though in themselves they never consisted of organic matter. Only

Figure 9 - 6. Strings of footprints, most of them 10 to 18 inches long, the tracks of dinosaurs on a surface of stratification of Triassic sandstone, Dinosaur State Park, Connecticut. More than 200 million years ago the sandstone was sandy alluvium. Arrows point to much smaller footprints probably made by the animal in Figure 14 - 2. (John Howard for Yale Peabody Museum.)

Figure 9-7. Tracks of insects and footprints of birds on siltstone, some 45 million years old, along the shore of an Eocene lake in Utah. (Bill Ratcliffe.)

those which were covered with fine sediment before erosion could destroy them are preserved. In some examples even the minutely detailed tracks of small insects (Fig. 9-7) (and the impression of pattering raindrops too) are as sharp as on the day they were made. Such prints are not just curiosities. They have a lot of information to give about size, length of leg, presence of claws, hoofs, and the like, and about whether the animals walked, ran, or hopped. Another unlikely source of good information is the fossilized feces of animals. Sliced with a rock-cutting saw and studied under a microscope, fossil feces give information about the food of some kinds of animals, and have shown unmistakably that some were cannibals: they ate their own kind.

These, then, are the chief ways in which fossils are created and preserved: deep-freeze of bodies, petrifaction of bones and shells, natural molds, footprints—all of them accompanied by burial of some kind, usually very rapid. The story of a fossil is told in a nutshell in Figures 9 - 8 and 9 - 9.

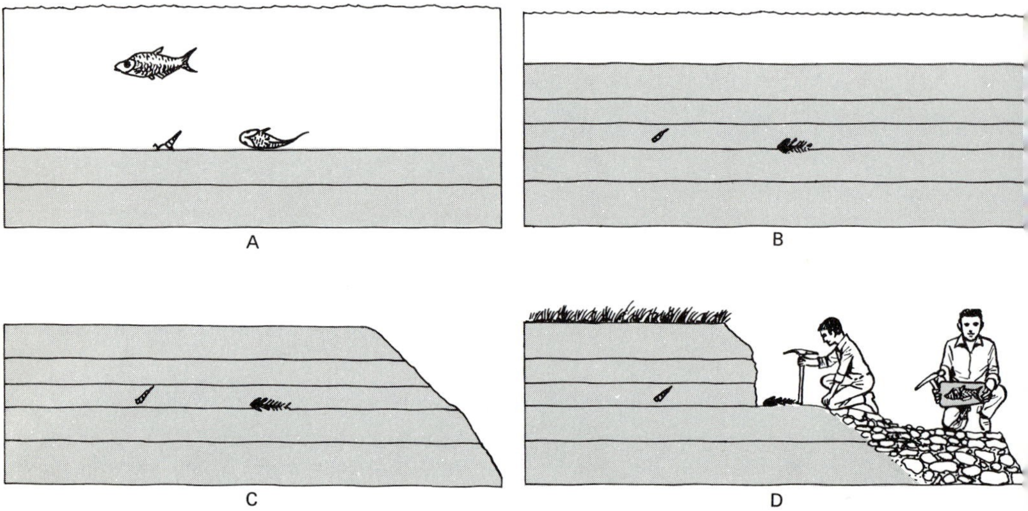

Figure 9 - 8. A fossil is made and discovered.

A. In an ancient lake a fish dies and sinks to the silty lake floor, on which a water snail is crawling.

B. The snail has died and, with the fish, has been buried by gradually deposited silt. The sediment begins to be cemented to form siltstone.

C. The lake has disappeared and the remaining layers of siltstone are eroded to form the slope at the right.

D. A geologist in search of fossils has exposed the fish, lying on what was once the lake bed. At right he displays the fish embedded in a slab of siltstone.

Reconstruction. To be properly studied by scientific specialists, a fossil must be reconstructed. Some fossils, especially marine invertebrate animals such as clams, snails, or oysters, require little or no reconstruction—perhaps nothing more than the fitting together of two halves of a complete shell. But reconstruction of a fossil vertebrate animal is likely to be a more complicated job. More often than not the individual bones are displaced or scattered. They are identified by numbering or in some other way and are carefully photographed in the position in which they were found. Then, removed to a laboratory, they are freed of all their stony matrix and reas-

Figure 9 - 9. A scientist and his assistant uncover the bones of the huge dinosaur *Diplodocus* in Jurassic sandstone exposed in the Bighorn Basin, Wyoming. (Courtesy of the American Museum of Natural History.)

sembled by technicians trained in anatomy. Figure 9 - 10 shows a scrambled mass of bones just as they were found in the rock, and skeletons restored from bones found in such disarray.

Reconstruction of the skeleton of an extinct animal is largely a matter of anatomy, but restoration of flesh, skin, and hair cannot be wholly accurate because it can only be based on analogy with some related animal that still exists. Even though the less obvious aspects of the anatomy of an ancient animal can only be guessed at, more information can be extracted from fossils than one might at first suppose. Fossils of land animals include not only bones, teeth, horns, and armor plate, but also excellent impressions of skin and footprints

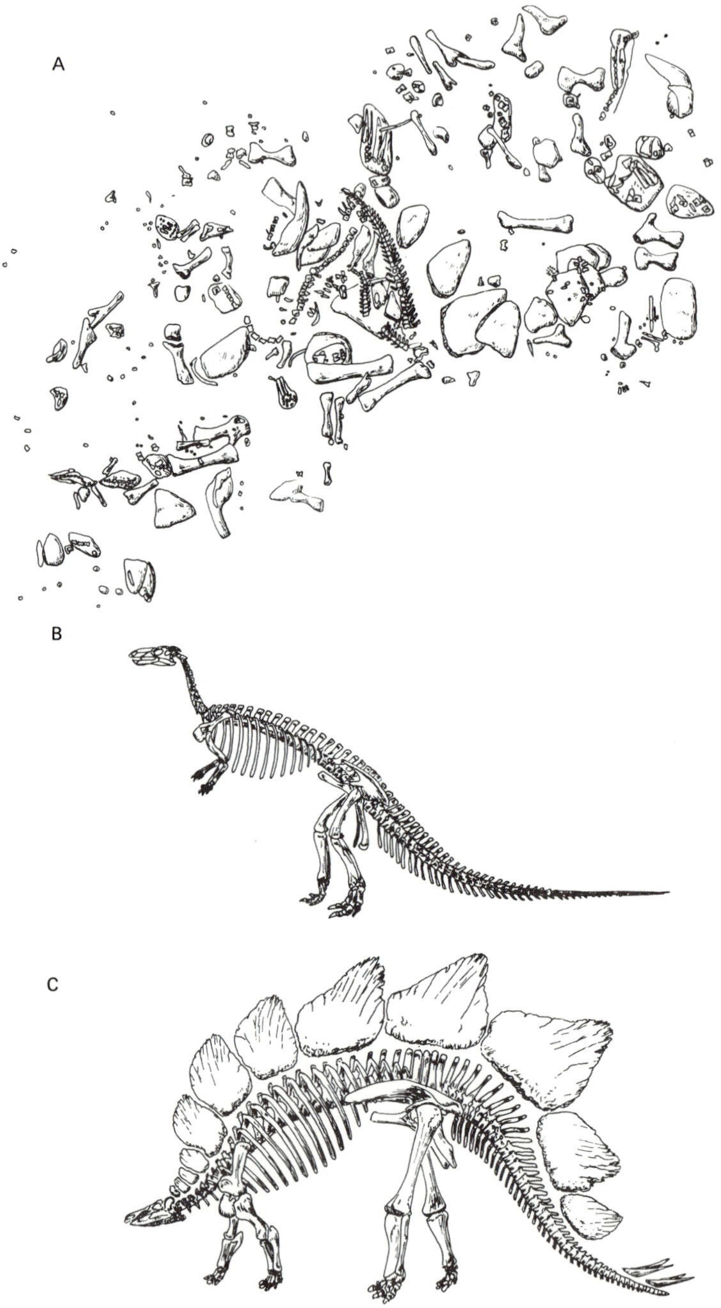

Figure 9 - 10. Scattered bones (A), are photographed at the site where they were found and are expertly reassembled in the laboratory to form reconstructed skeletons. These skeletons, of *Camptosaurus* (B) and *Stegosaurus* (C) are dinosaurs collected from similar bones in Late Jurassic strata in Wyoming. (C. W. Gilmore.)

preserved in detail. In some cases relative intelligence can be estimated from the size and shape of a plaster cast of the hollow within the skull that contained the brain. Skull configuration likewise can give information about senses, such as sight, hearing, and smell, and can even suggest whether an animal might have been capable of making sounds. Even the effects of disease can sometimes be discerned in fossil bones.

In restoring extinct mammals that lived during the latest of the glacial ages, modern scientists have been assisted (in advance) by Stone Age artists who sketched or painted on the walls of European caves the animals they hunted, just as they observed them (Fig. 9-11).

Classification

Species. Living things began to be classified long ago, when it was believed that each kind had been created separately. Later, when the process of evolution was understood, classification became far

Figure 9 - 11. Extinct glacial-age mammals accurately engraved by Late Stone Age artists.

A. Woolly rhinoceros. (Henri Breuil.)

B. A charging mammoth engraved on a piece of tusk ivory in western France some 10,000 to 16,000 years ago. The artist failed to finish the tusks, trunk, and forelegs, but did add some doodling (not shown). (After E. Lartet.)

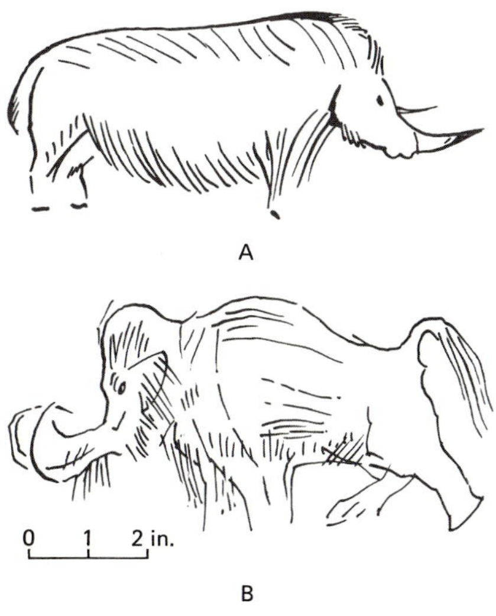

more necessary because of the huge cumulative number of organisms that were continually being discovered and that had to be catalogued for study. Without systematic classification, fossils would be as useless as a great library that lacked a card catalog.

The basic unit in classification is the *species,* a group of organisms that interbreed among themselves but not with other groups and that possess characteristics of form and structure common to the group. In living organisms species can be identified on the basis of whether interbreeding is possible. In extinct organisms, which include the majority of all known fossils, species can be defined only by analogy with living groups, or else by characteristics of form or structure, some of which are unrelated to reproduction.

A species is recognizable only because its members have been reproductively isolated in some way from neighboring populations, so that its genes cannot mingle with those of the neighbors. If isolation continues long enough the genes become incompatible with those of neighboring groups. When that happens, the isolated group has become a separate species by definition. If, on the other hand, the members of a group were not isolated, they would be free to interbreed with neighbors, and so in time the recognizable characteristics of the group would disappear. Species, then, are strictly a product of an isolated environment. Throughout Earth history, geologic activities have been causing environmental changes, and many kinds of isolation have resulted. Hence it is easy to understand that for nearly as long as life has existed there have been species. But owing to the slow geologic changes that are always in progress, species have become extinct while others have moved, merged, become isolated, and given rise to new species.

A convincing example of the process is found in the sea cliffs of chalk in England, facing the English Channel. The chalk is a white sedimentary rock consisting mainly of calcite in the form of millions of microscopic shells. It accumulated on the sea floor during a part of Cretaceous time. Besides the tiny shells the chalk contains abundant fossil sea urchins of a kind called *Micraster.* Many specimens were collected from a stratum at least 500 feet thick, apparently representing some 4 million years of deposition. The sea urchins at the base of the stratum are very different from those at the top, but the change from base to top is gradual. Those at the base consist of one group of species, whereas those at the top belong to distinctly different species derived by evolution from those below. This complete gradation illustrates the difficulty of setting boundaries to a fossil species, whose capabilities for interbreeding cannot, of course, be tested. Fossil species, therefore, as we have indicated, are defined on a basis of similarities of form and structure.

Linnaeus' catalog. In the middle of the 18th Century, a hundred years before evolution had been generally accepted, a Swede, Carl von Linné (who wrote in Latin and signed his name *Linnaeus*) spent long years making a catalog of living things. He classified plants and animals according to their obvious anatomical characteristics. Like others in that period, Linné thought of the many kinds of living organisms as unchanging products of Creation. Whereas that static concept is now interesting only as history, Linné's catalog has great scientific value. It is the basis of our modern classification of living things. Although improved upon in details, it was so well designed that its basic character remains unchanged, and it is still written in Latin, a language that for scholars at least is nearly universal. It gives each kind of organism a name consisting of two words. The first word names the broader *genus*, the second word the more limited species. Examples are *Pinus strobus* (white pine), *Felis domesticus* (house cat), *Felis leo* (lion), *Felis tigris* (tiger), *Felis pardus* (leopard), *Homo sapiens* (modern man). Successively larger groupings have been superimposed as umbrellas over the categories used by Linnaeus. Thus two or more related species are grouped into a *genus* (plural: *genera*), two or more related genera form a *family*, two or more families form an *order*, two or more orders a *class*, and two or more classes a *phylum*. Two or more *phyla* constitute a *kingdom*, the widest category of all because our three kingdoms embrace protists, plants, and animals respectively.

Modern classification. For our purpose in this book there is no need for a complete, comprehensive classification of organisms.* Therefore we will neglect protists, do a bit of simplifying, and so arrive at Tables 9-A and 9-B, our short-form classifications of animals and plants respectively.

* An illustrated, rather detailed classification of the three kingdoms is given in Ford & Monroe, 1971, p. 399-427. (See References at end of this chapter.)

Phylum *Protozoa* (Single-celled, mostly microscopic animals.)

______ *Porifera* (Sponges.)

______ *Coelenterata* (Corals, sea anemones, jellyfishes, etc.)

______ *Brachiopoda* (Lamp shells.)

______ *Bryozoa* (Sea mosses.)

______ *Echinodermata* (Starfishes, sea urchins, sea lilies, etc.)

______ *Mollusca* (Mollusks: snails, clams, oysters, squids, octopuses, etc.)

______ *Arthropoda* (Shrimps, crabs, lobsters, trilobites, spiders, insects.)

______ *Vertebrata* (*Chordata*) (Animals with backbones.)

Class (In reality 4 classes.) *Pisces* (Fishes.)

______ *Amphibia* (Frogs, toads, salamanders.)

______ *Reptilia* (Crocodiles, lizards, snakes, turtles, dinosaurs, etc.)

______ *Aves* (Birds.)

______ *Mammalia* (Warm-blooded animals that suckle their young.)

Table 9 - A. The Animal Kingdom.

Non-vascular plants:

Algae. (Simple plants, mostly aquatic.)

Fungi. (Mushrooms, molds, etc.)

Bryophytes. (Mosses, the simplest land plants.)

Vascular plants.

1. Primitive vascular plants.
2. Scale trees, club mosses ("ground pines").
3. Horsetail rushes.
4. Ferns and seed-bearing plants.

Ferns.

Gymnosperms. (Seed ferns, cycads, ginkgos, conifers.)

Angiosperms. (Flowering plants.)

Table 9 - B. The Plant Kingdom.

Numbers of species. Now that we know just what a species is, we can hardly help asking how many species, how many different kinds of animals and plants, there are. According to the best estimates, there are about 1.5 million species alive today. Of these, 860,000 are insects, 350,000 are plants, 8600 are birds, and only 3200 are mammals. Most of the remaining species, nearly 300,000 of them, are marine invertebrates. The total of 1.5 million living species consists only of those that have been described and published by scientists. It is estimated that twice that many still remain to be described. If that estimate is good, there are 4.5 million species living today. This number, of course, excludes extinct species known only as fossils. On the basis of the numbers of fossil species already described from particular strata it has been estimated that the total number of now-extinct species that lived through any part of the more than 3 billion years since life began ranges between 50 million and 4 billion. Although huge, these numbers refer only to species. When we remember that each species can embrace up to many billions of individuals at any one time (in 1972 there were nearly 3.4 billion people, members of the species *Homo sapiens*, alive in the world), we can begin to realize, however faintly, the size and complexity of the biosphere.

Fossils and stratigraphy

Changing sea level and marine strata. We have now seen how living things become fossilized and how fossils are classified so as to be of maximum use to science. With this information we can say something more about the relation of fossils to a sequence of sedimentary strata. Let us begin with the strata themselves and then deal with their content of fossils.

Figure 9-12, Block 1, represents a smooth, gently sloping coast consisting of layers of rock (it doesn't matter what kind). The layers have been bent by gentle pressure and then eroded so as to form a rather smooth surface. The surface has been partly submerged beneath the sea. Washing against the shore, surf has built a pebble beach. The pebble gravel of the beach grades seaward into sand, then silt, then clay, and with increasing distance from the shore this sediment becomes thinner.

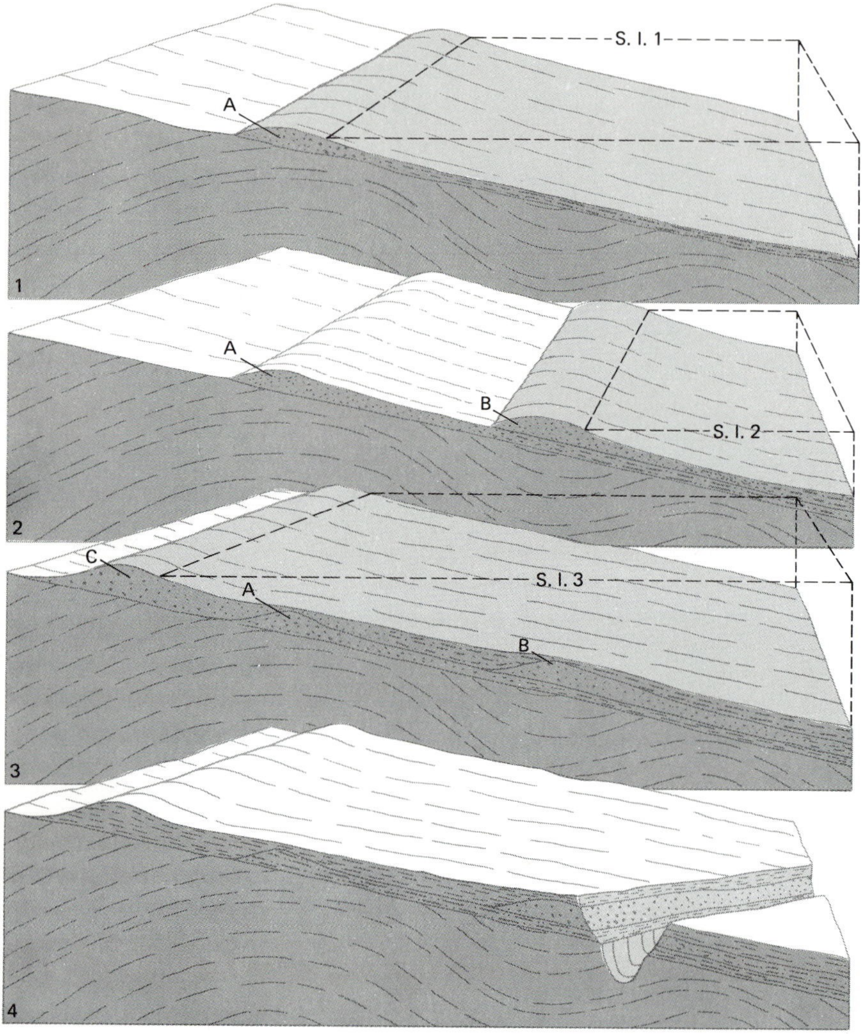

Figure 9 - 12. Model showing how a sequence of strata is built up in shallow coastal water.

Suppose the sea level (S. l. 1) sinks rather quickly relative to the land. This could happen for various reasons, but we are not concerned here with the cause. The shoreline moves seaward down the slope of the floor, just as it does when the tide falls, but through a much greater distance. This converts part of the sea floor into land, exposing it to weathering, stream erosion, and other processes that act on the land.

At a much later time (Block 2) the sea level rises again (S. 1. 2) but not as far as before. Along the new shoreline, surf builds a new pebble beach seaward of the older one and a new carpet of sediment begins to cover the sea floor, again grading in the offshore direction through sand and silt into clay. After this deposition of sediment the sea level subsides a second time relative to the land, once more exposing the bottom. The newly exposed surface is now underlain by two distinct layers of marine sediment. Toward the right-hand edge of the block the younger, upper layer consists of sand and silt, whereas the layer beneath it consists of clay. The coarser grain size of the upper layer reflects the fact that in that vicinity that layer was deposited closer to shore than was the lower layer. Also, the interface over which the two layers are in contact (like the interface of contact formed by any two successive leaves in a closed book) retains traces of chemical weathering and the scars made by erosion before the upper layer was deposited. Because of this evidence of interruption, geologists say that two such layers are not *conformable* with each other and that the interface is therefore a surface of *unconformity*. As might be expected, the geologic column is full of such surfaces.

After a further long lapse of time during which the surface of the younger layer of sediment becomes eroded, the sea rises once more, this time to a greater height than before (Block 3). Along the new shoreline another pebble beach forms, and the new layer of sediment grades seaward from it through the familiar changes of grain size: sand, silt, clay. This new sediment constitutes a blanket that conceals the older layers, including the two older beaches.

Now we are ready to expel the sea for the third and last time. When we have drained it off and can walk across it, we see (Block 4) a monotonous plain underlain by marine sediment that conceals everything beneath it. But if we wait a couple of million years or so —at least long enough to permit a stream to cut a sizable valley down through the sediment and into the rock beneath it—what we then see will verify the whole physical history of our coast.

The three layers, unconformably related to each other and to the underlying rock, show the three different grain-size arrangements appropriate to their distances from the former shores. The interfaces between them reveal the intervals of erosion, when the area was land rather than sea floor. And finally, by going toward the left we can find the latest and youngest beach, still at the surface because it was never covered up. The one feature we cannot see is the oldest beach, because that one is still not exposed. If we are really anxious to find it (for we can infer from the exposed strata that it should exist) we can bore some test holes. If we are not that anxious, we can leave the beach for some future generation of scientists to discover after it has become exposed in a future valley.

Changing sea level and evolution. The foregoing recital may appear monotonous to some readers. Yet to follow it closely is probably the easiest way to visualize the basic principle involved in the building up of a sequence of marine strata. In order to establish the physical happenings first, we have avoided saying anything about fossils in our strata. Now, however, it is legitimate to ask what effect all these comings and goings of seawater will have on the organisms that live in the sea off our model coast.

Our model is like other coasts in that the water is just as full of marine animals and plants as it can be. That is, as many individual organisms as the available food can support are living there. Any individuals that cannot be fed will either die or go elsewhere. We must remember, too, that the available food consists of local organic matter; in other words, the organisms are busy consuming each other. In the process they constitute long *food chains* like the short chain familiar to us: cow eats grass→man eats cow→human wastes nourish grass.

In order to take advantage of every bit of available food, organisms have become adapted through natural selection to every variation, however small, in their living space. For example, along our model coast different kinds of living things occupy (1) water of different depths; (2) such different kinds of bottom as rocks, clean sand, and mud; (3) exposure of different degrees, such as open coast with surf, inlets with calm water, and so forth.

When our coastal water is thus filled to capacity with teeming life, any change of sea level relative to the land will upset the very delicate balance that exists and will force changes upon the populations. Along a gently shelving coast such as the east coast of North America from Massachusetts to Florida, say 1000 miles long, relative lowering of sea level by only 20 feet could convert into land a strip of sea floor 5 to 7 miles wide. The area of such a strip would be at least 5000 square miles. With the population already at the point of saturation, such a change would affect billions of individuals large and small. Much would depend on rate of emergence, but even slow emergence would eventually eliminate at least 5000 square miles of living area. The surplus inhabitants, those least able to cope with the new, unbalanced conditions, would either die or move elsewhere. If they moved they would of course face competition with the occupants of any territory they invaded.

Such competition would intensify the evolutionary process. So if the successive changes of sea level along our model coast were spaced widely in time—if they were, say, a million years or so apart—we could expect that as the sea rose and opened up additional sea floor for settlement, the organisms that invaded it would include successful new species evolved since their ancestors had been excluded from the region. These new species add something different to the assemblages of fossils that become part of the fossil record. They should be of great help to the future scientists who will have the task of identifying the strata in Figure 9 - 12.

Remembering our discussion in Chapter 3, we can conclude that changes such as those we have just sketched are the main reason why distinctive groups of fossils are peculiar to the strata in which they occur, making it possible to correlate strata from region to region.

Extinctions. The path of evolution through Phanerozoic time is littered with extinctions—with the dying-out of species, genera, and groups of even larger magnitude. One important cause of extinction is, as we have just seen, change in the area of a sea floor brought about by relative change of sea level. But many kinds of land life have become extinct; so there must be causes other than sea-level changes. Change of temperature and other factors of climate, radical changes in food supply, and the evolutionary development of especially efficient predatory organisms are likely causes.

A basic cause of extinctions, suggested very often and still a matter of theory, is change of climate or other factors in the environment brought about by reversal of the polarity of the Earth's magnetic field. Although our calendar of reversals (Fig. 6 - 1) covers only the last few million years, reversals have been identified in Mesozoic and Paleozoic time as well and probably have been happening since early in the Earth's history. The largest numbers of extinctions revealed by the fossil record occurred toward the end of the Paleozoic Era and at the end of the Mesozoic. Those two times were marked by comparatively frequent magnetic reversals. The idea that the two kinds of events are related is now being explored.

With the information about fossils that we have gathered in this chapter, we can turn now to an aspect of the Earth's former populations to which we have paid little attention thus far. In the next chapter we will see how it is possible to perceive former environments by drawing inferences both from fossils and from the strata in which they occur.

References

Beerbower, J. R., 1960, Search for the past: Prentice-Hall, Inc., Englewood Cliffs, N.J.

Fenton, C. L., and Fenton, M. A., 1958, The fossil book: Doubleday & Co., Inc., New York.

Raup, D. M., and Stanley, S. M., 1971, Principles of paleontology: W. H. Freeman & Co., San Francisco.

Romer, A. L., 1966, Vertebrate paleontology: University of Chicago Press.

Stirton, R. A., 1959, Time, life, and man: John Wiley & Sons, New York.

10

Ancient Environments

North America today

In Chapter 1 and again in Chapter 6 we looked thoughtfully at the surface of the spinning Earth. Now we can profit by looking at it again, this time in terms not of structure or history but of the surface of a continent as it exists today: the surface with its plains, mountains, rivers, and the teeming organisms that live on and close to it. We could have said equally well " . . . the teeming organisms *appropriate* to it." They are "appropriate" because of the adaptations forced upon them through countless generations by their environment. If we neglect details, most of North America (Fig. 10 - 1) has had its present general aspect—Appalachian Mountains in the East, Cordilleran Mountain System in the West, broad plains and lowlands in between, draining to the Mississippi River and to Hudson

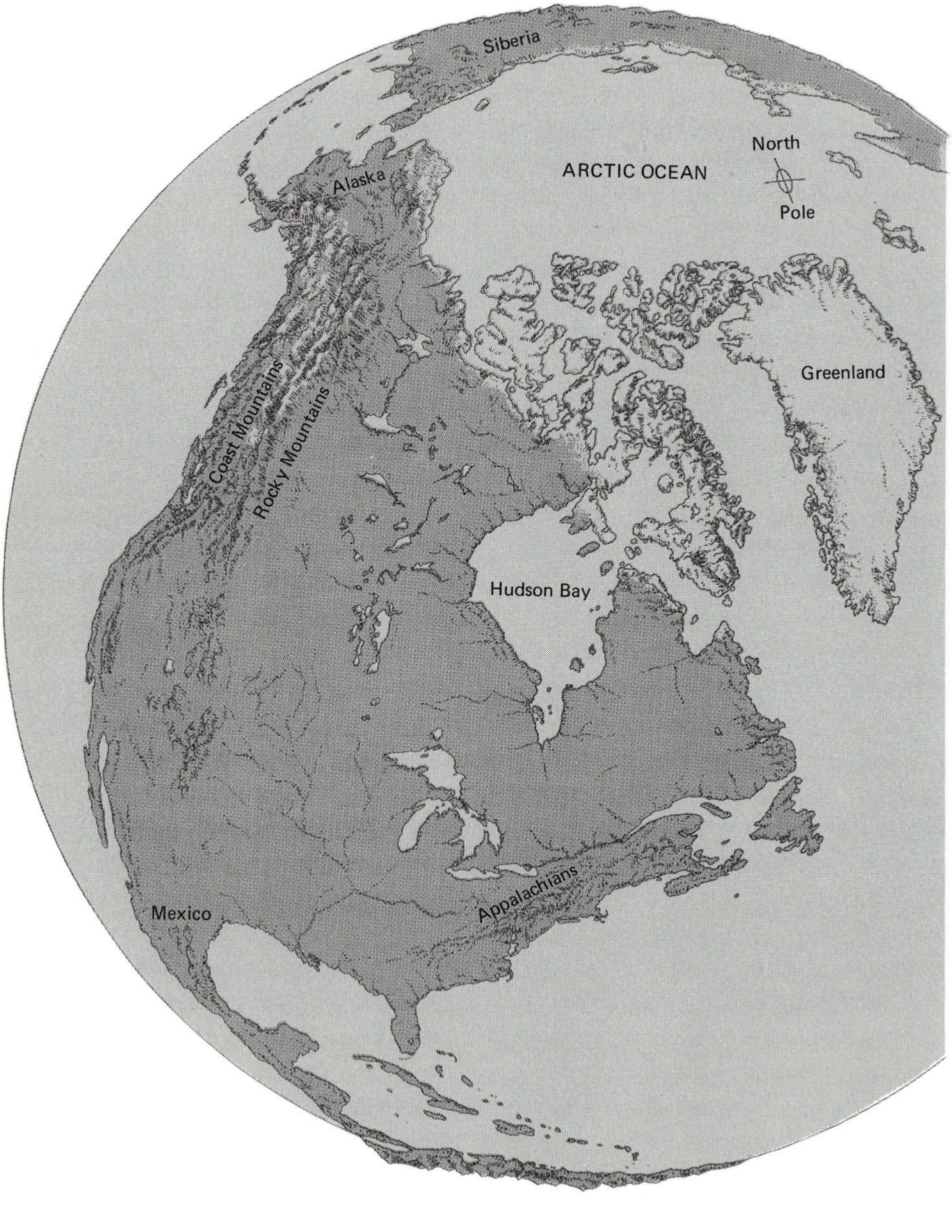

Figure 10 - 1. The face of North America.

Bay—for the last 50 million years or so. That time has been ample for adaptations, and for re-adaptations as well, in places where significant geologic changes have occurred. So North America today has thousands of environments, each with its own collection of appropriate plants and animals living together, preying on one another, or even in some cases cooperating.

The word *environment* embraces all the physical, chemical, and biological aspects of a given area, small or large: the configuration of the land, the depth, chemical content, turbidity, and movement of water, the climate, the soil, the vegetation, the animals. Environment is made up of so many different factors that we usually study it in terms of the influence of one factor on another or of the whole environment on some one factor in which we are interested. Thus we might be concerned with how temperature, depth, and salt content affect the organisms that live in a certain part of the sea. Or we might want to know how increasing altitude, as we climb up a high mountain range, affects the kinds of plants that grow on the mountainside. Again, interest is widespread today in the effect of smoke and gases in the atmosphere over a city on the health of the city's human population.

Pattern of vegetation. The distribution of vegetation on a continent gives a good general idea of the principal environments available to land animals, including man. Figure 10 - 2 shows the zones of natural vegetation (neglecting the artificial effects of farming) in North America. Adjacent zones are separated by sharp lines although actually, of course, the zones grade into one another. Comparing this map with Figure 10 - 1, we can see at once that the distribution of mountains strongly influences the vegetation pattern. In the East a long, narrow tongue of northern forest extends more than 600 miles southward along the tops of the Appalachians, encouraged by the cooler and slightly wetter climate. In the West, similar long, narrow areas of specialized vegetation are seen on the Rockies and other high mountains. As we scan the United States from Appalachians to Rockies, our eyes pass successively across forests, prairies, and dry grassy plains. These successive belts of vegetation are controlled mainly by rainfall, which decreases westward. Again, as we scan central and eastern Canada from south to north, we see northern forest succeeded by subarctic forest succeeded by Arctic tundra in well-defined belts. These belts are controlled chiefly by temperature and also in part by rainfall, both of which decrease

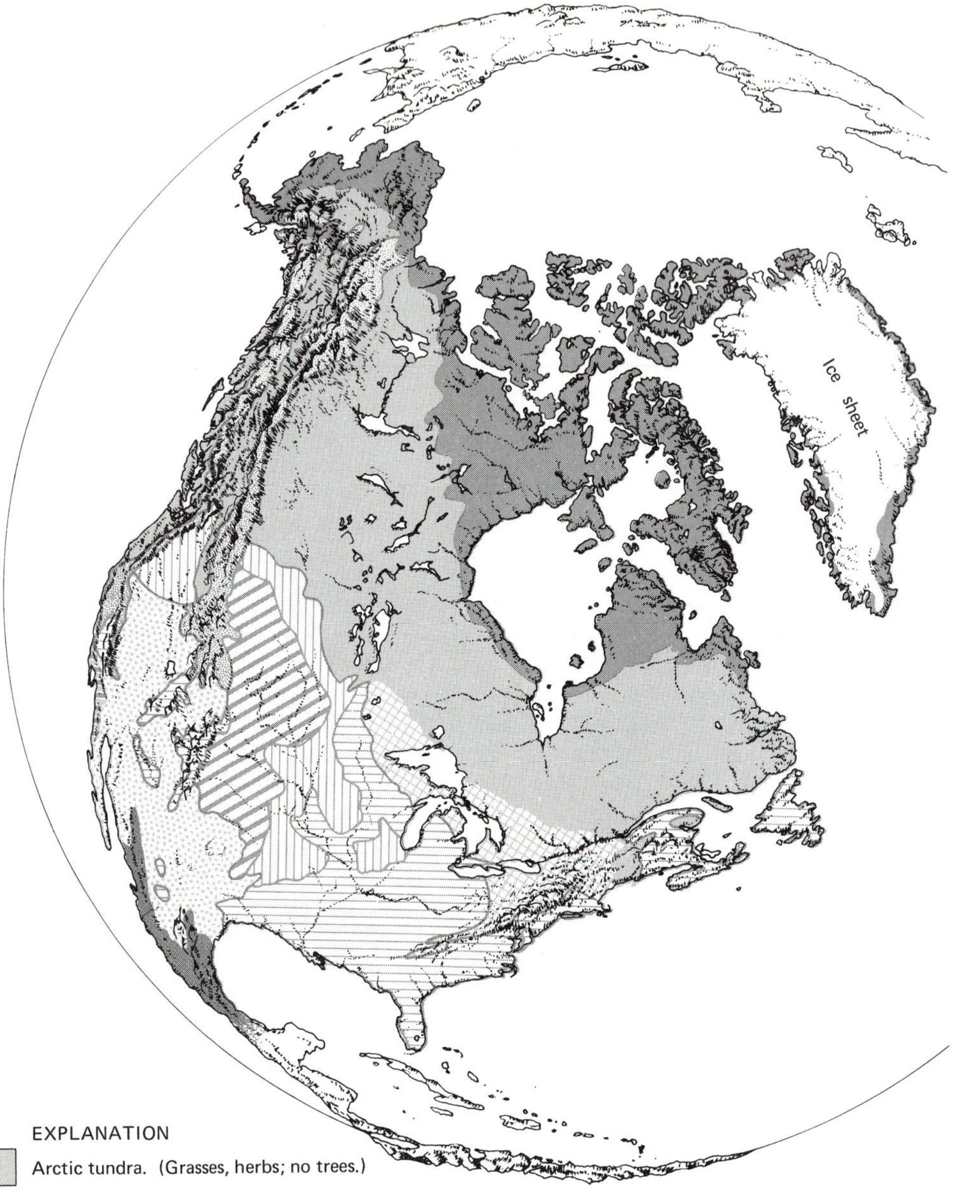

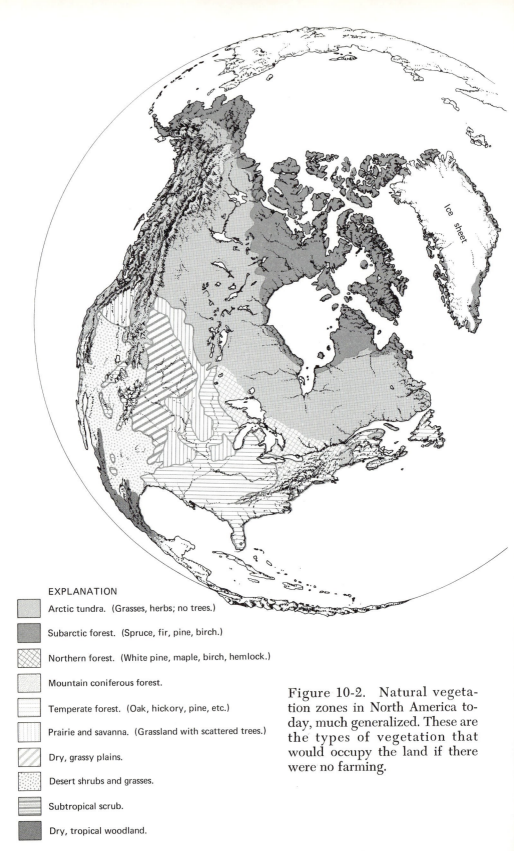

Figure 10-2. Natural vegetation zones in North America today, much generalized. These are the types of vegetation that would occupy the land if there were no farming.

toward the north. Along the Arctic coast of Canada and Alaska, rainfall and snowfall are slight. In most places they are no greater than the rainfall over large parts of Arizona. These examples show that the broad distribution of plants is easy to understand in terms of basic factors of climate.

Local assemblages of living things. But when viewed closely, each of these broad belts is seen to consist of many subdivisions. When we look at some small part of a continent we immediately see a variety of smaller areas, each with its own kind of vegetation growing there in response to local differences of altitude, slope, rainfall, and soil. Not only vegetation but also animal populations differ from one area to the next. Although it does not represent any actual tract of country, Figure 10 - 3 gives an idea of the possible variety of local environments. Far to the left is shallow seawater with a sandy bottom. In it is a varied group of invertebrate animals, fish, and seaweeds. Crabs, birds, and insects are common on the beach. Hardy grasses, insects, and perhaps rodents inhabit the belt of dunes built by sand blown inland from the beach, where it has already been sorted by the swash and backwash of the surf.

Inland from the belt of dunes and separated from each other perhaps because of differences in the soil, are grassland and scrub, the latter consisting of a variety of bushes. Through the grassland winds

Figure 10 - 3. Sketch suggesting a few of the land environments that are expectable on part of the west coast of North America and in the country inland from it. (Not to scale.)

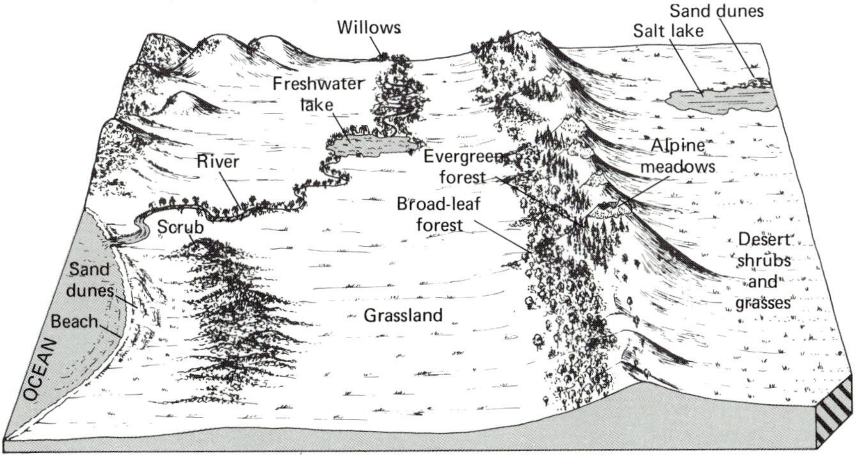

a stream, its banks lined with water-loving trees such as willows, its water populated with fish, freshwater clams and snails, and smaller creatures. In a lake are fish, invertebrates, and weeds, rushes, and reeds that attract water birds. The land animals here might include deer, antelope, coyotes, and rodents.

Still farther inland are mountains with grass-covered basal slopes, a belt of broad-leaved trees merging upward into evergreen forest, and at the top Alpine meadows with grasses and herbs. In the forest we might expect deer, fox, and mountain lion, and above, marmots and rabbits. Beyond the mountains, which draw rainfall on themselves but act as a barrier to rain on the terrain in their lee, is desert, clothed with discontinuous brush, shrub, and tuft grasses. Its inhabitants would include antelope, jackrabbit and other rodents, coyotes, lizards, snakes, and scorpions. In its salt lake one might find invertebrates and minnows adapted to salt water, and of course water birds. All these varied plants and animals have adapted to their surroundings.

In places the patchwork of contrasting environments exists on a still smaller scale. For instance, the south-facing slope of a single valley may be clothed with grass while the opposite, north-facing slope is covered with woodland, simply because moisture evaporates from the soil on the sunny side much more rapidly than from the shady side, on which more moisture therefore remains to promote the growth of trees. As a result, the animal populations of the two sides also differ, especially, perhaps, the insects and small mammals. Indeed anyone who traveled slowly and thoughtfully enough across a stretch of terrain could make a list of several, and perhaps many, local environments in which the assemblages of living things are adjusted nicely to various differences in living conditions.

Environments in the past

Because of pressures caused by overpopulation of the Earth by our own species, man is only now becoming widely aware of environments and their interrelations. But natural scientists have long known about the importance of environmental pressures in the development of all organisms. They came to understand that environment shapes the distribution of plants and animals and shapes the irregular paths of their evolution as well.

The Principle of Uniformity tells us that mountains and hills, plains and plateaus, rivers and lakes, valley floors with alluvium, deltas, beaches, sandy deserts, and other features of the lands as well as those of shallow-sea floors must have existed in the past as they do today (Fig. 10-4). The fossil record tells us that at any earlier moment in geologic time plants and animals were adapted to those various environments much as their modern counterparts are adapted. This relationship is so clearly evident that even where physical indications (described near the end of this chapter) in the

Figure 10 - 4. Ripples in the Dakota Sandstone, exposed in the Rocky Mountains in Colorado. Although possibly 100 million years old, the ripples are identical with those forming along seashores today. These ripples were made by a gentle current moving from left to right in very shallow water, in the sea shown in Figure 10 - 7. Modern aspen leaves indicate the size of the ripples. (From U.S. Geol. Survey Bull. 1291.)

strata are lacking, the fossil form of an organism now extinct is confidently believed to indicate a former environment broadly analogous to that in which its nearest living relatives exist now.

Environments inferred from strata. Both the fossils and the physical features of the strata in which they occur give information about the former distribution of mountains, shallow seas, and other major features on the continents. True, the continents have been moving, slowly and either intermittently or continuously, across the face of the Earth, borne by the moving crust plates on which they float. The farther back in time we look, the less certain we are of where each continent was at that moment. But despite this disadvantage, we can still identify the positions of former shallow seas, mountains, and the like on any particular continent such as North America. We can even sketch a series of crude maps of the continent, showing how the main features changed through time.

Most of what we can do is limited to the last 600 million years or so, the span of Phanerozoic time as shown in Table 3-A. As we have said, in the strata earlier than Phanerozoic too many of the fossils and too much of the physical evidence they formerly contained have been altered or destroyed by metamorphism. It is the same with human history. Historical data for the last 2000 years are comparatively good, for the preceding 1000 years data are far less abundant, and for the preceding 7000 years (that is, on back to 10,000 years ago) data are very spotty indeed. Much has been eroded, much has been covered over by sediment, and even more has been destroyed by human hands.

What we have just said adds up to this conclusion: throughout Phanerozoic time the *kinds* of environments present on the continents and their shelves have been much the same as the ones we identify today. But the *patterns* they make have changed continually. The changes have been brought about by local uplift or subsidence, by the drifting of a crust plate into a higher or lower latitude with its different climate, by possible collision of two continents, or by the outpouring of magma at the surface, forming volcanic cones or widespread flows of lava.

Ordovician sea, coast, and hills. Let us now examine the very different pattern that existed in a familiar part of North America at a particular time toward the end of the Ordovician Period, some 465 million years ago, in order to see why we know what the pattern was like. The strata that were being deposited as sediments at that time in New York, Pennsylvania, Ohio, and southern Ontario are well preserved today, and are exposed at the present surface in many places where we can examine them without difficulty. As we follow them from west to east, the strata change character (Fig. 10-5). In Ohio and western Ontario these rocks are rather thin, nearly horizontal layers of limestone, rich in fossil invertebrate animals characteristic of a shallow sea floor. The strata tell us that that part of North

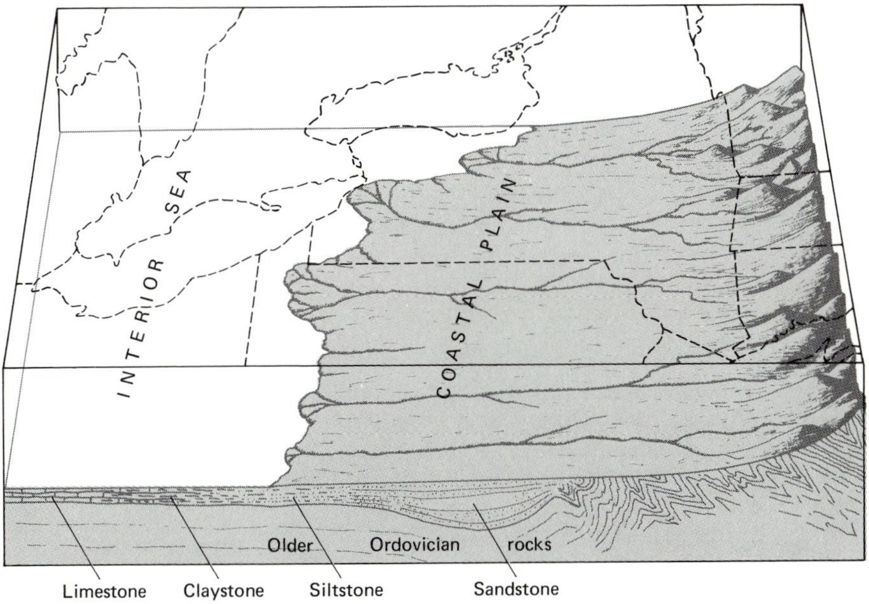

Figure 10-5. Environments in what is now the southeastern part of the Great Lakes region, at a particular moment late in Ordovician time. The canopy (broken lines) above the block indicates the positions of state boundaries and Great Lakes today. Front of block is about 700 miles long. (Suggested by reconstructions by R. C. Moore and by Dunbar and Waage.)

America was a shallow sea inhabited by a great variety of animals and plants, creating a scene much like that shown in Figure 10-6. Evidently the water was clear, because the limestone consists mainly of calcite with few of the clay particles that would have made the water muddy.

However, as we follow these strata eastward, we find that their clay content increases, eventually to such a degree that the rock

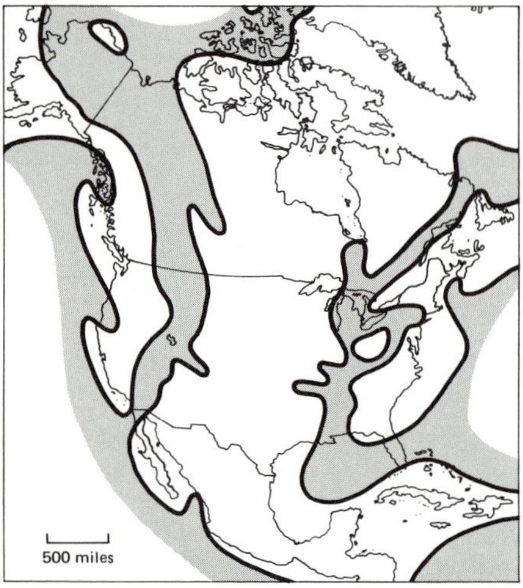

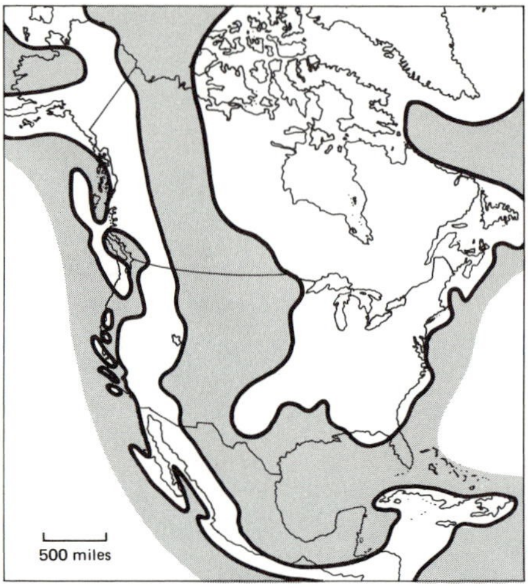

Figure 10 - 6 *(left)*. Generalized map of North America in mid-Devonian time, showing seaways (shaded) and lands (blank). The continent as it is today is represented as a faint background for comparison. (Ancient coastlines from Schuchert, 1955.)

Figure 10 - 7 *(right)*. North America late in Cretaceous time. Conventions and source same as in Figure 10 - 6.

becomes claystone. Also, silt and then sand begin to appear, so that eastward the claystone grades into siltstone and then sandstone. Along with these changes, the rock layers themselves become much thicker. The changes in kind of rock are accompanied by changes in stratification and by the disappearance of the marine fossils.

Our interpretation of the changes is that in going eastward we have left the former marine environment and have crossed a former shore onto a low-lying, gently sloping former coastal plain. The plain was built up by the deposition of alluvium by rivers that were flowing from east to west. The increase in grain size toward the east means that the headwaters of the rivers lay among hills or low mountains somewhere east of the coastal plain. Stream gradients must have been steep enough to enable the streams to carry particles as coarse as sand.

Chemical weathering of the hill country was sorting the rock material in the ways described in Chapter 5, and quantities of the quartz grains that are residual from such weathering were being washed westward toward the sea. Nearly all such grains, whether of sand size or silt size, were being deposited on the coastal plain, as was much of the clay. The rest of the clay particles were carried on into the sea and deposited in the coastal water. Farther seaward, having no more clay particles to make it turbid, the water was clear and formed an attractive environment for many animals and some marine plants.

We see in the strata little evidence of sandy beaches along the shore at the time when these particular sediments were being deposited. This is not surprising, because the parts of the strata that consist of sandstone lay east of the shore. So, near the coast, there was little or no sand to form a beach. Instead, the coast was low-lying and very muddy, somewhat like the present Mississippi delta as we observed it in Chapter 4. Indeed, the coast from Ontario southward seems to have consisted of a whole series of large and small deltas. As for vegetation, there is no evidence that any plants were growing on the world's lands at this early time. Hence, presumably, the Ordovician coastal plain was a wet expanse of bare mud and sand crossed by turbid, slow-moving streams. Could man have been there, he would have found the environment unattractive.

The hills or low mountains whose weathered rocks furnished all the mud are long since gone. Today they are represented by metamorphic rocks whose general trends are northeast-southwest, more or less parallel to the coast of the Ordovician sea. The hills, which are ancestors of part of today's Appalachians, occupied part of the New England region. While the metamorphic squeezing was in progress, a belt of the crust farther west was slowly subsiding, forming a pocket in which, as Figure 10 - 5 shows, the accumulating sediment was especially thick. This is an example of the subsiding pockets we found occurring in continental shelves (Chapter 5).

Maps of ancient geography. The ancient geography we have just reconstructed cannot be accurate in detail, but it is based on enough facts to be a fair representation of what the former terrain was like. With enough layers of sedimentary rock to work with, we can make a similar interpretation of former environments in any land area for the time or times represented by the available strata. The combined interpretations of many modern geologists have resulted in maps of the geography of North America and of other continents as they must have been at many of the ancient times represented in the geologic column. The maps vary in accuracy because the amount of information that can be extracted from the layers of rock varies a good deal. For some areas and time intervals there are few data or none at all, either because sediments were not deposited there or because erosion has subsequently destroyed strata that are needed for accurate interpretation.

What the maps show, through time, is a continually changing pattern of shallow seas, plains, and mountains of various kinds, as well as outbursts of volcanic activity in one region or another. Each successive map pattern reflects the continual interplay between the internal and the external processes we described in Chapter 1. The details of the maps are useful mainly to Earth scientists. For our purpose it will be enough to compare two maps (Figs. 10-6, 10-7) that show the reconstructed geography of North America in the middle of Devonian time and again in a late part of Cretaceous time. Those two geologic periods were separated from each other by a lapse of more than 200 million years.

Paleozoic North America. If we neglect details and think only of its general pattern, we can take the mid-Devonian map (Fig. 10-6) as broadly representative of Paleozoic North America. Its principal features are two interior seas, each both narrow and shallow, one near the present eastern coast and the other much farther west. The eastern sea on this map was one of several successors of the Ordovician sea shown in Figure 10-5. Although details of its shores kept changing, and although at times the shallow water drained completely out of the narrow seaway, it returned persistently to the same general position, at least partly because the crust beneath the seaway gently but repeatedly sagged downward, maintaining the pocket shown in Figure 10-4.

At times the pocket, or at least part of it, was squeezed so that sediments that accumulated in it were crumpled (Fig. 10-5), perhaps because of collisions between the continent and another one along what is now the eastern coast. But until the end of the Paleozoic Era, more than 300 million years long, the sea kept returning, until a final squeeze crumpled the pocket of sediment to form the Appalachian Mountains and lifted the entire region above sea level. After undergoing erosion through the 250 million years of Mesozoic and Cenozoic time, during which a huge volume of rock waste was carried away from them as sediment, the mountains still remain as stumps of their once imposing form. Their uptilted layers of sandstone and siltstone resist erosion and stand up as long ridges, while between the ridges the less resistant claystones and limestones underlie the lowlands (Fig. 10 - 8).

Figure 10 - 8. Appalachian Ridges south of Harrisburg, Pennsylvania. Although the ridges in the background are nearly straight and parallel, the ridge in the foreground represents a layer of rock that has been sharply bent. Erosion has cut it diagonally to create a zigzag form. (John S. Shelton.)

Returning to the seas that flourished before the Appalachians were created, we note that through the more than 300 million years of the Paleozoic Era, marine life evolved and fishes became abundant. The surrounding lands, although bare in Ordovician time, became clothed with vegetation of a primitive sort, and land animals appeared and evolved into many kinds. These events are recounted in later chapters; here we are only sketching the physical background of former environments.

The history of the other narrow sea in North America was similar. The sea occupied a belt that repeatedly sagged down, seawater came and went, and thick sediments accumulated. The seaway persisted into the Mesozoic Era for a short time only and in the Triassic Period it disappeared.

Cretaceous North America. Having looked closely at the mid-Devonian map, we can turn to the Late Cretaceous map (Fig. 10-7) with the confidence of some experience. On it we see the North America of about 70 or 75 million years ago. The sea in the Appalachian region had gone, having disappeared before the end of the Paleozoic Era. The Atlantic coast lay east of its present position, except for the Florida peninsula, most of which was submerged. A large part of the eastern half of the continent was land. The striking feature of the map is the rather broad seaway connecting the water of what is now the Gulf of Mexico with that of the Arctic Ocean. It cut North America into two pieces. Along the Pacific coast the sea covered much of what is coastal land today. Much of the land that existed then was low and rather close to sea level, and swamps were abundant over broad coastal regions. Climates then were milder than they are today. The lands were fully clothed with vegetation that was still exotic in character compared with what we see around us at present. The dominant land animals were reptiles, chiefly dinosaurs, which flourished in the mild climate of the lowland environment. The seas were full of fishes and a great array of invertebrates, and the air above them was populated with flying reptiles and at least some kinds of birds.

Reconstruction of environments

We have now seen some reconstructed environments in Ordovician eastern North America. We have also looked at maps of the reconstructed ancient geography of North America at two different times, one in the Devonian and the other in the Cretaceous. On what kinds of evidence within the strata have these reconstructions been based? The maps represent history, and the geologists who make them—even very little parts of them—are as much historians as the scholars who reconstruct the human affairs of the past.

Historians make maps of former situations and conditions. Their maps show that political boundaries in the ancient world are as different from the political boundaries in Europe and the Near East today as the map of Cretaceous geography is from the familiar wall maps hanging in a schoolroom. Historians also make maps of ancient military campaigns, trade routes, the spread of culture, and many other things. Where did they obtain the information from which maps of past conditions could be developed? Partly from the library, from documents written by people who lived in earlier times. The Magna Carta and the Declaration of Independence are written documents that represent specific events known in this way. But there are historic documents of other kinds, such as paintings, sculptures, and buildings. The Mona Lisa and the Parthenon are documents from each of which something has been learned about the culture of a former time.

The historian of man is lucky if he can find written documents that will take him back more than a few thousand years. Although the scientist-historian of the Earth has no written documents at all, he has an abundance of natural documents in strata, many of them with fossils. And he has at his disposal the techniques of radiometric dating with which, as we have seen, he has constructed a reliable calendar.

Examples of environmental interpretation. The fossil mollusks in Figure 10 - 9, when collected, identified, and compared with related living mollusks, prove to represent the kind of animals that live, not

Figure 10 - 9. Layered silt containing fossil mollusks of several different kinds. Pencil (right center) is 5 inches long. In the drainage system of the River Tiber downstream from Rome. (R. F. Flint.)

in the open sea but in the protected, brackish water where rivers enter the sea. The thin, nearly parallel layers of silt, likewise, are compatible with an estuary environment. By correlation, the fossils are known to belong to an early part of Pleistocene time, possibly as much as 2 million years ago. This rather young age is compatible with the state of the silt (loose and not cemented to form siltstone) and of the shells (not petrified). Although the locality is at least 10 miles from the sea and stands nearly 200 feet above sea level, clearly it was standing in tidewater at the time when the mollusks were living, very likely in the then-drowned valley of the Tiber. Since then this part of the west coast of Italy has been tilted up, causing the sea to recede from what was formerly a partly submerged coastal region.

In contrast, the locality of Figure 9-9 is northern Wyoming, far from the sea in the middle part of a continent. The fossil forelimb being painstakingly uncovered is so huge that it must belong to a dinosaur. The genus is *Diplodocus,* an animal somewhat similar to the one in Figure 14-7. Being a reptile, the dinosaur must have lived in a fairly mild climate. The anatomy of his skeleton shows that he walked on all fours and fed on vegetation, apparently soft stuff not unlike modern water weeds. The enclosing strata are of Late Jurassic age (perhaps 140 million years old) and consist mostly of fine-grained alluvium and sediments characteristic of swamps and ponds.

From these data it has been concluded that the Wyoming region at that time was a series of broad river plains with connected swamps and ponds, luxuriant vegetation, and a warm climate, a combination that suggests comparison with the basin of the Amazon River today, and that its altitude was very low. The low altitude is easily understood when we note that late in Jurassic time an interior sea, like that in Figure 10-7 but much less extensive, lay over western Wyoming not far west of the locality of Figure 9-9.

If the strata in which our dinosaur bones are wrapped were deposited on a low plain close to sea level, in a mild, moist climate, the environment has changed a lot since then. The fossil locality is now nearly 8000 feet above sea level and the climate is semi-arid, with sparse vegetation and very cold winters. The internal processes have been at work, lifting up broad areas in western North America in a series of arching movements that began at the end of Cretaceous time and have continued through most of the Cenozoic Era.

So our picture of gradually changing environments throughout the last 600 million years of Earth history is gradually being put together from bits and pieces of information extracted mostly from sedimentary strata. Some of the information is read from physical characteristics of the strata; more is read from the character of the fossils. Each reconstruction makes a picture that may be clear or dim, depending on the amount and the reliability of the data on which it is based. Gradually, the series of pictures is increasing in clarity, but a long time will have elapsed before we can cease to rely on a liberal amount of speculation to fill in our pictures.

In the chapters that follow we will attempt to sketch many former environments as we trace the history of plants and animals through Phanerozoic time.

References

Darling, Lois, and Darling, Louis, 1970, A place in the sun. Ecology and the living world: William Morrow & Co., New York.

Farb, Peter, and others, eds., 1963, Ecology: Time, Inc., New York. (Life Nature Library.)

Laporte, L. F., 1968, Ancient environments: Prentice-Hall, Englewood Cliffs, N.J. (Paperback.)

Raskin, Edith, 1967, The pyramid of living things: McGraw-Hill, New York.

Storer, J. H., 1956, The web of life; a first book of ecology: Devin-Adair Co., New York.

11

Geologic History of Plants

Our history of the development of the biosphere through Phanerozoic time begins logically with plants. Not only are plants the direct or indirect food of animals, but the fossil record suggests that plants emerged from the sea, the original home of all life, onto land before animals did so. Consequently, we can trace the history of animals more smoothly if we first depict the clothing of the continents with vegetation and the evolution of that vegetation to its present state. The record of plant fossils, especially the early part of the record, presents us with some pretty wide gaps. This isn't surprising, because land environments, to which most fossil plants are related, afford poorer conditions for preservation than the sea floor does. But the gaps make it difficult to determine the evolutionary relationships among some plant groups. Despite this fact, if we avoid going into much detail we can outline a consistent story based on the plant fossils that have been discovered up to now.

Structure and physiology of plants

Vascular system. A brief look at the characteristics of today's plants—at their basic structure and physiology—shows what we are aiming at when we try to outline the steps that led up to the plant world of today. Most plants, especially most of the higher plants, have a distinctive structure. This consists of (1) a long stem, commonly vertical, commonly branching, and in most cases rigid; (2) a base with roots firmly anchored in the soil; and (3) an upper surface marked by an array of green leaves. The primary action takes place in leaves (whether broad leaves, needles, or something else), and consists of photosynthesis. Leaf cells synthesize *chlorophyll*—"plant green"—the green pigment that is a receptor of sunlight and that, using the energy it receives from sunlight, can build food from carbon dioxide taken from the atmosphere and water drawn from the ground. The carbon dioxide enters through pores in the leaves, the photosynthesis reaction (chemical reduction of oxygen) occurs, and the result is the production of sugar, water, and free oxygen.

The water needed for photosynthesis is drawn upward from the ground through the root-stem-branch system. At this point we can appreciate that that system is not merely a structure to support leaves. It is also a system of plumbing, of ducts through which water is pulled up in quantity even to the top of the highest tree. The tallest of the Douglas firs in the Pacific Coast region of North America can pull water up as high as 400 feet above the ground.

Actually what is pulled upward is not merely water; it is *sap,* a watery solution. The woody tissue of a living tree contains vertical strings of dead cells. In the most specialized kinds of trees the floors and ceilings of the cells have been dissolved away. It is as though in a tall building one were to remove the floors of a series of rooms one above the other in order to make an elevator shaft. The strings of dead cells in a tree are filled with sap. They form tubes through which the liquid, which enters from the ground, rises from the tree's roots up to the active leaves. Water in the leaves is continually being lost to the atmosphere by transpiration through pores. This loss reduces the pressure within leaves and twigs and causes sap in the tube system to be pulled through cell walls into the leaves to make up the deficiency. Thus the upward pull affects the whole system of tubes, including the roots, which then absorb more moisture out of the ground. The whole *vascular* system is a kind of pump, a very efficient one.

Considered in a broader way the vascular system is just a segment of the water cycle (Fig. 2 - 1). In the cycle, part of the rainfall percolates through the ground, is taken up by the roots of plants, pumped upward to their leaves, and transpired to the atmosphere, from which it will fall again as rain. But part of the water stays in the plants. In the process of photosynthesis it has been combined with the carbon dioxide that is taken from the atmosphere through leaf pores, to form food and substance for the plants.

The manufacture of plant substance requires a lot of water and carbon dioxide to keep going. For instance, one acre of corn in a farmer's field consists of about 10,000 plants. These will yield about 100 bushels of grain (the "seeds" of the corn). To build 100 bushels of seeds, apart from stalks, roots, leaves, and corncobs, which we will neglect, the plants need 20,000 pounds of carbon dioxide, from which they will sort out and use 5500 pounds of carbon and return the liberated oxygen to the atmosphere. To do this the plants must process 21,000 tons of air—more than two tons per plant. At the same time the plants must take from the ground and pump up to the leaves 500,000 gallons of water, equal to 50 gallons per plant.

One acre of corn accomplishes this activity every year; the resulting seeds, stalks, and leaves go mostly to feed animals. An acre of wheat or any other grain accomplishes a similar task. An acre of trees does likewise, though more slowly, creating fruit, nuts, firewood, or lumber. When we think of the number of acres of forests, orchards, pastures, and cultivated fields on the Earth's lands, whether the plants are useful to man or not, we get some idea of the huge part the plant section of the biosphere plays in the water cycle as well as in the weathering, both chemical and mechanical, that forms a segment of the rock cycle.

Reproductive systems. We must never neglect the way in which a living thing, animal or plant, reproduces, because its success in reproducing itself is one of the major factors in its evolutionary success—its ability to survive through the thick and thin of environmental changes. So we can learn something about geologic history by looking at the ways in which the plants of today operate their reproductive systems.

A small proportion of the kinds of plants living today reproduce vegetatively. In vegetative reproduction a plant develops shoots, bulbs, or suckers, which take root and become separate plants. The offspring are replicas of their parent. We write the word *parent* in the singular because there is only one parent. In consequence there is no recombination of genes; the offspring possesses nothing new with which it can, on its own, take advantage of new possibilities offered by the environment.

Sexual reproduction, however, permits the combination of different characteristics received from two different individuals, and if each new characteristic is favored by the environment, evolution can progress much faster. Basically, sexual reproduction in seedless vascular plants consists of a cycle with two phases. Phase 1 takes the form of a small leafless plant that produces sperms or eggs (or both). The sperm from a male plant swims through a body or film of water to fertilize an egg on a female plant nearby. The fertilized egg then develops into a new plant, in this case a plant larger than that of phase 1, with leaves. In phase 2 this larger plant produces *spores*, reproductive cells that need no further fertilization. These are carried off and sown by wind, take root, and develop into the small leafless plant of phase 1, thus completing the cycle. As is evident, at one point in the cycle the presence of water is essential, and in the early history of land plants this requirement restricted the environments in which reproduction could occur.

The invention of seeds eliminated this difficulty and thereby improved the process. With seeds it was not necessary for a sperm to move through water to unite with an egg. A seed is not just a single cell. It has many cells that together form a root, a stem, and one or two leaves, all wrapped up tightly in a package of food with a physically and chemically protective cover. Also, the seed begins its existence as a structure that contains an egg, usually fertilized while still attached to a parent, and is nourished by that parent through the early part of its growth. So by the time it leaves home it is already a miniature organism with a built-in survival kit. It can keep alive until it succeeds in rooting itself and starting its career of independent growth. With this kind of start the seed has a chance of survival vastly better than that of seedless plants, and this in turn means a wider choice of environments and a better competitive position for the species. As we shall see, the history of plants demonstrates the value of these advantages.

The early marine plants

As we recall by looking back at Table 7 - A, the evolution of plants begins with marine algae. The world's earliest known fossils, 3 billion years old, include blue-green algae, the simplest of plants. For almost 2 billion years such plants inhabited the sea along with protists. But by about one billion years ago, as we know from fossils found in Australia, green algae were also in existence. The presence of green algae implies that plants had already accomplished two important improvements over their simpler ancestors: more complex structure and the capability for sexual reproduction. The latter improvement we have discussed already. Because each offspring developed from two parents rather than only one, it possessed a new combination of genetic material, with which evolution could proceed more effectively.

Although perhaps a less obvious improvement, the greater complexity of structure was also of basic importance for the future. When the original single-cell plants added a second cell and then still more cells, their form changed from nearly spherical to linear or irregular. A possible explanation of the change lies in the fact that any shape other than a sphere has a greater ratio of area to volume than a sphere has. The change in a plant's shape to linear or irregular would therefore have meant increased proportional area of contact with the seawater that contains carbon dioxide and sunlight, the things a plant needs for nourishment. Once irregularities of form were established, they could be molded into organs having diverse functions. This had to happen before algae could possibly have existed on land.

Establishment of vegetation on land

Up to the present, only two finds of very early fossil land plants have been made. One is from upper Silurian strata in central Europe. It consists of fragments of naked, leafless plants a few inches long, the earliest known land plants. The other find was made in Australia in strata of early Devonian age. It comprises fossil stems, as much as 10 inches long, of a plant (Fig. 11 - 1) very similar to the living club moss *Lycopodium* ("ground pine"), with a stem that creeps along the ground and sends up vertical shoots. *Lycopodium* is not only a land plant; it is a vascular plant as well. The presence of a fossil plant of this kind implies not only a substantial advance in evolution but a great geographic change, the emergence of plant life from the sea onto the land.

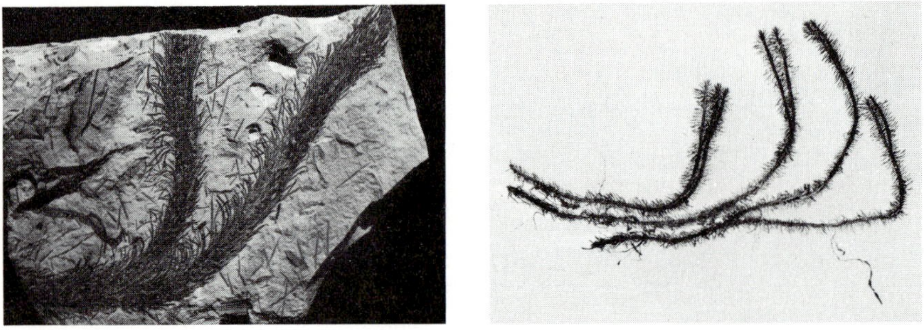

A B

Figure 11 - 1. A. The oldest known land plant. A fossil impression of the genus *Baragwanathia* from lower Devonian strata in southeastern Australia. (R. J. Tillyard, courtesy of C. O. Dunbar.)
B. The very similar living relative, the club-moss *Lycopodium lucidulum,* whose shoots are 4 to 5 inches long. (Theodore Delevoryas.)

Clearly the emergence had been accomplished by late Silurian time. How much time, then, was taken up in the process? If we count from the date of the earliest known green algae in the sea, approximately one billion years ago, to the date of the early land plants in the late Silurian of Europe (roughly 430 million years ago), we trace a span of around 570 million years. That is indeed a long time, nearly equal to the entire length of the Phanerozoic. So even though we admit that conquest of the land and "invention" of a vascular system were indeed a spectacular feat for limp, sea-living green algae to perform, those plants nevertheless had a huge amount of time within which to perform it. There was ample time for millions of trials and errors. Remembering from our discussion in Chapter 7 that with so much time available a seemingly slim chance becomes almost a certainty, we are justified in wondering whether it really did take algae that long—570 million years—to conquer the land. Perhaps somewhere there are strata of Cambrian or even greater age that contain fossil land plants. If so, someone will probably find and recognize them some day. If this should happen, a wide gap in our knowledge of the history of plants will have been narrowed.

The journey from sea to land. Trying to fill the wide gap with logical guessing, we might speculate on how the great transition from seawater to dry land took place. As we reasoned in Chapter 7, the presence of algae in the sea liberated free oxygen, which in time leaked into the atmosphere, filling it to its present concentration. The oxygen created a parasol against lethal solar radiation, and so made it possible for organisms to invade the land without danger of instant death. Plants, apparently green algae, were the invaders. They themselves had paved the way for invasion and conquest by liberating, through countless years of photosynthesis, the huge quantities of free oxygen that converted the Earth's atmosphere into a parasol.

We can reasonably suppose that green algae drifted along coasts and into estuaries. After millions of years they could have spread up the rivers themselves, passing from salt seawater through brackish water into fresh river water. We can imagine some slow rise of land that blocked rivers and created lakes, and in time the fluctuation and even drying up of lakes, over and over again. The ceaseless interplay of internal with external Earth processes could have created a thousand different configurations.

Of course we don't know how many kinds of algae may have got as far as this. But at least one kind survived the struggle against the drying-up of its environment. While in the water, algae had absorbed their food through the entire surface of the plant. Once they began to emerge they had to abandon that simple practice. The first step may have been stimulated by the experience of a drying environment, perhaps at low tide in some estuary. They developed a skin-like covering to protect themselves from drying out, and water could still enter and leave the plant by diffusion through the new skin.

Meanwhile the irregularities of body shape, the protuberances, had to be developed into primitive organs for special uses. One such protuberance, perhaps threadlike, might have been used as an anchor, and also for absorbing water and dissolved nutrients from the ground. Again, more vascular tissues would have to be added as a support on which the plant body could be lifted upward. In other words, the plant was no longer a limp alga. It was becoming a primitive land plant with a sort of root, a stem, and an elementary vascular system. Yet with a naked green stem and slender branches lacking leaves, it might still have looked rather like an alga.

By late Silurian time at the latest this alga-like stage had been reached and passed, and no wonder. Once the difficult process of emergence had been accomplished, the natural result—spreading over the land and adapting to many local environments—must have been comparatively easy. The basic necessities—soil, water, minerals in the soil, and carbon dioxide in the air—were there ready to be used. The surface of the land, for a long time as bare as the surface of the Moon, began to acquire a green mantle.

Early vegetation on land; seedless plants. We get glimpses of this early vegetation from petrified plants in strata here and there, for example in strata that represent sediments deposited by streams in the first half of Devonian time. The plants are of several different genera, but they all belong to primitive groups that later became extinct. Most of them crept along the ground, sending up shoots a foot or so high that were bare or had only rudimentary leaves. All were seedless. From these finds reconstructions have been made. To one of them (Fig. 11-2) we have added a flight bag 400 million years younger than the plants. The bag is incongruous, yes, but without something familiar to serve as a scale the vegetation gives the impression of being taller than it was.

As they continually adjusted to slight differences in the environments they encountered, the new plants gradually diversified into more kinds. The result was that by the end of Devonian time, 360 million years ago, the vegetation had a different aspect. Conspicuous in it were fernlike plants, ancestors of true ferns. These too were seedless. Some kinds grew to heights as great as 40 feet and some had trunks more than 3 feet thick at their bases. Visualiz-

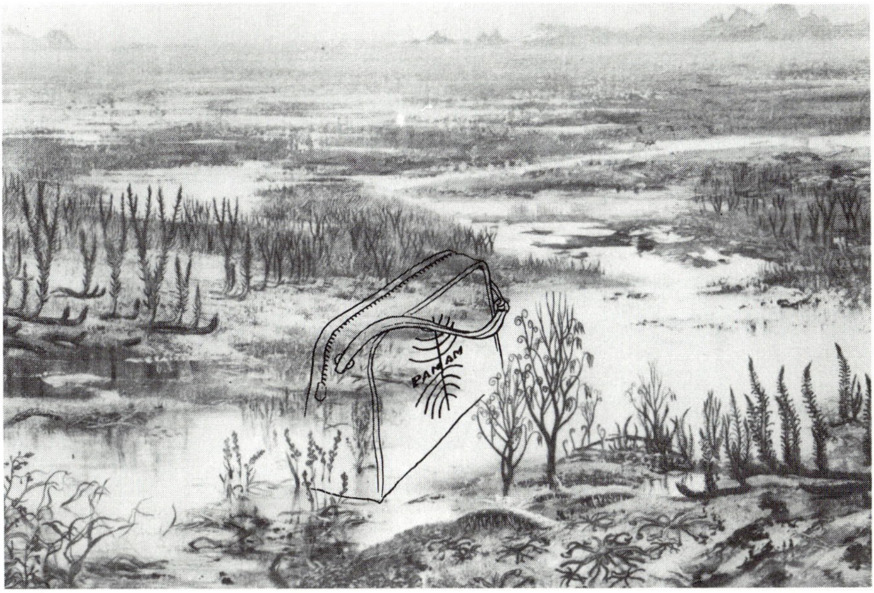

Figure 11 - 2. An early Devonian landscape, in which several genera of fossil plants from strata in various parts of the world have been brought together. Bag indicates scale. (After Augusta and Burian, Artia, Prague.)

ing these large dimensions, we can say that the Earth's first woodland began to appear in late Devonian time. Besides these tree-size plants there were large horsetail rushes and large club mosses, descendants of the small earlier Devonian kinds. Figure 11 - 3, another composite scene, suggests the look of a late-Devonian landscape.

Late-Paleozoic floras; seed bearers. The next important event in the history of plants was the advent of seeds. This is evident in fossils collected from Mississippian and Pennsylvanian strata. Like the early models of many other inventions, the seeds were rather primitive. They were bare, naked bodies that formed on leaves. A sperm cell fertilized the egg of a seed attached to the same plant. The fertilized seed matured and either dropped to the ground or was blown away. Among the early plants with seeds were seed ferns. Modern descendants of other early seed-bearing plants, such as the conifers pine, spruce, and hemlock, still reproduce in this way. Together the seed ferns, conifers, and similar plants belong to a

Figure 11 - 3. Composite landscape of late Devonian time, showing true ferns (F), primitive scale trees (S), and smaller vegetation. These trees were about 30 feet high. (After Augusta and Burian, Artia, Prague.)

group named *gymnosperms* (the word means "naked seeds") all of which possess seeds of a primitive kind.

Once they had come into existence, gymnosperms had no need to cluster close to wet places, for their style of reproduction did not depend on the presence of even a film of water. They began to invade uplands, mountains, and dry places that earlier had been denied to plants, and in consequence the green mantle of vegetation spread more widely over the Earth.

Obviously, since gymnosperms have survived through 350 million years and are very numerous today, their system of reproduction works. But to make it work, gymnosperms have always had to produce a huge number of seeds per plant, and an even huger amount of pollen (the powder-like male sex cells) to pollinate them. They depend on the winds to scatter their seeds, which are thereby subject to dangers because many land in the wrong places: in water, in deserts, in places too cold for the germination of seeds. Because of this helter-skelter method, gymnosperms waste a good deal of energy in producing seeds and pollen, most of which cannot survive.

Figure 11 - 4. A frond of the fern *Pecopteris dentata,* preserved in delicate detail on a slab of claystone that was once clay on the floor of a Pennsylvanian coal swamp. The frond grew from the top of a tall trunk; so the plant as a whole was shaped somewhat like a palm tree. The fossil, about 30 inches long, came from a coal mine near Danville, Illinois. (Courtesy of the American Museum of Natural History.)

Subsequently the plant world overcame that difficulty by inventing a more economical method, but not until Cretaceous time, nearly 200 million years later.

Meanwhile let us look at three representative scenes from within that long intervening time. The first is a Pennsylvanian landscape. It is richly detailed because it is based on an enormous number of fossil plants (Fig. 11 - 4) collected from beds of coal (compressed plant matter) and from claystone strata that were deposited with the coal. Extensive lowlands and a mild climate prevailed widely in several continents in Pennsylvanian time. Lush vegetation flourished in huge swamps whose abundant water bodies covered dead plant matter, protected it from decay, and preserved it for later compression into coal. Our picture of Pennsylvanian plants is therefore one of luxuriant lowlands. About the hilly uplands of the time we know little, because the steeper slopes of uplands and the general absence of standing water on them do not favor the preservation of a plant once it has died.

Figure 11 - 5. Reconstruction of a coal swamp in the Pennsylvanian Period around 300 million years ago, showing the luxuriant vegetation that prevailed in such places. F: true ferns; H: horsetail rushes; S: scale trees; SF: seed ferns; also D: a large dragon fly. Some of the horsetail rushes were as much as 100 feet tall. (Courtesy of the American Museum of Natural History.)

The swampy lowlands may have looked like the scene in Figure 11-5, which portrays a wet forest. In it were large numbers of insects and salamander-like animals, as we shall see in the next chapter. This sort of swamp was by no means peculiar to the Pennsylvanian. It has existed since then and is common today in places along the Atlantic and Gulf Coasts of the United States. The swamp in Figure 11-6 is basically similar, though of course most of the plants in it are of kinds radically different from those of the Pennsylvanian Period.

Figure 11-6. Modern swamp with trees and vines, on the Texas coastal plain. (Jim Bones.)

Figure 11 - 7. Great logs of Triassic conifers, now consisting of quartz, exposed in the Petrified Forest National Park, Arizona. (Alan Pitcairn from Grant Heilman.)

Mesozoic vegetation. The second scene we want to examine dates from the Triassic Period and lies in the region of the Petrified Forest National Park in eastern Arizona. That "forest" is not really a forest at all. It is the wreckage of a forest that stood elsewhere, and consists of a great accumulation of logs embedded in claystone from which they are now gradually emerging by weathering (Fig. 11 - 7). During their long sojourn in the stratum that enclosed them, the logs were petrified by having their cellular structure filled with silica, and so were converted to agate, a variety of quartz. They represent conifers (gymnosperms of course) that are relatives of today's pine trees. When living, they stood 100 to 200 feet tall, with trunks as much as 7 feet thick. We know that they did not grow at the site where they are now seen, because their surfaces are all abraded and all branches and roots have been broken off. They are driftwood, washed downstream by floods from some upland environment and buried in fine alluvium. These logs tell us that as much as 220 million years ago upland groves of great conifers in western North America were fully as majestic as those which were growing there much later, when European man saw them for the first time.

Figure 11 - 8. Composite view of reconstructed plants (gymnosperms and seedless plants only) that were common in the Mesozoic Era. 1. Cycad relatives; 2. Horsetail relatives; 3. Ferns; 4. Conifers. Other Mesozoic landscapes are seen in Figures 13 - 5,-7; 14 - 4, 14 - 6, 14 - 7, 14 - 12. (Augusta and Burian, Artia, Prague.)

The third scene is another composite view, this time of plants characteristic of the Mesozoic Era (Fig. 11 - 8). Dominant among them are two sorts of gymnosperms: conifers and cycads and their relatives. The conifers included pine-like and cypress-like trees, as well as ginkgos. The cycads probably evolved from seed ferns. With their crowns of graceful fronds they resembled palms in general form, although they were not related to palms in an evolutionary way. Some had stubby, globular or barrel-shaped bases; others had long stems. A few still survive today. Less conspicuous in Mesozoic time than these gymnosperms were the seedless plants, led by ferns and horsetail rushes.

Angiosperms

Flowers and fruit. The long history of plants is marked by one more major evolutionary development, invention of the flowering plants or *angiosperms*. The name, which means "encased seeds," is applied because the seeds are not carried as naked bodies on the leaves. Instead, one, several, or many seeds are enclosed together in a special container that forms the outer part of the plant's fruit, as in an apple. Also, angiosperms have not only fruit but flowers. These two possessions insure better reproduction. For instance, the colored flowers attract animal pollinators such as bees, which transport pollen (the male sex cells) that in a more primitive plant would depend on the wind for transportation. The fruits also attract other animals to eat them and so to transport the seeds they contain to other places.

Flowers, fruits, and other improvements permitted angiosperms to enter the few exceptionally difficult environments that remained without life, just as Mount Everest could not be climbed by man until after a whole array of special equipment, including the oxygen bottle, had been invented or adapted for the task. Today angiosperms constitute more than 95 percent of the existing species of vascular plants.

We do not know when angiosperms first appeared. The fossil record indicates that they are certainly present in Cretaceous strata and possibly in Jurassic and even Triassic strata (the fossils in the latter look like angiosperms but lack the parts necessary for certain identification). At any rate, by the middle of Cretaceous time, about 100 million years ago, 90 percent of the plant species identified from fossils were angiosperms. This group includes most of today's deciduous trees, fruit-bearing shrubs, common vegetables and flowers, grasses, and cereals. This is an impressive list; the appearance of angiosperms marks the transition from a medieval to a modern plant world. The late Cretaceous floras already had a modern look, and by the beginning of Cenozoic time the trees and other vegetation (Fig. 16-3) would have been not merely recognizable but familiar to human eyes, if human eyes could have been there to look at them.

Angiosperms and climate. A conspicuous difference between angiosperms and the vegetation of earlier times is that except in the tropical zone angiosperm trees and shrubs are deciduous. After their seeds ripen they shed their leaves, photosynthesis stops, and they become dormant for a time, in some cases for months. This fact has suggested to scientists that the invention of angiosperms was at least partly an evolutionary response to a change in the climates of the

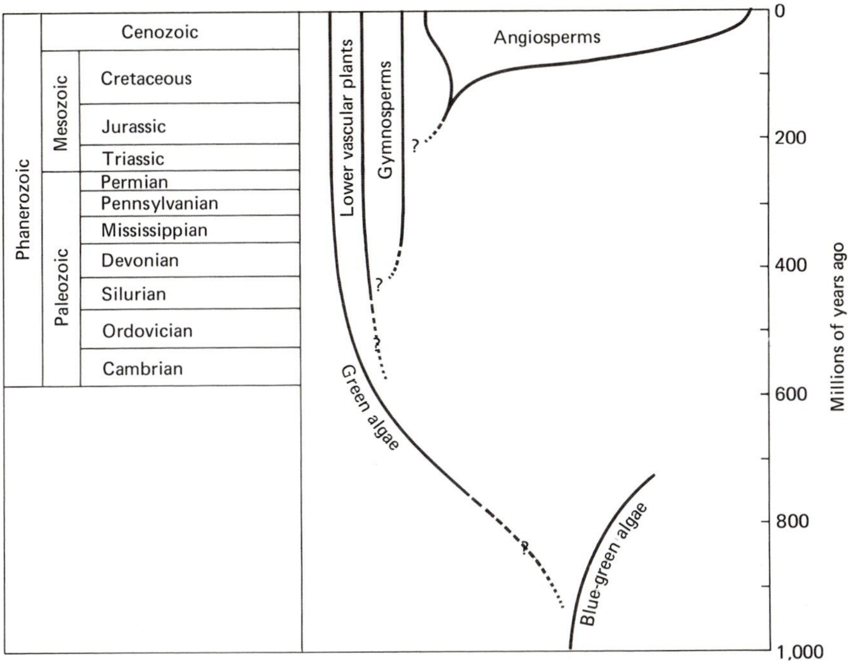

Figure 11 - 9. Highlights of the geologic history of plants.

continents. If climates in Mesozoic time were becoming more markedly seasonal, with seasons that were colder and drier than before, the shedding of leaves and the interruption of activity would make good sense. Thinking about this suggestion, we can hardly avoid looking again at Figure 6 - 10 to see where the movements of crust plates were carrying the continents during the Mesozoic Era. North America and the European part of Eurasia, at least, were moving northward into higher, cooler latitudes. Just possibly this idea explains the development of deciduous kinds of plants in Jurassic and Cretaceous time. It is only an idea, but by trying to fit together historical evidence of quite different kinds we do make progress. This is how we have learned the great relationships among events in the Earth's long history.

Significance of grasses and cereals. Although in terms of the dominance of angiosperms the world's vegetation became modern before the end of the Mesozoic Era, the history of plants does not end there. Important changes *within* the angiosperm group began before the end of the Cretaceous and continued through much of the Cenozoic. Of these, the most interesting for our story is the development of grasses.

Many city dwellers today think of grasses only in terms of lawns and golf courses. But grasses, a distinctive group of angiosperms, are specialized for life in a dry or seasonally dry climate. The natural home of grasses is on dry plains and on the tree-dotted grasslands called savannas. The vegetation map (Fig. 10 - 2) shows that nearly all of west-central North America is natural grassland, to say nothing of the great areas of sparse grass in the desert and semidesert country farther west, from Washington State to Mexico.

Probably because of the spread of dry climates, grasses became more and more widespread during Cenozoic time. With their stems, leaves, and seeds, all of them edible, grasses are an ideal food, rich in energy, for animals equipped to eat it. And as we might expect, the rapidly evolving mammals of early Cenozoic time did eat it in large measure. Through natural selection they became better and better adapted to eat it; their teeth and even their limbs took shape under the dominating influence of their diet of grass, as we will see in Chapter 16.

Cereals are special kinds of grasses in which the concentration of food is even greater than in more ordinary grasses. Men discovered the high food value of cereals 10,000 years or more ago. Thereafter it was not long before men began to cut down and clear forests so that they could cultivate cereals. Still later they began to improve crop yields by selecting seeds and breeding plants artificially, and by the use of fertilizers and irrigation. Today most of the world of men lives mainly on cereals in the form of rice, wheat, corn, other grains, and bread. Ten thousand years is too short a time for this cereal diet to have affected the structure of the human body. But that diet made it possible for man's numbers to increase to a spectacular degree, a fact of which we are awesomely aware today. In this fact we see, as we have seen in other parts of the Earth's history, the tightly woven interrelationships, not only among various parts of the biosphere but also among elements of each one of the Earth's four spheres.

Before we turn to the evolution of animals, let us recount the major events in the history of plants, remembering that without plants there could have been no history of animals on the lands. The historical milestones revealed by our record of fossils are these:

1. Emergence of marine algae; primitive land plants.
2. Seedless plants. Ferns, rushes.
3. Invention of the seed. Gymnosperms.
4. Invention of angiosperms, with their flowers, fruits, and other improvements.
5. Evolution of grasses and then cereals.

The evolutionary relations of the principal groups of plants to each other and a calendar of plant history are contained in Figure 11 - 9.

References

Andrews, H. N., Jr., 1964, Ancient plants and the world they lived in: Cornell University Press, Ithaca, N.Y.

Darrah, W. C., 1960, Principles of paleobotany: The Ronald Press Co., New York.

Delevoryas, Theodore, 1966, Plant diversification: Holt, Rinehart and Winston, New York. (Paperback.)

Scientific American Staff, eds., 1957, Plant life: Simon and Schuster, Inc., New York. (Paperback.)

12

Invertebrates and Fishes

Basis of animal history

The geologic history of plants makes a good background for the history of animals because of the close biochemical relationship between the two groups of organisms, with plants providing food and free oxygen for animals and animals furnishing carbon dioxide for plants. By reason of that relationship we would expect both groups to have evolved more or less in step with each other. In fact they did; we read in the last chapter that the evolution of some animals was influenced by the kinds of plants that evolution created—always under the remote control of the conflict between the Earth's internal and external processes.

Our study of animal history, like that of plant history, is based on inferences drawn from fossils. But the succession of fossil animals is broken by many gaps, which we have to bridge by speculation until the key fossils are found. As more fossils are discovered and studied, some of the speculations made in earlier years are proved right and become part of our firm knowledge. Others turn out to be wrong and are replaced by better ideas.

Differentiation of animals from plants

As we said much earlier, the difference between plants and animals is fundamental and extends far back toward the beginning of the history of life. It lies in the difference between the ways by which the two groups are nourished. Basically a plant needs carbon dioxide and sunlight, with which it photosynthesizes the organic compounds that are both its food and its body substance. Animals are a different matter. Whether single celled or many celled, animals take their food from the substance of other organisms, either directly from plants or at second hand from other animals. Also they breathe oxygen, a source of energy provided to the atmosphere and hydrosphere by plants. Without both food and oxygen, animals could not exist. Evidently, then, they could not have evolved until after plants had come into existence and had begun to produce free oxygen by photosynthesis.

As Table 7 - A indicates, we have fossil evidence that early in the last billion years of the Earth's history, before the Phanerozoic began, both plants and animals were living in the sea, complementing each other's life processes. Plants were freeing the oxygen from CO_2 while animals were breathing the freed oxygen, oxidizing plant food, and creating CO_2 to be used again by plants. This is biochemical cooperation.

However, although the cooperation continued, animals began to look more and more different from plants. The structures of the two groups gradually diverged because of differences in their ways of life. Plants, first in shallow seas and later on land, had no need to go in search of food because food was all around them. They could simply stand still and absorb it. On land all they needed was leaves, each with exposure to sunlight, a system of roots to collect water solutions from the ground, and a connecting structure to support the leaves and transmit the solutions. And that is what most plants have. But animals are different. Except for certain simple kinds that live fixed on the sea floor, animals, because they prey upon the bodies of plants and other animals, must go in search of food. They are hunters. For this reason, right from the beginning of their evolution they had to be able to move about. Early animals that lived in the sea solved the problem of movement in various ways. The most successful of these was also the most interesting because it laid the foundation for the structure of the human body. It consisted of developing:

First, a firm solid framework to which muscles could be attached;

Second, a tail to act as an oar;

Third, a framework of limbs (an equal number left and right) which muscles could move, in order to balance the body and later to propel the animal through the water;

Fourth, eyes, placed in front like the headlights of a vehicle, to see the prey being pursued and obstacles that must be avoided;

Fifth, a mouth (also in front) to take in not only food but also water to be passed along to a set of gills that can extract oxygen from it;

Sixth, a central control box (a brain, again in front) connected with muscles by a nerve system.

If we add to this list, seventh, an elongated streamline shape to reduce the friction of motion through water, we end up with a resulting object that is basically fishlike.

Of course these inventions did not appear all at once. They developed little by little through tens if not hundreds of millions of years. No single one of them was brand-new. It was only a gradual extension or remodeling of some existing feature. Because of the way natural selection works, it can never start from scratch. It can only change, by slow degrees, something old and put it to a new use.

The basic fishlike form was successful in the sea. It was later carried over onto the land and became a standard pattern for remodeling into land animals. We can now see why modern animals differ in appearance from modern plants. The animal form resulted from the need to pursue food.

Environments within the sea

It is not difficult to imagine that early animals would have evolved into many kinds; the ocean as we see it today is very far from being monotonously uniform. It harbors a large number of different environments, which through natural selection mold organisms into many forms and many specializations. The ocean of a billion years ago, like that of today, would have provided environments for animals floating at and near the surface, for dwellers on the cold, dark bottom, and for some animals in between. In the shallow waters along coasts there would have been, as there are today, differences of temperature, of the kind of rock or sediment on the bottom, and of intensity of agitation by waves or currents. Each slight difference in the environment, however subtle, imposes its own influence by means of natural selection.

Sea animals specialized accordingly. Some were herbivores, eating marine plants; others were carnivores, eating other animals. Some of the carnivores were merely scavengers, eating dead bodies. Others were predatory, devouring live prey. No wonder, then, that

Figure 12 - 1. Restoration of a Cambrian sea floor. C: Sea cucumbers; J: Jellyfishes; S: Sponges; T: Trilobites. The trilobites are two to three inches long. (Prepared by George Marchand under the direction of I. G. Reimann, University of Michigan Exhibit Museum.)

shallow seas were lively places even in early times, the principal activities being the search for food, escape from predators, eating, and being eaten.

Figure 12 - 1 gives some idea of how a shallow sea floor in Cambrian time looked. Against a background of waving seaweed are cabbage-like clusters of branching sponges, sea cucumbers, parachute-like jellyfishes, and the now-extinct relatives of modern crustaceans called trilobites. This group of living things has been reconstructed with unusual accuracy, thanks to the way in which it was preserved. Near Field, British Columbia, in the heart of the Canadian Rockies, a geologist discovered, almost by accident, small fossils in a layer of black shale. The shale was carefully quarried and became the source of 130 species of Middle Cambrian invertebrate fossils, their limbs and even their antennae preserved in minutest detail (Fig. 12 - 2), carbonized like the fossil fish in Figure 9 - 3. Because edible parts of animals on the sea floor are almost invariably destroyed by scavengers, the occurrence at Field is most unusual. It represents an area of former sea floor from which oxygen-bearing water was cut off. The bottom water, lacking oxygen, became deadly to all life except certain bacteria. Organisms perished and their

Figure 12 - 2. Two fossil trilobites (of the species *Olenoides serratus*) in Cambrian black shale from British Columbia. Not only antennae but long appendages trailing from the tail are preserved. (Smithsonian Institution.)

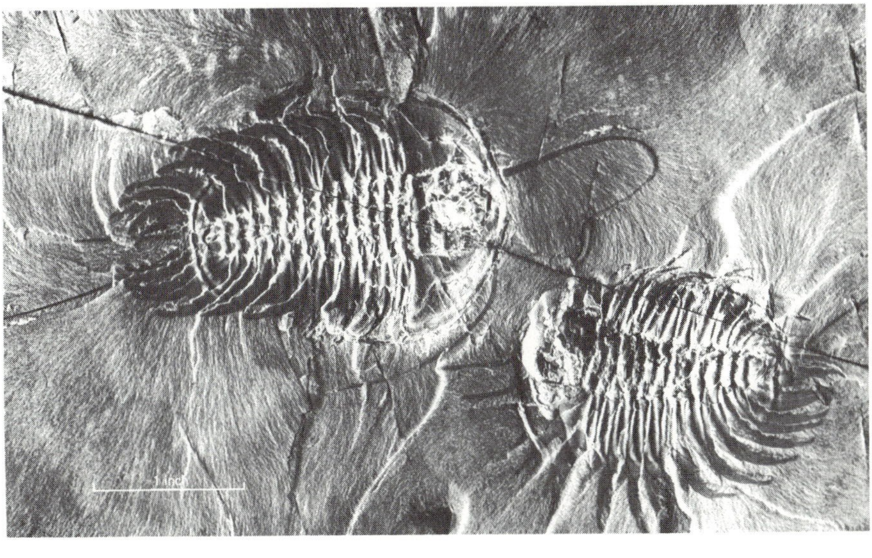

bodies were buried in the fine black mud of the bottom. Because scavengers too cannot live in an environment deprived of oxygen, bodies were preserved down to the finest detail. Areas of water that lacks oxygen exist today, one of them in the Black Sea.

Among the 130 species collected from the black shale there are no brachiopods at all, despite the fact that, worldwide, nearly one-third of all Cambrian species are brachiopods. The absence of these animals is hardly likely to have been an accident. We might surmise that it was a matter of environment. The brachiopods, which lived attached to rocks, were not mobile and did not float. Therefore, they would not have inhabited a mud bottom, nor would they have been light enough to drift with the currents into the lethal area.

The lethal character of that patch of sea floor has added immeasurably to our knowledge of the degree of evolution reached by marine invertebrates in Cambrian time. It was far advanced. Yet the roster of Cambrian invertebrate animals was very different from that of invertebrates today. Fossils collected from Cambrian strata consist overall of about 60 percent trilobites, about 30 percent brachiopods (Fig. 3 - 2), and only 10 percent other kinds. Today trilobites are extinct, brachiopods are restricted to a small area of the ocean, and "other kinds" have taken over.

The sea-floor scene in Figure 12 - 1 may look crowded, but it is no more crowded than the scenes along modern coral reefs, easily visited by skin diving or snorkeling. Sampling and counting fossils from a single cubic yard of sedimentary rock shows that sea floors were crowded everywhere, and they probably became so long before Cambrian time. Living things tend to fill the available space, and if invaders drive the inhabitants out, the same space is then fully occupied by the invaders.

Shells and skeletons

One feature of the scene in Figure 12 - 1, the presence of animals with shells and skeletons, is noteworthy because those structures had far-reaching influence on the path of evolution. Sponges had skeletons of silica, and certain tiny floating animals and plants not shown in our scene had shells of the same material. Trilobites possessed shells of calcium phosphate interlayered, plywood-like, with films of chitin, a nitrogen compound found also in human fingernails.

Although not present in this particular scene, most of the numerous brachiopods, and other animals as well, had shells of calcium carbonate. Shells are the hard parts that are readily preserved as fossils. They are also external skeletons, armor that serves a useful purpose. The sea animals of much earlier time were soft-bodied and flabby, lacking the hard parts that lend themselves to preservation. Yet here, in the Cambrian sea, more than half a billion years ago, were animals that had managed to acquire shells of three different kinds.

The fossil record offers few clues as to how sea animals acquired hard parts; so we have to fall back on speculation. As an example, we might imagine that the calcium-carbonate shell developed in some such way as this. Calcium ions are present in seawater and in much of what a sea animal eats. Thus calcium could penetrate into the eater's body. Much of it would be excreted and lost, but some (in amounts depending on the individual's biochemistry) would remain in or at the surface of the body and accumulate as a sort of stiffening. On a sea floor with many predators about, it was probably safer to wear armor, and greater stiffening was favored by evolution. As a result skeletons and shells gradually developed. Whether or not this was the way it happened, the process of stiffening in sea animals parallels the process of the developing vascular system in land plants.

Through the 60 million years or so of Cambrian time invertebrates with an external skeleton in the form of a shell were the most complex kind of organisms. Higher organisms with the internal skeleton like that possessed by vertebrate animals had not yet been invented.

Glimpses of invertebrate evolution

Nearly all the major groups of marine invertebrates are present as fossils in Cambrian strata. Most of them have survived (in evolved forms) right up to the present day. These facts are summarized in Figure 12 - 3. Definite evolutionary changes took place within each group, but most such changes were less conspicuous than those which affected many land animals. In most cases this difference was the result of a smaller range of changes in environments within the sea than in land environments. Although the evolutionary changes in marine invertebrates are clearly recognizable to experts, they are less so to the ordinary visitor in a museum. Nevertheless we must

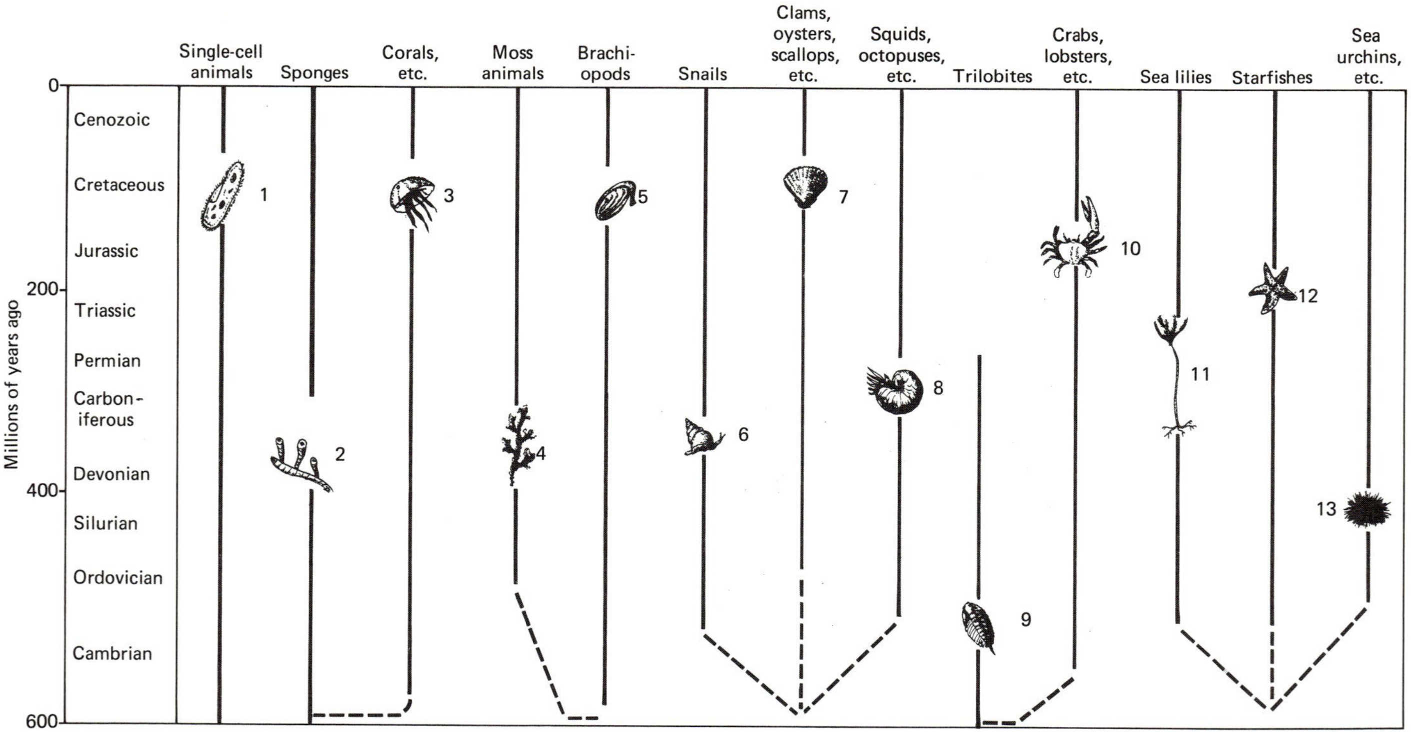

Figure 12 - 3. The main groups of marine invertebrate animals and their positions in the geologic column. Widths of white bands suggest approximate abundance of each group from one stratum to another. Dashed lines suggest likely evolutionary relations among some of the groups.

Figure 12 - 4. Fossils of invertebrate sea animals on a slab of mid-Devonian sandstone from Hamilton, New York. For identifications see text. The reconstruction in Figure 12 - 5 is based on fossils of this age and region. (National Museum of Natural History.)

not forget that it is mainly the evolutionary differences and similarities among invertebrate fossils that have made possible the correlations of strata through which the geologic column (Table 3 - A) became possible.

Therefore, in order to be able to give more attention to the history of vertebrate animals, we will content ourselves with looking at two samples of the life on shallow sea floors at two times later than that of the Cambrian scene already displayed in Figure 12 - 1.

A Devonian sea floor. One is a collection of fossils from a sandstone stratum of Devonian age exposed at Hamilton, in the central part of New York State. A sample of the rock is seen in Figure 12 - 4. In it the big central figure is *Devonaster eucharis* ("graceful Devonian star"), a starfish that was already pretty much like a modern species.

Figure 12 - 5. Scene on the floor of a shallow Devonian sea in what is now the central part of New York State. A: algae; B: brachiopods; C: cephalopods; CL: corals (some branching, some with horn-shaped shells, some in colonies resembling a bundle of organ pipes); S: snails; SL: sea lilies; T: trilobites. (Courtesy of the American Museum of Natural History.)

In the "northeastern" angle formed by the star is a thinner relative, a "brittle star." In the "northwestern" angle, out at the edge of the sample, is a cylindrical hollow with rings around it. This is the external mold of part of the stem of a crinoid or sea lily, the complete form of which appears near the right margin of Figure 12 - 3. Most of the other fossils are of three or four different kinds of brachiopods, a group having hinged shells somewhat like the shell of a clam, and also represented in Figure 12 - 3.

It was from a fauna of this kind that the scene depicted in Figure 12 - 5 was restored. If you imagine yourself a Devonian skin diver, you could have seen something like this, in a field of view about 2 feet wide along the base of the photograph, on which the visible animals are named. Especially noteworthy are the two cephalopods, one kind with a long straight shell and the other with a frilly, coiled shell. Both belong to the same group of carnivores as do the squids and octopuses of today; all have tentacles equipped with sucking disks to seize and hold their fleshy prey.

Figure 12 - 6. Scene on a Cretaceous sea floor not far from what is now Memphis, Tennessee. Base of photograph is 6 feet long. A: algae; B: belemnites; cephalopods with (CC) smooth, coiled shell like the modern nautilus, (CS) straight shell, and (CSP) spiral-coiled shell; S: snail. (Courtesy of the American Museum of Natural History.)

A Cretaceous scene. The other sample is a sea-floor view representing Cretaceous time in an area that is now southwestern Tennessee (Fig. 12 - 6). The straight-shell cephalopods common in the earlier part of Paleozoic time had been eliminated in the evolutionary struggle; instead there were coiled shells, in the form of a flat disk like a watch spring or a roll of Scotch tape. In the wide mouth of the coil was the head of the owner with his tentacles ready to seize small fishes and other prey. Also characteristic of Mesozoic but not of Paleozoic seas were squid-like cephalopods called *belemnites.* They had internal shells resembling backbones, built of calcium carbonate, most of them about the size and shape of very thick pencils or thin cigars, which are preserved in great numbers as fossils. One kind was much bigger, with a length of more than 5 feet. Belemnites, like Paleozoic cephalopods, moved rapidly by jet propulsion. Of course the jets were water, not gas, but the basic principle was similar to that of a rocket. The cephalopod's jet represented the invention of a means of movement quite different from the use of feet or fins. In addition the belemnite, like the modern squid, had a sac

Figure 12 - 7. A carbonized fossil "sea scorpion" (named *Eusarcus scorpionis*) 4 inches long, from a layer of Silurian rock exposed near Buffalo, New York. The jointed body, legs, and tail (the tip of which has been bent sideways) are clearly visible. (Courtesy of the American Museum of Natural History.)

containing "ink," which it could squirt out through its jet, dyeing the water behind it, presumably so as to foil pursuit. Both the jet and the ink illustrate once again the originality of the evolutionary process in contriving adaptations to particular environments.

"Sea scorpions." Mostly because of their exceptional size, we ought to mention a group of arthropods (Fig. 12 - 3) that are distantly related to king crabs and are perhaps ancestors of modern scorpions. These are "sea scorpions," which existed through much of the Paleozoic Era but reached a peak in Silurian seas. Breathing with gills, they possessed a segmented shell built of chitin, two pairs of eyes, and six pairs of jointed "legs" (Fig. 12 - 7). Although many kinds were small, one was more than 7 feet long. Some kinds swam, while others crawled along the muddy bottoms of estuaries seeking worms or small fishes (Fig. 12 - 8). Bigger than any fishes of the

Figure 12 - 8. The largest of the known "sea scorpions," more than 7 feet long, restored with snails and trilobites on the floor of a late Silurian estuary near Buffalo, New York. At right, above, is the smaller kind represented in fossil form in Figure 12 - 7. (Courtesy of the Smithsonian Institution.)

time, "sea scorpions" were the largest animals that had yet appeared in the biosphere.

Insects. Like plants, invertebrates emerged from the sea onto the lands. Land snails evolved from marine snails; insects and spiders evolved from marine arthropods. But these events have so far left only a very spotty record. Small size and lack of hard parts do not lead to the preservation of abundant fossils except under unusual conditions. Therefore we know little about this scramble among plants and various animals to climb out of the sea. Fossils establish that scorpions, probably descended from one of the smaller "sea scorpions," existed as early as Silurian time, and that insects (still lacking wings) and spiders existed at least as early as Devonian time. Then, as we know, the lands were already partly covered with primitive plants. To some extent the marine contenders were already selected for adaptation to land life, for marine arthropods already had a protective shell and were accustomed to moving about. The principal evolutionary changes they still had to accomplish in order to become insects were to develop a system of breathing air and to learn to fly. They did the first by evolving quantities of pores in their shells, with tubes leading from them into the body proper. Hence their breathing, their absorption of oxygen, took place over their whole body surface. They accomplished the second change before Pennsylvanian time by extending parts of the body wall outward to form wings.

Compressing these events into a thumbnail sketch, we have probably made them sound like the story of a plucky little arthropod determined to succeed. But wait. Before we begin to think that some kind of free will must have been involved, we must remember about time. The span of time from mid-Silurian to the beginning of the Pennsylvanian was approximately 120 million years—probably more than ten times greater than the time it took to create the Grand Canyon. Time for the lives of at least 120 million generations of arthropods to run their course; plenty of time for two principal adaptations (breathing air and flying) to be selected from a host of trials and experiments.

The widespread Pennsylvanian coal swamps, with their shallow bodies of water in which clay was slowly deposited, were unusual conditions that made possible the preservation of great quantities of insects as fossils. Those collected and described so far include *eight hundred* species of cockroaches alone, one of which was four inches long. Other fossilized insects include kinds more than a foot long, and one fossil dragonfly has a wingspread of 29 inches, about that of a sizable modern hawk. One of these giants is shown in Figure 11 - 5. Evidently the mild climate that favored the luxuriant coal-swamp vegetation also favored insects.

It isn't surprising that these new insects, recently emerged from the sea and still wingless, gradually developed wings and turned to flying as a way of life. For a small creature, flight through the air offered the best means of escape from enemies (other insects, because birds did not materialize until more than 200 million years later) and of pursuing prey. We must admit their action was successful (if success is defined as ability to occupy territory and to reproduce) because there now exist three times as many species of insects, most of them fliers, as of all other animals combined.

Figure 12 - 9. A fossil *Glossina* (tsetse fly) from strata deposited in a lake of Oligocene age. With it is a fossil leaf of a tree related to the modern birches. (Courtesy of the American Museum of Natural History.)

After insects had taken to the air and had adjusted to the many possible varieties of aerial living, the pressure to evolve rapidly slackened and evolution slowed. Early in the Cenozoic Era, after some 250 million years of evolution, many kinds were already very similar to those around us today. In Figure 12 - 9 we see a fossil insect that was preserved in the fine sediment deposited at the bottom of a lake in central Colorado about 30 million years ago. It is a *Glossina,* the same genus although not the same species as the tsetse flies of today (one of which carries the parasite that causes sleeping sickness in man). So for *Glossina,* at least, the rate of evolution has been rather slow.

Origin of fishes

Earlier in this chapter we described the basic fishlike structure and form. As yet we have found no fossil representatives of a sea animal that might have been ancestral to fishes. The best guess (and a guess is all it still is) is that what we seek might have been an early small, elongate, soft-bodied ancestor of modern sea urchins and starfishes, lacking a skeleton and eyes, but with a jawless mouth that sucked up food particles from bottom mud. One of several reasons for the guess is the character of the earliest (Silurian) *whole* fossils of fishes. (Bony scales but nothing more have been found in Ordovician strata). These whole fossils indicate that the fishes had sucking mouths, no jaws, and no fins. The bodies had a somewhat flexible covering of bony plates, and they occur in deposits of streams and lakes but not of the sea, an indication that primitive fishes followed the algae out of the sea into rivers and lakes. These facts have led to two speculations. First, the elongate, flexible body might have developed in response to environments characterized by steady river currents. Second, the armor plates, admittedly clumsy, served not as protection against predators (perhaps the "sea scorpions"?) but as a barrier to the entrance of too much fresh water

Figure 12 - 10. *Dinichthys,* a 30-foot carnivorous fish with bony armor, pursuing sharks through a shallow Devonian sea. His jaws were armed not with teeth but with the sharpened, zigzag edges of upper and lower jaws. (Augusta and Burian, Artia, Prague.)

into the body through pores. Such a barrier might have been necessary to an animal whose ancestors had always lived in salt water and whose body chemistry was adjusted to salt water.

Be that as it may, Devonian fossils include marine fishes with jaws (remodeled from one of the arches that support the gills) and more efficient equipment for swimming. These early marine fishes were of several kinds, the most spectacular of which was a predatory creature of the genus *Dinichthys* ("terrible fish"), with a body length as great as 30 feet (Fig. 12 - 10). The sharks in the picture possessed skeletons of cartilage, like all sharks right up to the present. But in *Dinichthys* a change in composition was occurring; its fossil skeleton shows the presence of calcium. More fully evolved in this respect were the *bony fishes,* which likewise branched off in Devonian time from more primitive freshwater ancestors. Their skeletons were fully calcified and therefore harder and firmer. Also, their fins were adapted for faster swimming and the streamlining of their outer form had been perfected. These advantages fitted them for a dominant position among fishes, one they still occupy today. We call them successful because they evolved into more kinds and produced greater numbers than any other group. But because nearly all of them continued to be confined to the sea, in terms of evolutionary change they were in a blind alley that lacked the challenges to be met with on the land.

Transition to life on land

In the last sentence we said "*nearly* all." The exception was one inconspicuous group of bony fishes that did not return to the sea but stayed on in fresh water and was thus already, as it were, halfway on land. Because they stayed, these fishes had a much more interesting history than did their relatives in the sea. They were obliged to meet the challenge of land life. They accomplished what plants and then insects had accomplished before them. They succeeded in freeing themselves from a water environment, even a freshwater one, and emerged onto the land. In so doing they became the ancestors of all vertebrate land animals, including, at length, man.

Through these freshwater representatives, then, the race of bony fishes bequeathed their serviceable skeleton of bone to man. The human skeleton was developed in basic form by fishes in Paleozoic waters. The same fishes also bequeathed to us the essentials of our breathing apparatus. This came about because of peculiarities in the anatomy of our freshwater-fish ancestors, an obscure group of lobe-finned fishes that are ancestral to the modern sea-dwelling lobe fins, coelacanths (Fig. 8 - 2). They were also ancestral through another line of descent to modern lungfishes that now live in streams in dry countries of the Southern Hemisphere. As their name implies, lungfishes possess an auxiliary lung, built from a former swim bladder, with which they can draw oxygen from the air. When the dry season comes, the lungfish wallows in the mud of the stream bed, forming a sort of cocoon, which it lines with mucus. Curled up inside, the fish becomes dormant, its body activity reduced to as little as 10 percent of the wet-season rate. When the rains come again, the mud softens and the fish emerges and swims away.

The lobe-finned fishes of Devonian time had a lung of the same kind. We shall see in a moment what this meant. First let us note that in Devonian strata, consisting of sediment deposited by streams, there have been found skeletons of *Ichthyostega,* which we mentioned in Chapter 8. This is one of the amphibians, whose group name means "ones that lead a double life," that is, a life partly on land and partly in water. They are the group of animals to which modern salamanders and frogs belong. Now when we compare *Ichthyostega* with a Devonian lobe fin (Fig. 12 - 11), the two appear strikingly similar in size, shape, and structure, particularly the basic structure of the fish's fins compared with the amphibian's feet. Looking back at Figure 8 - 5, we can see the same comparison between the Devonian fish (a lobe fin, of course) and, next to it, an amphibian. Obvious, though of less importance, is the fact that at the head end both fish and amphibian possessed an armor of bony plates. Overall, the comparison leaves little doubt that the Devonian lobe-finned fish evolved into the Devonian amphibian.

The auxiliary lung and the structure of the fin are surely what made the change possible. But what was the push that touched it off? Two reasonable ideas have been suggested. According to the first, which is based on analogy with modern lungfishes, we have a climate with pronounced dry seasons, converting streams into separate shallow pools. Left high and dry, the lobe fin responds by using its peculiar fins to drag or "row" its body toward the nearest pool still remaining, meanwhile breathing with its primitive lung.

Figure 12-11. Two steps in the evolution of fishes into land vertebrates. A. A Devonian lobe-finned fish with a primitive lung. B. The Devonian amphibian, *Ichthyostega,* about 3 feet long. Note the feet, with separate toes. But the membrane in the former fin has not yet been lost.

Those individuals with the less efficient fins, or lungs, or both fail to make it to the next pool and are "selected out" by the environment; only the possessors of the more efficient equipment succeed. In time some of the successful fishes become amphibians.

The other idea does not involve the drying up of streams. It suggests that the lobe fins took to very shallow water in order to escape the predators that infested the deeper water. In the shallows the fugitives used their fins first as props to support their bodies, and then gradually for pushing themselves along. With fins strengthened by natural selection they could then emerge to the adjacent land, where in Devonian time there was still no threat of predators, and where the evolution of fins into weak legs could be completed.

Of course both these explanations are greatly oversimplified, but between them they probably indicate essentially what happened. Remember that the Devonian Period was a generous 50 million years long—long enough to permit many millions of trials. With that much time available we could almost have predicted that *some* group of pond or river animals would get out onto the adjacent land.

In converting themselves into amphibians, Devonian fishes set up another milestone along the path of evolution. They founded a new dynasty consisting of four-legged, vertebrate land animals.

In this chapter we have recounted the conversion of a soft, flabby sea animal into a fish having these attributes: an elongate body supported by a strong, jointed skeleton, a head with eyes, a mouth with jaws and teeth, fins with a vigorous thrust, and a tail. With these possessions already developed in a fish, it remained only for the environment to press the fashioning of a lung, and the pathway from water to land was assured. In the next chapter we will see how amphibians fared in their new continental environment.

References

McAlester, A. L., 1968, The history of life: Prentice-Hall, Inc., Englewood Cliffs, N.J., p. 39–81.

Smith, H. W., 1961, From fish to philosopher: The Natural History Library. Anchor Books, Doubleday & Co., Garden City, N.Y., p. 71–101. (Paperback.)

Swinton, W. E., 1948, The corridor of life: Jonathan Cape, London, p. 67–122.

13

Amphibians and Reptiles

Amphibians

Settling in on the land. In the foregoing chapter we left the first amphibians, newly evolved from lobe-finned fishes that had dragged themselves out onto dry land, looking around at the new territory they were to inhabit. They certainly could not have had a broad view of the landscape, because they were sprawled on the ground, almost like fish out of water, with legs still weak and extremely short. They could only have moved slowly and awkwardly, but at least they were threatened by nothing more serious than insects, which we can suppose gave them little or no trouble.

But even without threat of predators or serious competitors, any land environment presents far more difficulties for its inhabitants than does water, whether sea, lake, or stream. In the sea, environments are comparatively uniform. Temperature, light, and food supply change little, and marine vertebrate animals therefore have not been compelled to develop as great a sensitivity to incoming impressions as have land vertebrates. On land, vertebrate animals are faced with more and greater immediate stimuli: changes from warm to cold and from sunshine to rain, sudden high winds, floods, and fires, to name a few.

In their new environment the late Devonian amphibians, only recently evolved from fishes, were faced with such changes, just as the shipwrecked Robinson Crusoe was suddenly confronted with a completely new and different environment. Crusoe ingeniously found or manufactured means of coping with all the difficulties in the path of his survival. In other words, he adapted to life in the wilderness.

Faced with life on land, the new amphibians adapted to some of the changes in their environment. But they seem to have compromised to a degree by restricting themselves to those environments in which the changes were smallest. In so doing they acted as conservatives would act. In terms of the history of evolution as a whole, their reaction was quite normal. An organism stays put until it is forced out. Then, if it can change at all under the laws of evolution, it does change just to the extent necessary to bring itself into line with its new surroundings.

Evidently this policy—we would call it a policy if it referred to people today—of evolution tempered by compromise enabled amphibians to overcome the land's hazards because, as their fossil record clearly tells us, they branched out and continued through the remainder of Paleozoic time. For at least 70 million years they were unquestionably the world's dominant land animals. A good many more than 100 species of fossil Paleozoic amphibians have been collected. This large number means that the animals themselves had adapted to many varieties of the near-water environment (Fig. 13 - 1). Some lived along stream banks as frogs do today, while others lived in the water itself. Some were lizard-like, some were salamander-like, some were legless and snake-like. Many looked like stubby-tailed crocodiles, but the resemblance was only superficial, for crocodiles are reptiles, a more advanced group of vertebrates.

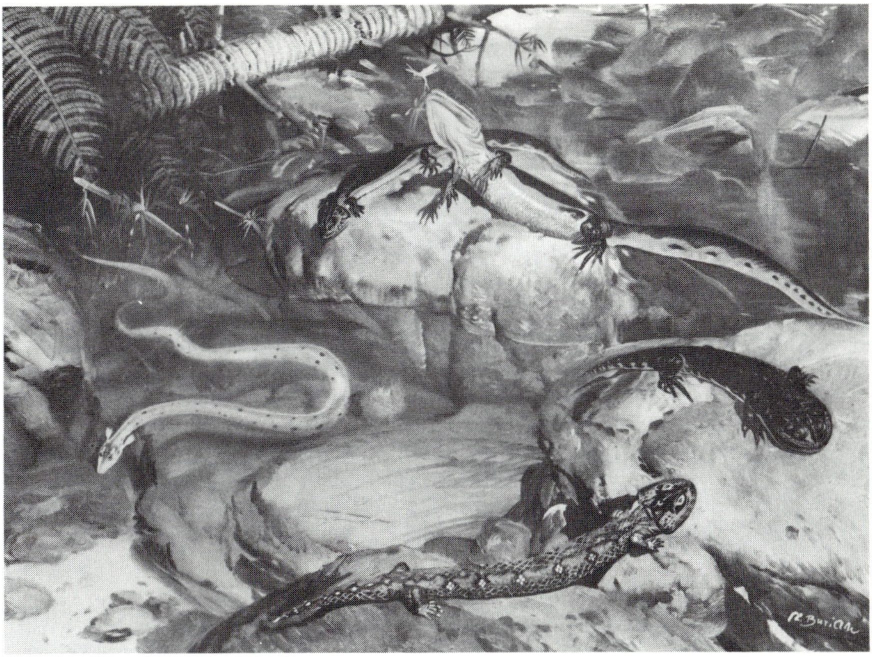

Figure 13 - 1. Four kinds of amphibians in the environment of a clear stream in a Pennsylvanian coal swamp. Most of them were carnivores. The patterns on the animals' skins of course cannot be learned from fossil skeletons. They are based on analogy with amphibians living today, and so are uncertain. (Augusta and Burian, Artia, Prague.)

The amphibians' sizes varied; body lengths ranged from a few inches to as much as 10 feet. Many seem to have had scales or bare skins; others, like their lungfish ancestors, had heads encased in bony-plate armor. They seem to have been mostly carnivores, eating insects, worms, or small fishes.

Evolutionary changes. Studying the many species of Paleozoic amphibian fossils, we can perceive several basic changes in the anatomy of amphibians, all of which took place during the first tens of millions of years following their emergence from water. It is worth our while to make a list of the changes, because some of them played a part in the ancestry of our own human skeleton.

1. *Development of a rotary limb joint.* A fish's fin is a paddle. To be effective in the water its stem need only work back and forth against the rest of the fish's skeleton. But in a land environment sidewise movement is needed also, and this demands more flexibility at the points where limbs join shoulder and hip. The envi-

ronment created the necessity, and natural selection responded with the evolution of universal joints, ball-and-socket joints. These proved so useful that they were passed on by amphibians to later land vertebrates, and have ended up as standard equipment in the skeleton of man.

2. *Evolution of the limb with five fingers.* The fin of the lobe-finned fish was already supplied with "wrist" and "hand" bones, but these were extended outward by a thin membrane strengthened by many thin, rib-like bony supports, like part of an umbrella with its fabric and ribs. As the fish evolved, "wrist," "hand," and their thin extension were gradually bent into a tight curve (Fig. 13 - 2), the number of "ribs" was reduced to five, and the membrane was eliminated. The result was a five-fingered limb, a very useful contrivance for support on irregular terrain. The amphibians kept

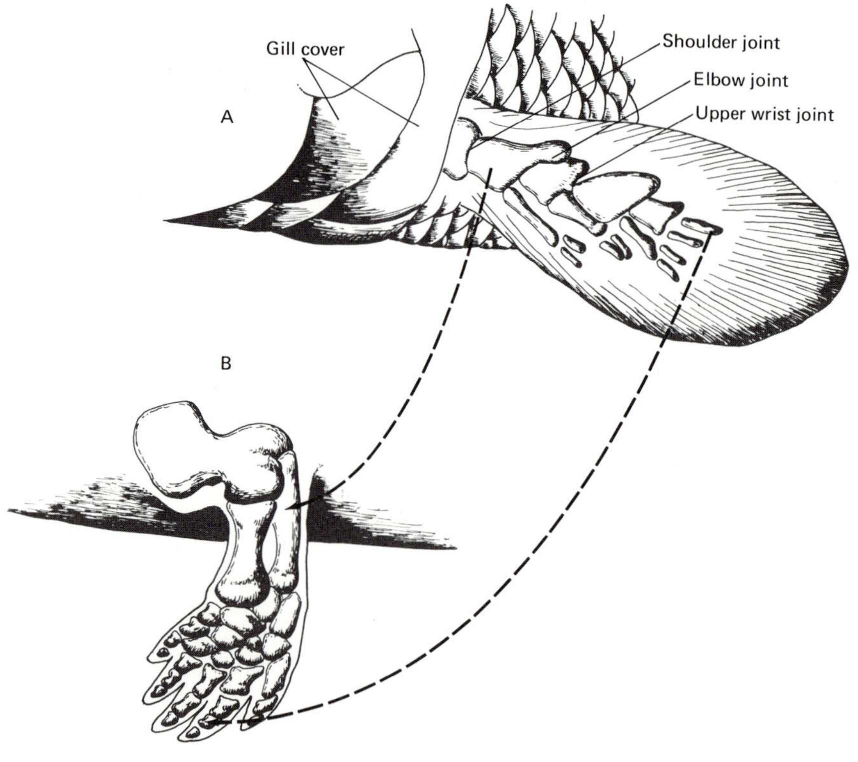

Figure 13 - 2. Evolution of fish fin to five-fingered limb (not to scale).
A. Left front fin of Devonian lobe-finned fish.
B. Left foreleg and foot of a Paleozoic amphibian. (Broken lines indicate bones that correspond with those in A.) (After W. K. Gregory.)

it and passed it along to their descendants including, eventually, ourselves.

3. *Lengthening of the limbs.* When the lobe-finned fish first began to use its fins to prop up or steady its body, it could not lift its body off the mud. Any locomotion was mere wallowing. In contrast, the amphibian, although its limbs were short and inefficient, could hold its body a little above ground at least temporarily, so that wallowing became waddling. This was the beginning of the process of walking.
4. *Loss of armor.* The armor of bony plates that encased the head of the lobe-finned fish seems to have been needed as protection against predators in the water. But with emergence onto the land, where there were as yet no predators, the need disappeared. The armor became a useless and cumbersome weight, like the armor of European soldiers in the 16th Century after firearms came into use. As a result, the evolutionary process eliminated it. In one group after another the armor disappeared because the environment no longer demanded it.
5. *Increase in size of the brain.* Enlargement of the brain case (the hollow in the skull that contains the brain) is shown clearly in the fossil skulls of amphibians. But even if there were no fossils as evidence, we would still feel sure that the brains of amphibians must soon have become larger than those of lungfishes. This would have been a necessity. Because a land animal receives many more stimuli from its surroundings than does an animal surrounded by water, the land environment exerts strong pressure upon it to improve its mental equipment so as to be able to respond. Without improved responses, amphibians might not have survived.

These five changes were important ones. But even as a group they were not great enough to hide the obvious basic resemblance between fossil fish and fossil amphibian. Look again at Figure 8 - 5 and see that the resemblance is strong.

Apart from the five improvements that are visible in fossils, we are sure that during the 50 million years or so that elapsed between late Devonian and early Pennsylvanian time amphibians were upgraded in one significant respect not visible in fossils. This was improvement of the still rather primitive lung into a larger and more efficient breathing organ. A greater supply of oxygen was essential for the more rapid movement implied by longer legs. More oxygen would have been essential also for use while the animal was submerged for short periods, as the legless, snake-like amphibian seen in Figure 13 - 1 must have been.

Despite their history-making move from water to land, during which as lobe-finned fishes they accomplished the dual feat of breathing air and beginning to move about on land, and despite their having invented internal fertilization, amphibians progressed no further. They never succeeded in developing a kind of reproduction that would allow them to leave the water environment for good and all. They always remained tied to water because they had to return to it in order to lay eggs. Limp and fragile like those of fishes, the eggs of living amphibians must be deposited in water so that they can be fertilized externally, and must remain there until they hatch. Exposure to air dries them out quickly and they do not develop. After hatching in the water, the new individual goes into a larval phase (the modern tadpole is an example), breathing with gills like a fish.

Because of this sort of reproductive process, amphibians are roughly analogous among animals to the seedless group among plants. Both depend on water for reproduction. In consequence amphibians were never able to colonize high or dry environments, and to this day they tend to remain in places where, seasonally at least, there is water in which they can deposit their eggs.

Although amphibians had these disadvantages, we cannot simply write them off, for they occupied a truly important place in the evolution of living things. They not only established vertebrate animals firmly on the land, but also handed onward the evolutionary torch. They did this through an invention that made reptiles possible and through them led to a new wave of improvements. Just as one kind of fish in Devonian time had invented a lung, so also one kind of amphibian in Pennsylvanian time invented a better egg and thereby became a reptile by definition. The egg was a big success, and became another milepost on the path of animal development. Amphibians are far from dominant today, but for at least 50 million years of Paleozoic time they were both dominant and inventive. Fifty million years is a respectably long time, far longer than the more recent time during which all the races of man and all man's ape-like ancestors have inhabited the Earth. It was time amply long for all the statistical trials and errors that led to the invention of a truly remarkable egg. Let us look first at the egg and then at the background behind its invention.

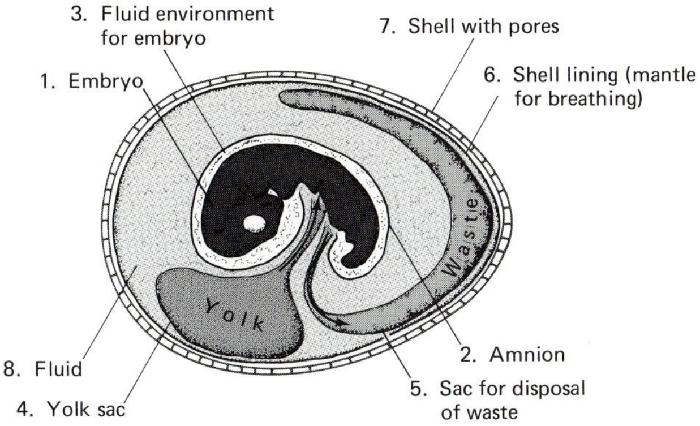

Figure 13 - 3. Essential features of the amniote egg.

Reptiles

The amniote egg. This newer-model egg (Fig. 13 - 3) was epoch-making because it did not have to be laid in a water environment. It is built so as to contain its own water environment even though it is surrounded by air. It is something like the space vehicle hurtling toward the Moon, containing its own microenvironment, with oxygen that can be comfortably breathed by the astronauts inside the protecting shell of their vehicle. Like the space vehicle, the amniote egg had a familiar microenvironment and a supply of survival equipment inside its protective shell. Its inhabitant, or astronaut if you like, was an embryo reptile; because by inventing the egg, amphibians founded the line of reptiles. No modern amphibians have the egg, but all reptiles do. So do all birds and two of the mammals. The egg was so successful that its basic features appear to have remained unchanged throughout the 300 million years since it was invented.

The egg contains, among other things, three membranes that form three sacs. The first and innermost sac, the *amnion* (from which the egg derives its name, *amniote egg*), encloses the embryo, suspended in a fluid. The fluid is an artificial watery environment, a substitute for the pond or stream water in which amphibians and fishes had always deposited their eggs, and in which the amphibian egg hatched as a larva or tadpole, breathing with gills, ready to change later into an air-breathing adult.

The second sac contains the yolk, a big package of food for nourishing the embryo. The third sac connects with the kidney of the embryo and holds its waste, the urine. Around all three sacs as a group is a hard but permeable shell. Lining the shell is a membrane so constructed as to make it possible for the embryo within to breathe. As a kind of lung it conducts oxygen from the surrounding air into the egg and permits carbon dioxide to escape. These substances pass in or out through the permeable shell. The rest of the space within the shell is filled with fluid, which separates the three sacs and also acts as a hydraulic shock absorber to protect the embryo further.

The amniote egg was fertilized internally while still within the body of the female. The eggs of fishes are fertilized after they have been laid. To enable their kind to survive, fishes must lay many eggs. Like gymnosperms among the plants, they must be prodigal. Some fishes lay almost a million eggs each breeding season. So many eggs are destroyed, mostly by being eaten, that many are necessary in order to insure the survival of at least a few. Internal fertilization results in far more secure reproduction because the chances that any fertilized egg will survive are greatly increased. So, far fewer eggs need be produced.

So it is evident that the invention of the amniote egg was another link in the continued improvement of reproduction that began early in the history of the biosphere with the invention of sexual reproduction in plants. Internal fertilization and the amniote egg were important, but the improvements did not end with them. More improvements were to come, as we shall see shortly.

All this careful provision for the helpless embryo cradled inside its own sac both before and after the egg is laid looks extraordinarily like forethought and planning, but by now we can understand without difficulty that it was nothing of the kind. On the contrary, it was a response to environmental pressures (natural selection) on each new generation of amphibians throughout a very long time. What exactly were the pressures? They may well have been mainly the effect of continual increase, in the ponds and streams, in the kinds and numbers of fish and invertebrate predators, all with a large appetite for eggs and for the small tadpole-like larvae that were the amphibians' rather incapable young children. This would have been a situation full of danger, not for the adult amphibians themselves but for the survival of the race. Any device that would keep the eggs out of the water at all times and enable the young to pass through their vulnerable tadpole stage *inside* the egg instead of out in the dangerous world, would have been an enormous improvement.

The response can be stated in two different ways. We could say: "This danger made it seem attractive to amphibians to deposit their eggs in safer places, out of reach of the egg-eaters—namely, on stream banks and lake shores, above water level." More correctly, although less dramatically, we could say that those amphibians which varied genetically in the direction of harder eggshells, enclosing bits of apparatus having the rudiments of any of the features we have described, would have had a slight statistical edge in terms of survival over their relatives who did not. Apart from this, it was only a matter of time to enable enough successive generations to develop and perfect the improvement. And, as we always find when we look back at some particular event in Earth history, time was there in abundance. The 50 million years that elapsed between the first known appearance of amphibians and the first appearance of reptiles (who presumably had the egg) provided room for a very great amount of evolutionary experimentation.

The amniote egg was inevitable. Either it or some similar improvement that would have accomplished the same result would have been invented sooner or later. So probably the amniote egg was less a question of *what* than a question of *when.*

Although we can have modern amniote eggs for breakfast any morning we like, no one has yet found *fossil* amniote eggs in Paleozoic strata. The earliest known fossil ones have come from strata of Mesozoic age. But Pennsylvanian and Permian strata contain the fossil skeletons of reptiles in large numbers, and all reptiles possess the amniote egg. More than this, we have evidence of the gradual conversion of amphibians into reptiles, not from fossil eggs but from fossil skeletons. In these we see many different combinations of amphibian characteristics and reptilian characteristics. Taken together, they establish the transition firmly (Fig. 13 - 4). This fossil evidence of the transition implies that if we ask the question: "Which came first, the reptile or the egg?" we would have to answer it by saying: "Neither one; both must have evolved together, little by little through a long series of phases."

Just now we mentioned amphibian characteristics and reptilian characteristics. What are the principal differences between these two groups of animals? To begin with two differences visible in fossil skeletons, we can look at neck and limbs. The reptile has a real neck with a single ball-and-socket joint; so that the head can pivot on the backbone. An animal that can look this way and that can be more agile, both in pursuit of food and in escaping enemies. The first reptile, of course, had no enemies; but it wasn't long before enemies appeared from among the reptiles themselves. Reptiles began to eat each other; we will see this about to happen in Figure

Figure 13 - 4. *Seymouria,* a land vertebrate from lower Permian strata, with a nest of eggs at the base of a scale tree. Its skeleton, 2 feet long, is a 50-50 mixture of reptilian and amphibian structures. At first believed to be a reptile, it is now thought to belong more logically to the amphibians. Whatever its place in the classification, it is a perfect example of a halfway position in the evolution of one kind into another. (Courtesy of the American Museum of Natural History.)

13 - 7. As for the limbs, those short, weak amphibian limbs that had evolved from the fins of a fish grew both stronger and longer. In this way they became capable of holding reptile bodies up off the ground, enabling reptiles to take longer steps and to move faster. The advantages of faster movement are obvious.

These and other differences are visible in fossil skeletons. In addition to the skeletons, we see important differences between the soft parts of living reptiles and those of living amphibians. First of all the soft, permeable skin of the amphibian is replaced in the reptile by an impermeable hide, often covered with scales or plates made of tough, horny material. Evidently this developed in order to reduce loss of water by evaporation through the skin, but it could not have happened until the lungs had been greatly improved, because the amphibians breathed partly through their thin skins.

Next, the reptile has a better system of muscles and a better system of blood circulation, with a more efficient heart, the pump that keeps the blood moving. Finally, in part because of a better supply of blood, the reptile's brain, although small in size, possesses the beginnings of cerebral hemispheres, those high frontal parts of the brain, important for intelligence, that are large in the brains of mammals but not present in amphibians. On the other hand, despite these great advantages, a reptile lacks any good means of regulating the temperature of its body. It depends on the Sun for body heat; when the environment becomes cold or extremely hot the reptile becomes sluggish and inefficient. A reptile can be killed by extremes of temperature that higher animals can survive.

Results of the amniote egg. The fine new egg had two principal results, both very important for those who possessed it. The egg increased survival rates and increased the kinds of territory that could be inhabited. Survival improved because of where the egg was laid and because of the extent of development of the embryo at the time it hatched. The egg could be deposited out of reach of the egg eaters and the tadpole eaters that were numerous in the water. Also, thanks to all the equipment within the egg, the embryo could be fully developed and ready to fend for itself almost from the moment it emerged from its egg. A Pennsylvanian life insurance company would surely have charged lower premiums to its amniote customers. Further, the potential living space for land vertebrates became far more extensive for an obvious reason. Freed from the need to return to water in each breeding season, reptiles did not have to live near water. They were at liberty to invade drier regions, on which vertebrates had never before set foot, and where competition was yet nonexistent. The outward-spreading reptiles left the amphibians tied to ponds and streams, where most of them remain to this day. The departure of the reptiles from their amphibian ancestors and the old water-related environment recalls the way amphibians had earlier left their fish ancestors to the environment they had once shared, the sea. The conquest of the lands by vertebrate animals had progressed one more step.

Climate and the evolution of reptiles. Once the amniote egg was established as the symbol of a fuller life, another factor seems to have stimulated the further development both of the egg itself and of the reptiles that produced it. Comparing Permian strata with Pennsylvanian strata, we find great differences, indicating that the new Permian environments were unlike those of the Pennsylvanian. Climates became cooler and the great coal swamps gradually disappeared. At least some continents stood higher, and a good many high mountain ranges had been built. The causes are still wrapped in uncertainty but they probably had to do with the relations among moving crust plates and collisions between continents.

Whatever the cause, the result was that wet terrain, the natural environment of amphibians, became scarce. The spreading of amphibians, which had continued through the Mississippian and Pennsylvanian Periods and had been encouraged by the presence of widespread swampy lowlands near sea level, came to an end. There was no longer room for the number and variety of amphibians then in existence. For animals with great dependence on water, life had gradually become difficult. But the changing climate presented no such difficulties to animals with the amniote egg. The newly evolved reptiles could dispense with the numerous swamps and streams. However, lacking a means of conserving body heat, they had to avoid the higher and colder regions.

Sprawling reptiles. Once the amniote egg had brought them into existence, reptiles took full advantage of the freedom it gave them. They spread rapidly and adapted the forms of their bodies in accordance with the demands of a variety of environments. To make the story simple, we can bypass details and divide the reptiles into two groups, a conservative group of sprawling types and a radical group called dinosaurs. The latter were so numerous and spectacular that we must describe them in a chapter all to themselves, but the sprawlers can be mentioned here. When we say "sprawlers" we mean that these reptiles continued the tradition they had inherited from amphibians, with a long body kept close to the ground on short legs. Because of this body form the sprawlers were rather crocodile-like (Fig. 13 - 5). The way of life of many of them was crocodile-like too; they seem to have spent much time in streams and lakes, pursuing fish. The late-Paleozoic sprawling reptiles have been traced through fossils in younger and younger strata, evolving slowly along the line that leads separately to modern crocodiles and lizards.

Figure 13 - 5. *Rutiodon,* one of the many kinds of sprawling reptiles. The fact that it is found as fossils in lower Triassic strata in both Europe and North America suggests that perhaps in early Triassic time the two continents were still connected.

In the foreground are cycads; just behind the reptiles are horsetails. In the right background are ferns followed by conifers. (Courtesy of the American Museum of Natural History.)

One of the sprawlers found in large numbers in Permian strata is remarkable for a special adaptation, a showy-looking sail-like structure on its back (Fig. 13 - 6). In fossil skeletons the "sail" consists of long, thin, bony spines that project upward from the backbone. The bones were almost certainly connected by a membrane like that on the webbed foot of a duck. Measurements of the many fossil skeletons show that the area of this reinforced web was proportional to the volume of the reptile's body. Apparently the web carried many blood vessels and so, when exposed to sunlight, could quickly warm its cold-blooded possessor, much as the radiator of an automobile cools the water that circulates through its many tubes. This early though clumsy apparatus was invented independently by two different genera of sprawling reptiles. *Dimetrodon,* the genus shown in Figure 13 - 6, was a carnivore, as we can infer from its teeth. But a quite different sprawler, a plant eater, invented a similar web. The plant eater's web was different in that its long spines had short bony crossbars on them. But although it was a different model, its function evidently was the same.

Mammal-like reptiles. The tail end of the Paleozoic Era was a time of rapidly changing topography and climate. It is therefore not surprising that it was also a time of very active evolution among reptiles, then the most complex and successful land animals. Of the new kinds of reptiles that evolved then, one group played by far the most significant part because it later founded the line we call the mammals. The earliest known evidence of this group consists of Pennsylvanian fossils and others collected from Permian strata in South Africa and in Argentina. Probably those two regions were joined in a single land in Permian time, as the map in Figure 6 - 10 suggests; so it is not surprising that the same or closely related land animals are found in both, even though the lands are now separated by several thousand miles of ocean.

These fossils are the skeletons of reptiles, but reptiles that are already beginning to show unmistakable mammal-like characteristics. We call them mammal-like reptiles, and we find them again in Triassic strata, several genera of them, and even more like mammals. The critical features, the ones that can be seen in skeletons, are three: long legs, differentiated teeth, and an improved brain. All these features are suggested in Figure 13 - 7. The legs had become longer, with elbows moving rearward and knees forward. This

Figure 13 - 6. *Dimetrodon,* a Permian sprawling reptile more than 10 feet long, with an amazing blood-warming device on its back. In the foreground are horsetails; in the background, ferns and scale trees. (Courtesy of the American Museum of Natural History.)

Figure 13 - 7. Two of the many kinds of mammal-like reptiles found in Triassic strata in South Africa. Three carnivores (*Cynognathus,* left) preparing to attack a plant eater (*Kannemeyeria,* about 6 feet long). The vegetation consists of cycads (1) and a conifer (2). (Charles R. Knight, © Field Museum of Natural History.)

changed the reptile's posture from sprawling towards verticality, as it is in mammals (Fig. 13 - 8), in which the bones of the limbs firmly support the body, relieving much of the strain on limb muscles. The teeth, instead of being cone shaped and alike as in reptiles, were beginning to be differentiated into nipping teeth, canine teeth, and molars. The brain was larger and better developed in its upper-frontal part, the seat of intelligence.

Mammal characteristics that are not visible in skeletons include (1) mammary glands for suckling young; (2) hair, as insulation to reduce loss of body heat; (3) warm blood; and (4) a four-chambered heart, the most efficient pump yet devised for nourishing the brain despite changes in the external temperature. When the Sun sets, the reptile becomes cold and sluggish; the mammal remains comparatively warm. Probably the Triassic mammal-like reptiles possessed most of these characteristics at least in rudimentary form. As for

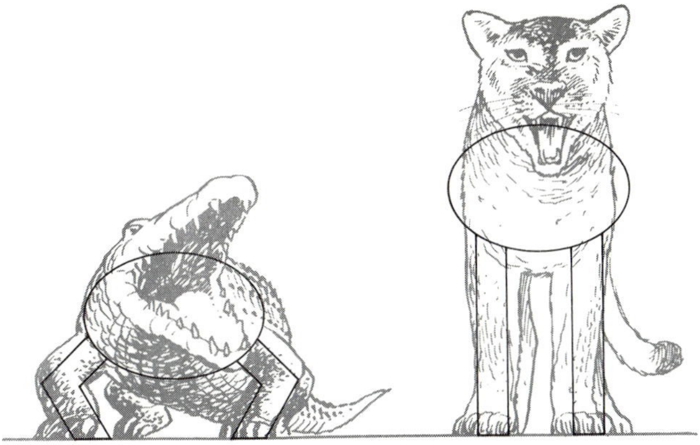

Figure 13 - 8. Postures of reptile and mammal.

reproduction, it has been argued that they probably laid eggs, for the reason that the most primitive living mammals do so, but there is no certainty. What we can be sure of is that the longer, less sprawling limbs and the improved brain go together. These two features must be related. The limbs mean greater speed, which requires greater muscular coordination, which demands a more complex control box. Any aircraft designer can tell us that the more operational capability you build into a vehicle, the more control mechanisms you have to install in it.

The mammal-like reptiles were many and varied. Some were more than 10 feet long; others were no bigger than mice. All became extinct. But before it disappeared, one of them—very likely the carnivorous *Cynognathus* ("dog-jawed one") seen in Figure 13 - 7—created the line of true mammals and so placed another milestone along the evolutionary road.

In summary, we can visualize four paths (Fig. 13 - 9) along which reptiles made their evolutionary way onward. The first is the path of the conservative sprawlers, which themselves never did anything spectacular and have remained sprawlers ever since. The second is the path of the dinosaurs (Chapter 14), which evolved into many dramatic kinds, yet met their end before the Cenozoic Era began. The third is the aerial path of birds (Chapter 14), which evolved out of a reptile at some time before the Jurassic Period. Fourth and last is the path that began with mammal-like reptiles, led quickly to mammals, and in the Cenozoic Era branched into a multitude of threads, one of which led to man.

Figure 13-9. Evolutionary paths traced or pioneered by reptiles. All were made possible originally by the amniote egg.

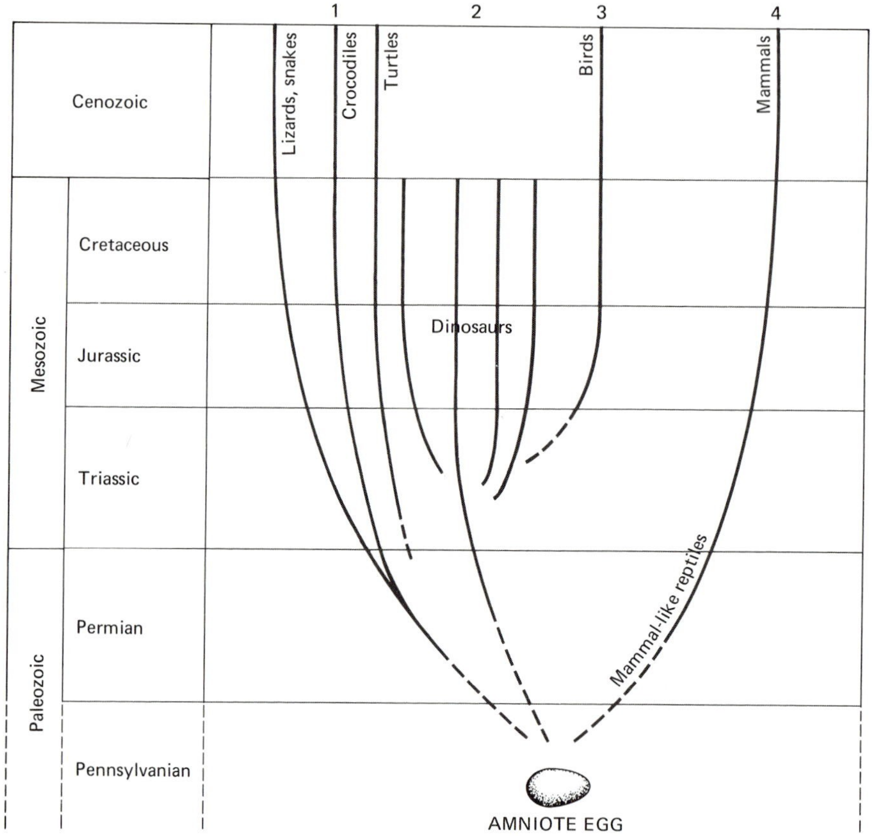

Many dramatic things happened along these paths. Among the most dramatic was the story of the dinosaurs, which will be told in the next chapter.

References

Gregory, W. K., 1951, Evolution emerging: The Macmillan Co., New York, vol. 1, p. 246–287.

Romer, A. S., 1964, The vertebrate story: 4th ed., University of Chicago Press, p. 87–175.

Smith, H. W., 1961, From fish to philosopher: American Museum of Natural History, New York, p. 85–101, 120–121. (Paperback.)

14

The Age of Dinosaurs

Reptiles on the lands

The end of the Paleozoic Era was followed by one of the most remarkable chapters in the history of life, the reign of Mesozoic reptiles. Within the Mesozoic span of about 190 million years a dramatic spreading of reptiles occurred. Thanks to the advantages of the amniote egg, reptiles, placed on the stage by evolution in the Pennsylvanian, spread across the lands, invaded the seas, and took to the air with newly invented wings. One line of reptiles evolved into birds that competed with the flying reptiles themselves. Another line, as we have already seen, evolved into mammals. Yet the most eye-catching characters in this reptilian drama, the dinosaurs, together with their swimming and flying relatives, did not survive beyond Mesozoic time. They died out to the last individual, leaving the stage to be populated by new kinds of animals, mostly descendants of Mesozoic mammals.

Hundreds of genera of dinosaurs, from chicken-size kinds up to lengths that approached 100 feet, and scores of different adaptations to many ways of life are represented in the fossil record. Yet, being reptiles, all were probably cold blooded and all depended directly or indirectly on the availability of great quantities of leafy vegetation. Yet fossil dinosaurs have been found in all continents save Antarctica. These facts lead us to believe that because such beasts could hardly have flourished in high, steep mountains or in cold climates, lowlands with mild climate and luxuriant vegetation must have been widespread in Mesozoic time. This belief seems to agree with our information about the movement of crust plates. Looking again at Figure 6 - 10, we remember that probably much territory now in middle or high latitudes lay in lower latitudes in mid-Mesozoic time. Very likely southern North America and southern Europe then touched the equator. If continents had been in their present positions in Mesozoic time, it seems improbable that reptiles would have been so numerous or that any of them would have been so huge.

A look at another map, Figure 10 - 7, helps in a different way to explain the profusion of Mesozoic reptiles. Toward the end of the Mesozoic the area of North America had far more shallow sea and much less land (most of it low-lying) than it has now, and the Gulf of Mexico connected with the Arctic Ocean. Such a pattern would have given the central and northern parts of the continent milder climates, especially in the winter season, than are possible today. Europe and other continents were also overspread extensively by Mesozoic seas.

So the great Mesozoic flourishing of reptiles, which might at first thought seem mysterious, appears after all to be explained satisfactorily in terms of surroundings that were very agreeable for cold-blooded beasts. In this explanation we see again, as in the earlier history of living things, the basic response to the dictates of environment that was made through the process of natural selection.

Kinds of dinosaurs

We have already noted that there were hundreds of different kinds of dinosaurs. But all the kinds belonged to two distinct stocks that diverged from a common ancestor at a time in the Triassic Period before dinosaurs as such had come into existence. The name *dinosaur* is popular rather than scientific. It means "terrible lizard," and when it was coined it referred to a very large, ferocious kind. But such kinds were rather few among the many sorts of reptiles we include in the dinosaur group today. In that group are a large number of reptiles that were neither large nor terrible.

The two basic stocks of dinosaurs are distinguished by scientists on the basis of their hip bones. One stock had hip bones shaped like those of lizards, and the other had hip bones like those of birds. This difference is important technically and is evident in the arrangement in Figure 14 - 1. But for our purpose we need say no more about it; we are concerned mainly with external appearance and ways of life. With this concern in mind we can describe some of the headliners in the dinosaur world without paying attention to the differences among Triassic, Jurassic, and Cretaceous kinds. All we

Figure 14 - 1. Relationships among dinosaurs mentioned in the text.

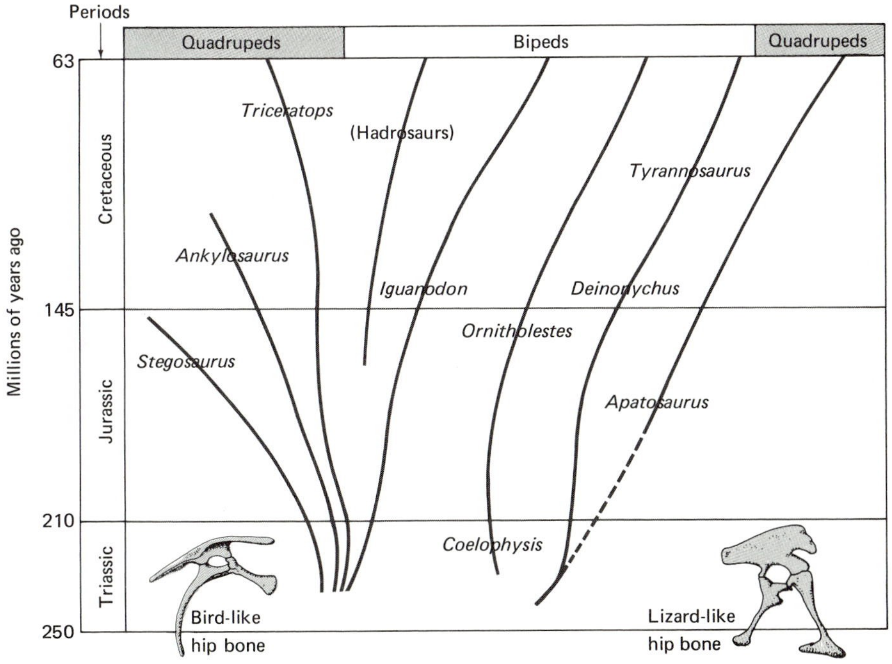

Figure 14 - 2. *Coelophysis,* a typical Triassic dinosaur. Very likely it made the small-size footprints seen in Figure 9 - 6.

need note is that (not surprisingly) Triassic dinosaurs were rather primitive and of only moderate size. All of them had got up onto their hind legs, with their forelegs smaller in size and well off the ground (Fig. 14 - 2). Necks had lengthened far beyond those of the Permian sprawlers. Yet although dinosaurs had become bipedal, they were not truly upright like bipedal man. When they walked or ran, their bodies were more nearly horizontal than vertical, although no doubt they could stretch to an upright posture at times, as a squirrel can. As for their feet, if we look at the many footprints they left in wet sand and mud (Fig. 9 - 6), showing three or four long toes and an additional short one that rarely touched the ground, we can understand why those who studied them first thought they were tracks made by birds.

Most of them were carnivores like their Permian ancestors, and the presence in Triassic strata of a few kinds with conspicuous armor, spines, and spikes suggests that they were already starting to protect themselves against predatory enemies in the form of other dinosaurs.

Of course the rather primitive Triassic dinosaurs as a group included the ancestors of all the later dinosaurs. It will be best to describe them in terms of diet, way of life, and special equipment. We can distinguish plant eaters from carnivores, bipeds from quadrupeds, and those having armor, plates, or conspicuous horns from those without such accessories. We will divide our examples into four groups.

Plant-eating bipeds. Although most early Mesozoic dinosaurs were carnivores, their successors included a great many plant eaters. Their tracks show that they rather frequently walked on four legs. Common among them was *Iguanodon* (Fig. 14 - 3), a rather heavily built animal with a maximum length of about 35 feet. In one locality more than twenty complete skeletons were found, together with those of turtles, crocodiles, and fishes, indicating a swampy environment. On their five-fingered "hands" the "thumb" consisted of a big, sharp spike, which was very likely an effective weapon of defense. Apparently they fed by pulling down branches and biting off the twigs. Their tracks show that they walked about (probably not very fast), with occasional short hops.

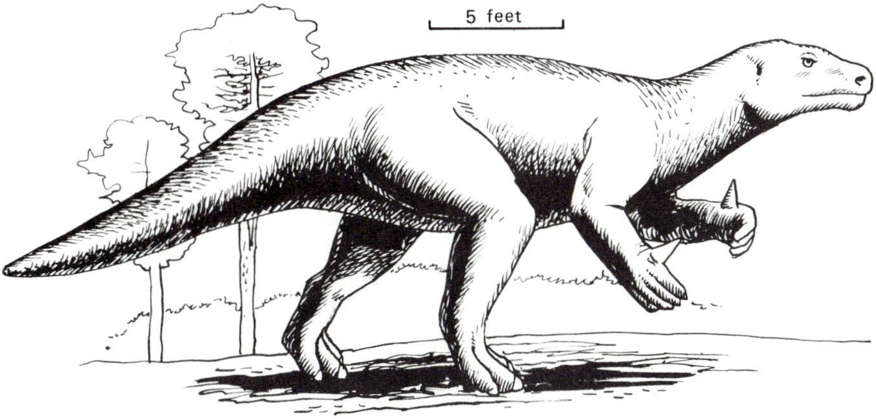

Figure 14 - 3. *Iguanodon,* a large plant-eating biped that was common in Europe.

Other plant-eating bipeds, a group 20 to 40 feet long called Hadrosaurs, were specialized to an amphibious life in and around swampy pools (Fig. 14 - 4). Their feet were slightly webbed, and

Figure 14 - 4. Hadrosaurs (1), an armored dinosaur closely similar to *Ankylosaurus* (2), and the carnivorous dinosaur *Struthiomimus* (3). The tree at left is an angiosperm. (Charles R. Knight, © Field Museum of Natural History.)

their tails thin like those of crocodiles, adapted for sculling their bodies through water. Nostrils were placed so that the body could be almost completely submerged. The mouth consisted of a horny bill, rather like a duck's, behind which were as many as 1000 teeth, long, very thin, and packed close together like a bundle of wooden matches. As the horny bill pulled soft plants from swamp mud, the tooth-shod upper and lower jaws slid forward and backward, like two wire brushes being rubbed past each other, tearing up the food.

Flesh-eating bipeds. Where there are plant eaters, there are always carnivores to prey on them. The dinosaur clan included a lot of running, bipedal beasts of prey that came in many sizes and styles. One, *Ornitholestes* (similar to Fig. 14 - 4, no. 3), was only 6 feet long and so lightly built that it may have weighed less than 50 pounds. It was agile and built for fast running; its grasping hands, equipped with three very long fingers, could snatch a very small reptile off the ground as it tried to escape. Another, *Struthiomimus* (Fig. 14 - 4, no. 3), was a little larger, in form much like an ostrich. It even had a toothless beak. A crushed skull of a related dinosaur was found in a fossil dinosaur nest filled with eggs. This, together with its light build and flexible hands, led to the thought that *Struthiomimus* was an egg eater and a professional robber of nests. Possibly the owner of the skull had been caught in the act of stealing.

Another biped, *Deinonychus* (pronounced *dye-nón-ikus*), about 8 feet long and possibly a descendant of *Ornitholestes,* had two remarkable adaptations to a carnivorous way of life. The second toe on each hind limb was armed with a claw much larger and sharper than the other claws. The toe with this claw had a peculiar joint on which the toe could be lifted off the ground and rotated through 180 degrees (Fig. 14 - 5). This enabled the reptile to attack its prey with a strong backward kick of its leg, a kick that should have been able to disembowel an animal as large as the attacker. Furthermore the long tail of *Deinonychus* was equipped with tendon-like rods that could instantly lock the bones together, converting the tail into a rigid counterbalance for the rest of the body. With this claw and tail, such a dinosaur must have been an extremely agile and dangerous animal.

Several carnivorous bipeds were much bigger than this, reaching lengths of more than 30 feet. One of them, *Tyrannosaurus,* between 45 and 50 feet long, 18 to 20 feet tall, and probably weighing 7 or 8

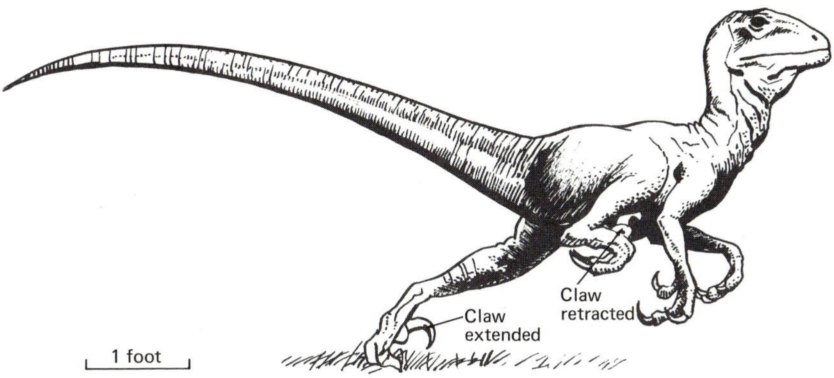

Figure 14 - 5. *Deinonychus,* the carnivore with the slashing claw. (Restored at Yale Peabody Museum under the direction of J. H. Ostrom.)

tons, was the biggest land carnivore on record (Fig. 14 - 6). Its mouth, in a skull 4 feet long, was armed with a great number of curved, sharp, 6-inch teeth. As its arms were extremely small, it must have done its killing and feeding without using its arms or hands. Probably its principal prey consisted of plant eaters such as Hadrosaurs and the horned dinosaurs we shall describe presently.

Figure 14 - 6. *Tyrannosaurus,* the biggest carnivore, attacking a helmeted *Triceratops,* which seems prepared to defend itself. The trees are palms, representative of angiosperms. (Charles R. Knight, © Field Museum of Natural History.)

Amphibious quadrupeds. Next we come to the giant dinosaurs, which have appeared so often in popular literature and even in advertisements that their form is well known far outside the realm of science. In the fossil record there are at least four different genera, superficially all rather alike; we need mention only two of them.

Figure 14-7. *Apatosaurus,* an amphibious quadruped 75 feet long, at the edge of a Jurassic water body. Two others are feeding in the water. Their huge size dwarfs the crocodile types in the foreground. Vegetation consists of cycads and horsetails. (Charles R. Knight, © Field Museum of Natural History.)

Although these two were quadrupeds, it seems odd at first sight that their forelegs were considerably shorter than their hind legs. But actually it is expectable, because they were descendants of Triassic bipeds with short forelimbs. Probably the best known genus is *Apatosaurus* (Fig. 14-7), a huge, slow-moving plant eater as much as 75 feet long, with a short center section supported on massive columnar legs with clawed feet. At the front of this section was a long, flexible neck with a small head, nicely balanced at the rear by a long, tapering, flexible tail. The whole assembly must have weighed more than 30 tons, four or five times as much as the largest African elephant.

As the ancestors of *Apatosaurus* increased in size and weight, evolution kept the weight of its skeletal structure down by developing vertebrae with holes and hollows in them, thereby reducing weight in places where great strength was not needed, but retaining it in the pillar-like legs where it was essential. The great feet left imprints nearly 36 inches long in the Mesozoic mud.

Diplodocus, another giant, was a plant eater similar in many ways to *Apatosaurus.* It differed chiefly in that it was longer (one is calculated to have measured more than 95 feet with a height of nearly 45 feet) but much less heavily built; so he is thought to have weighed in at 10 to 12 tons. No doubt both giants spent a great deal of time in swamps and rivers, cropping soft vegetation from above and below the water surface. When they were well offshore among swampy islands, they were least likely to be reached by big carnivores; so such places were not only dining rooms but safe havens as well. The giants protected themselves further by moving their nostrils to the tops of their heads, which enabled them to breathe comfortably while remaining out of sight, almost completely submerged.

These and some other dinosaurs swallowed their green food whole and ground it up after it had reached their stomachs. Like chickens, in whose crops we always find many tiny stones, the dinosaurs swallowed stones the size of potatoes, and with these as tools the strong gizzard muscles soon reduced the food to a mush. At times heaps of these stones, worn round and smooth in ancient gizzards, have been found along with the bones nearest the abdominal parts of a big dinosaur.

Probably these great beasts laid eggs, although none have been identified positively. Eggs could not have survived in water; so they must have been laid on land, perhaps on islands and in other places that predators could not easily reach.

The huge amphibious quadrupeds possessed even smaller brains relative to the weight of their bodies than the rest of the dinosaurs, a group not famous for mental capacity. In *Diplodocus* the true brain weighed only about $\frac{1}{4}$ ounce per ton of body weight. We say "true brain," because *Diplodocus,* in common with many other dinosaurs, had an auxiliary and much larger message center located in the spine close to the pelvis. Connected with the true brain by the spinal cord, that center directed the activities of hind legs and tail. Although this arrangement may seem clumsy, we have to admit that it worked, because many different kinds of dinosaurs possessed it through tens of millions of years. Surely, however, the mild, little-changing environments in which dinosaurs lived helped out by offering few problems and generating few mental strains.

Quadrupeds with armor or horns. Our roster of distinctive kinds of dinosaurs includes a miscellaneous lot which, although not very closely related to each other, were characterized by odd sorts of armor or horns or both. Despite the fact that their Triassic ancestors were bipeds, this lot had put their forelimbs down onto the ground again and become quadrupeds. Yet their forelegs, like those of *Apatosaurus,* were still shorter than their hind legs. Being plant eaters, they stood in need of protection against preying reptiles; hence the development of armor plate and defensive horns.

Outstanding among these protectionists was *Stegosaurus* ("plated reptile"). The skeleton shown in Figure 9-10 C, some 20 feet long and weighing perhaps 4 tons, shows the thick, triangular plates of bone that fringed the spine, to which they must have been attached by ligaments. Probably these plates, the largest measuring 30 inches, protected the spinal cord against carnivorous bipeds, who may have aimed at the back of the neck for a kill, as a terrier kills a rat. In addition, *Stegosaurus,* as the illustration shows, was armed with pairs of stout, sharp spikes two feet long near the tip of its tail. A single swipe of such a tail should have been capable of knocking over a fairly large enemy, and of inflicting a lasting injury besides.

Ankylosaurus and its relatives (Fig. 14-4) protected themselves in much the same way as armadillos do today. Although 15 to 18 feet long and 5 to 8 feet wide, they were less than 5 feet high. Behind their strong, thick, beaked skulls heavy plates of bone sheathed the upper halves of their bodies. Some also had great spikes along their sides from shoulder to tail, which was like a heavy paddle or club. With all this defensive armor these reptiles must have been slow moving. But at the approach of danger they could sink down with legs folded beneath them and prepare to wield their tails effectively if they were molested.

Still another sort of defense based on horns was developed by *Triceratops* ("three-horned-face") and its many relatives (Fig. 14-6). These stocky, short-tailed quadrupeds grew to lengths as great as 25 feet and heights of 8 or 10 feet. Their most striking feature was a huge, heavy skull that spread backward as a great shield and protected the neck. Forward, the skull was equipped with three horns that projected forward, over a narrow, almost parrot-like beak. Inside the skull was a brain that, although small by human standards, was large for a dinosaur. Such a brain suggests that these helmeted, horned creatures were agile, and this is borne out by their hind parts, which had neither armor nor weapons. Clearly these animals could swing around rapidly to face and perhaps gore an enemy. Frequent scars on the fossil neck protectors suggest the effects of ancient combats.

As we think of combats between dinosaurs we cannot help wondering whether those struggles were silent or, as with dogs and cats today, loudly vocal. A specialist in the anatomy of dinosaurs tells us the little that is known on this subject. It seems that the configuration of the small bones at the base of the tongue is similar to that of the same bones in some living animals. The analogy suggests that at least some kinds of dinosaurs could have been capable of croaking or barking as living crocodiles do. For this reason, although Paleozoic lands may have been silent save for the sounds made by winds, streams, and surf, Mesozoic scenes could have been punctuated by primitive sounds from the animal world.

Protoceratops, a primitive relative of *Triceratops,* was a smaller dinosaur that had a beak but lacked horns and that lived in Asia. It is famous for its eggs and nests, found by fossil-collecting expeditions in Mongolia in the 1920s. Late in the Mesozoic Era that region was dry as it is today, and the eggs had been laid in neat little hollows in sand that is now converted to sandstone. The female dinosaurs scooped out the hollows and filled them with as many as 15 eggs 6 to 8 inches long. Several such nests have been found, and in at least two of the eggs were the tiny bones of the embryos that had failed to hatch. Eggs of other kinds of dinosaurs, both larger and smaller, have also been discovered.

Reptiles in the sea

One of the most striking things about Mesozoic life is that nearly half of the known kinds of reptiles lived not on land but in water, in streams, estuaries, and the sea itself. We have already noted that the Mesozoic Era was generally characterized by the wide spreading of shallow seas over parts of the continents; so environments favorable for water-living animals abounded.

Mesozoic strata contain a large number of fossil reptiles adapted to life in the water. This fact can mean only that some reptiles had gone back to the sea, the home in which their fish ancestors had originated long before. Because this suggests a retracing of steps, it needs a bit of explanation. Simply because Devonian fishes had emerged from water onto the land and had evolved through amphibians into reptiles, we are not justified in concluding that the later movement of reptiles into the sea was a step backward in an evolutionary sense. On the contrary, it was an example of the principle that any vigorously expanding race tends to move into every possible environment that is habitable. Indeed the movement was not greatly different from the return of amphibians to streams and ponds in Pennsylvanian time (Fig. 13 - 1). Food was to be had in the water, and competition was not prohibitive; so first amphibians and then reptiles moved in. Before the end of the Paleozoic, some reptiles had already become water dwellers and had begun to adapt to the new life. The adaptations were mainly changes in the direction of more efficient swimming. Of course the reptiles continued to breathe air, just as a modern whale, a mammal, breathes air despite its fish-like form. Furthermore, Mesozoic marine reptiles did not evolve from a single land-reptile ancestor that decided early to take the plunge backward into the water. Their fossil skeletons show unmistakably that they originated from different ancestors at different times. Thereby the fossils emphasize the plastic way in which living things respond to the strong influence of an environment that offers them new, food-filled territory to conquer.

Many details have been learned from fossils in strata of marine claystone and chalky limestone, the fine grain of which has preserved not only bones but also impressions of skin and scales. Besides small primitive kinds, the common groups of marine reptiles, all of them carnivores, were three: ichthyosaurs, plesiosaurs, and mosasaurs. Characterizing them briefly, we note first that ichthyosaurs ("fish reptiles") had evolved to a fully streamlined, fish-like form (Fig. 14-8), and so were splendidly adapted to the fast pursuit of fish and squids. Smooth skinned and ranging up to 30 feet in length, they had developed a fin on their back and a fish-like tail, and had evolved their four limbs into seal-like flippers, which they used for steering. In the flippers, fingers and toes were closely pressed and more bones were added for greater strength. Their eyes were big and were effective for seeing through water. They made one improvement in the reproductive process that was essential. An air-breathing marine animal that lives in seawater cannot lay eggs because the embryos would drown. So the ichthyosaurs invented *vivipary*, the process by which the embryo develops within the body of the mother and when mature is born alive. This is not a guess; it is fact determined from marvelously preserved fossils of female ichthyosaurs pregnant with as many as 7 fully formed young ones.

The second group, plesiosaurs ("reptile-like"), so called because in contrast with the fish-like ichthyosaurs they retained their original reptilian form, were some 25 to 40 feet in overall length. Except for their tails they had the general appearance of gigantic swans. Of course they were descended from land reptiles different from the ancestors of the ichthyosaurs. Their feet had evolved into long paddles, and their heads, at the end of a lengthy neck, were equipped with sharp-pointed teeth that interlocked to hold a slippery fish securely. With such teeth chewing was impossible; the plesiosaur swallowed his fish whole and ground it up in the stomach by means of gizzard stones. An interesting sidelight on the diet of plesiosaurs is an analysis of the fossil contents of the stomach of one individual that apparently died before its gizzard stones had done a thorough job. It was found that the mess of bone and shell fragments resolved themselves into fishes, flying reptiles, and cephalopods (which had been eaten whole, shell and all).

The third group of marine reptiles are called mosasaurs because their fossils were first found in strata near the Moselle River in northeastern France. They might equally well have been called Johnny-come-lately, for they did not turn up until late in Cretaceous time, after ichthyosaurs had been splashing around for almost 150 million years. Mosasaurs were descended from lizard rather than dinosaur ancestors. They were some 30 feet long and scaly, with jaws jointed in such a way that they could open their mouths wide, as snakes do.

Mosasaurs and ichthyosaurs were not unique in responding to the demands of a water environment by streamlining their bodies. The same response was made by a variety of animals that lived before and since the Mesozoic as well as during that time, as can be seen in Figure 14 - 8.

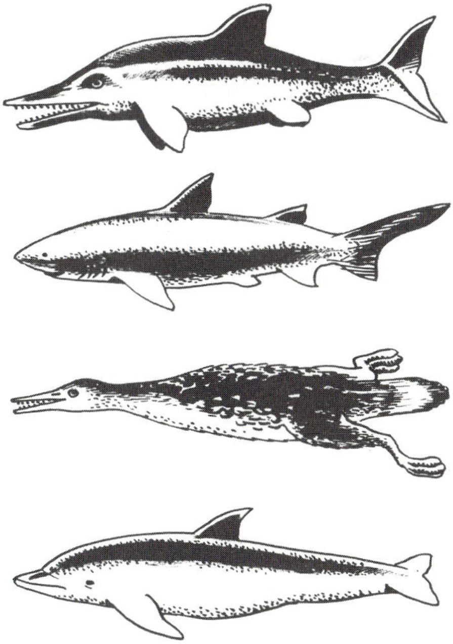

Figure 14 - 8. Adaptation of four groups of animals to aquatic life by streamlining the body: A. Reptile; B. Fish; C. Bird; D. Mammal. Starting from a variety of shapes, they end up looking alike.

Reptiles in the air

The story of the great thrust of reptiles during the Mesozoic Era is not finished yet. Reptiles not only spread across the lands and spilled over into the sea; they also took to the air by two different evolutionary routes. They learned to fly as reptiles, and in an entirely separate effort they also learned to fly as birds. As far as we can judge from the fossils, true aerial reptiles were not as numerous as were marine ones. Nevertheless they get credit for having been the first animals to invade the air since insects had performed that feat back in Devonian time. The credit is well deserved, for the air is a more difficult and more hazardous environment than is the sea. It requires more special equipment, energy, and skill (by which we mean agility and quick responses to stimuli) to navigate in the air or even merely to remain airborne than it does to get about on or in water. This is basically why man succeeded in inventing ships long before he invented airplanes. The time that elapsed between the two inventions amounted to several thousand years. But from the time they first appeared (Pennsylvanian) until they made it into the air (Jurassic), reptiles lived through some 80 million years.

We know a good deal about the structure and appearance of flying reptiles, thanks to an unusual kind of sedimentary rock that is quarried extensively in southern Germany. The rock is a layer of late Jurassic limestone so unusually fine grained that pictures for book illustrations were engraved on smoothed blocks of it in the days before steel and copper plates came into use. Thus the stone came to be called Lithographic Limestone. Its extraordinary fineness is thought to indicate that it was deposited in shallow lagoons, protected from the waves of the open sea by barriers of sand or by coral reefs. The soft sediment of the lagoon floors retained even the smallest details of the drifted plants and the animal bodies that sank and were covered by it. As a result the Lithographic Limestone has become famous for its fossil plants, invertebrate animals, fishes, and reptiles.

Many reptiles with wings have been collected from this stratum and similar fossils have been found elsewhere in other strata of Mesozoic age. Looking at one of the primitive Jurassic specimens preserved in all its fine detail (Fig. 14-9), we see at once that its body had been adapted for flight in at least three respects: (1) re-

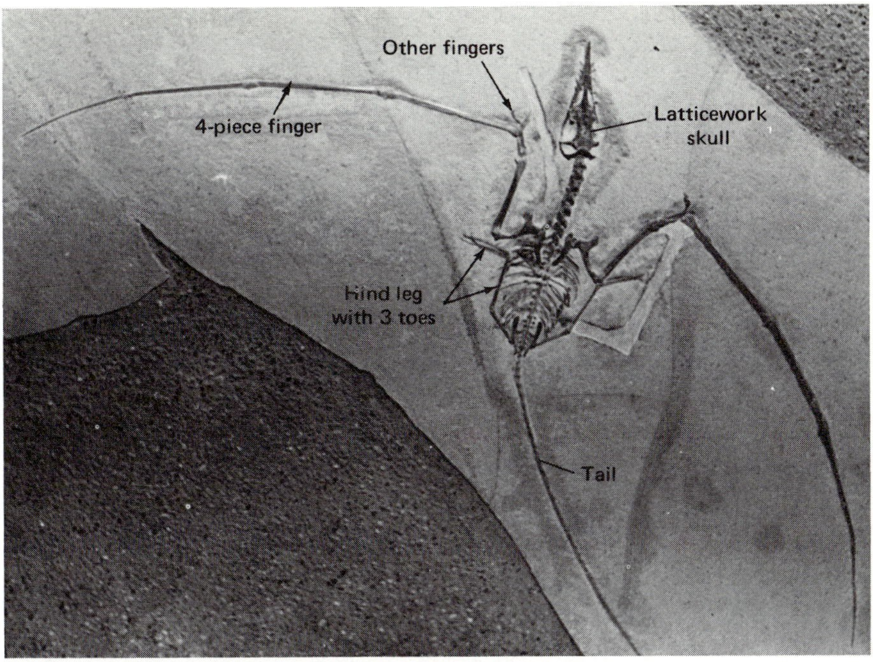

Figure 14 - 9. Skeleton of *Rhamphorhynchus,* a primitive flying reptile, just as it was found in the Lithographic Limestone in Germany. (Senckenberg Museum, Frankfurt a.M.; photo J. H. Ostrom.)

duction of weight, (2) development of features that could aid navigation, and (3) construction of a mechanism for flying. Here are some of these adaptations:

(1) Bodies were small; although some were nearly the size of a turkey, others were hardly larger than a canary. The skeleton was lightened through the development of thin, hollow wing bones, and in some kinds by reducing the skull to little more than a latticework of narrow bones.

(2) Eyes and that part of the brain that is connected with eyesight were unusually well developed.

(3) The most remarkable single feature was the wings. Looking at Figures 14 - 10 and 14 - 11 we can easily see that the fourth digit of the forefoot or hand—in other words, the little finger—was lengthened enormously whereas the other digits remained small. From the tip of the lengthened finger down to the hind foot and thence to the tail stretched a thin membrane of skin that made a wing.

The three adaptations taken together as an assembly formed a machine that was clumsy, but it *could* fly. Creation of the wing accompanied by improvement of the eyes and by reduction of overall weight made flight possible, but resulted in some remarkable proportions. For instance, one kind of flying reptile with a wing span of 3 feet weighed, when alive, only an estimated 16 ounces! Skins were bare and leathery, and jaws were equipped with numerous sharp reptilian teeth. Very probably these animals, like modern buzzards, were gliders rather than active fliers. Descended from ground-living carnivores, they apparently retained the carnivores' diet, skimming slowly over water in search of seafood or seeking large insects. The structure of their skeletons shows that these creatures could not walk. When they alighted, it was apparently not on the ground but on the branches of trees or on overhanging ledges of rock, from which they hung suspended, somewhat as bats hang today.

Later flying reptiles found in strata deposited in Cretaceous shallow seas had evolved by shedding their teeth in favor of a long beak, surely a better adaptation to the kind of life they led. One genus developed a peculiar flange or keel at the back of its skull (Fig. 14-10) that may have counterbalanced the long beak, making it easier for the skimming reptile to head into the wind. But the chief change consisted of enlarging the wing area, obviously for better support in the air. One flying reptile supported a body thought to have weighed less than 25 pounds, with a wing spread of some 25 feet. Because of this spread it can be called the largest winged creature in the history of life. Although they remained fragile to the end of their existence, flying reptiles survived through more than 100 million years.

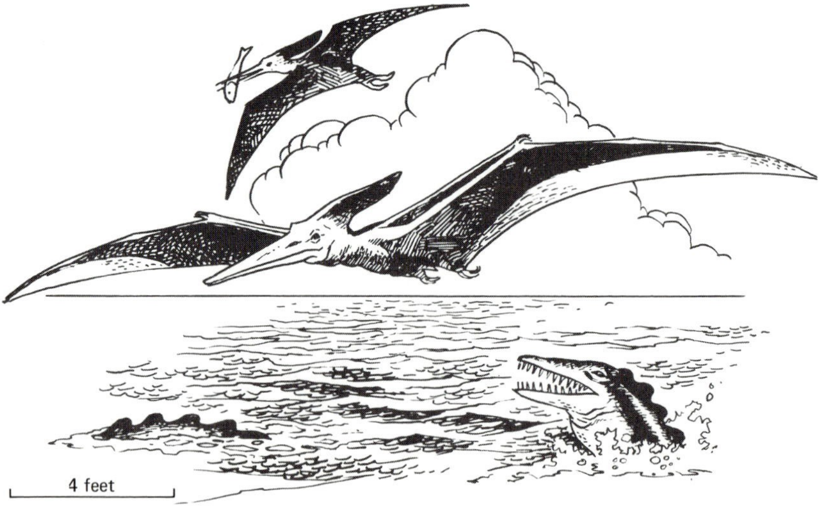

Figure 14-10. *Pteranodon,* a flying reptile with a keeled skull. It glided through great distances over the wide Cretaceous sea in Kansas and Nebraska.

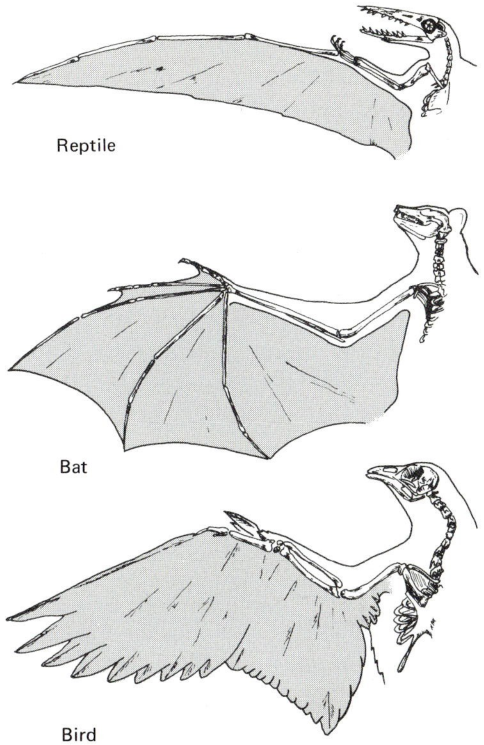

Figure 14 - 11. Schematic comparison of wing in flying reptile, bat, and bird. Each wing was invented at a different time. In the reptile the entire wing is supported on one finger only. In the bat the outer part of the wing is strengthened by four finger supports. In the bird more of the wing is supported by the upper and lower arm bones, while the wing surface consists of lightweight, stiff feathers. This is the most efficient of the three wings.

Even though the reptilian wing did work and did survive that long, it was a less efficient invention for flying than the wings invented independently, first by birds and later by mammals in the form of bats. In Figure 14 - 11 these three kinds of wing are compared, and one can see that the birds' wing is the most efficient.

Birds

Along the warm seashores in Jurassic time reptiles were experimenting with flight in different ways. We have already seen that more than one kind of land-living reptile got into the air in the form of the leather-winged flying reptiles we have just described. But still another kind went farther. In the middle of the 19th Century, in one of the quarries of Lithographic Limestone, was found the fossil skeleton of a reptile smaller than a crow, with clawed fingers on its forelimbs, reptilian teeth, and large eyes. Astonishingly, the fossil showed finely detailed prints of feathers attached to the forearm and to the vertebrae in the long tail. This was a bird, without the slightest doubt. It was given the genus name *Archaeopteryx* ("ancient wing") and the species name *lithographica* (for the rock stratum) (Fig. 14 - 12). Two other fossil skeletons and, separately, the imprint of a single feather have been found in the same stratum.

Of course these fossils are of immense interest and have been studied with great care. We can sum up the result by saying that *Archaeopteryx* is basically a flying reptile, but that because by definition birds are feathered and reptiles are not, it rates as a bird. Its features tell us surely that this earliest-known bird was descended from a reptile biped that lived on the ground. Its feathers strongly suggest it had warm blood, for efficient thermal insulation is one of the chief functions of feathers, as a down-filled comforter indicates. Most birds have even warmer blood than man has. Their normal temperatures, as high as 103°F, are made possible by their feather coats and the high level of their body activity.

Feathers are made of chitin, the same tough horny substance of which scales are made. Some scientists believe that the small reptilian ancestor of our primitive bird had scales, and that the scales first developed frilled edges, perhaps because such edges helped prevent the skin from absorbing too much solar heat. The frills proved useful in reverse, because they also reduced loss of heat from the body, and gradually evolved into feathers. The light weight and stiffness of feather substance were, of course, ideal aids to flying.

Despite its feathers, however, this first bird, like its cousins the leather-winged reptiles, must have been a poor flier. Its build suggests that it would have been a fairly capable glider. Probably it lived on land and as a carnivore ate either tiny living animals or carrion. The fact that it has been found as fossils only in marine limestone probably means only that some of them were blown or drifted to sea and were entombed in the soft bottom mud. The delicate bodies of those which perished on land were simply not preserved.

By Cretaceous time birds had ceased to be clumsy fliers, and many had substituted beaks for teeth. A number of them had specialized to life on the water. One of these was the rather loon-like swimmer and diver *Hesperornis* (Fig. 14 - 8), nearly 6 feet long, still retaining its teeth but with wings much reduced in size and strength. The existence of birds that had all but deserted the air in favor of swimming demonstrates that early in their history birds took up the pursuit of fish, just as reptiles had done repeatedly since Mesozoic time began.

Figure 14 - 12. *Archaeopteryx,* the first known bird. Having caught a tiny lizard, he is sitting on a branch of a conifer to eat it. In the right foreground are cycads; in the background, conifers and another similar bird. (From a watercolor by Rudolf Freund under direction of K. C. Parkes. Carnegie Museum, Pittsburgh.)

The end of the great reptiles

The close of the Cretaceous Period, and with it the end of the Mesozoic Era, marks a crisis in the history of the biosphere, because it was accompanied by the dying-out of many groups of animals.

Most conspicuous among the losers were reptiles. All the dinosaurs, all the flying reptiles, and all the marine reptiles except turtles became extinct, leaving lizards, snakes, and tortoises to carry on the reptile line. Among invertebrates most kinds of cephalopods, including all belemnites, and some lines of marine bivalves and snails died out.

The pattern of extinction is an odd one because mammals and land plants were affected little or not at all, and fishes and many invertebrates remained untouched. Because of this pattern, attempts to explain the extinctions in terms of a single cause have been unsuccessful. Before the events of Earth history were provided with radiometric dates, the close of the Mesozoic used to be called the time of the "great dying." Today, however, we realize the phrase is misleading. It was not as though some great plague had swept over the Earth, wiping out life, for at least two reasons. First, the extinctions were selective, striking down some and sparing others completely. They were not confined to any single environment but embraced land, sea, and air. Second, although extinctions were most evident near the close of the Cretaceous, overall they were spread through a considerable length of time. Groups of reptiles, particularly, had become extinct at various times throughout the entire Mesozoic Era. So whatever the cause, apparently it was not sudden, at least as we use the term "sudden" in human history. Even the most conspicuous extinctions, those that came at the close of Cretaceous time, must have been spread over millions of years.

Looking at the geologic evidence of widespread events near the end of the Cretaceous, we can see that continents in general rose higher. At the same time and perhaps mainly as a result of such rises, the broad, shallow Mesozoic seas drained away from continents, and the swampy lowlands that widely accompanied those seas disappeared. Also temperatures decreased, partly because of the higher lands and more restricted areas of seawater.

We have to acknowledge that the cause of the extinctions has still not been identified. Causes suggested formerly, such as disease, loss of food supply, and the vaguely stated "loss of racial vigor" seem wholly inadequate to explain the selective wiping out of *some* of the inhabitants of land, sea, and air but not *all* of the inhabitants of any one of them. The mammals seem to have escaped scot-free from all the destruction.

Recently it has been suggested that the end of the Mesozoic was marked by a conspicuous series of the reversals of the Earth's magnetic field described in Chapter 6, and that the reversals might have seriously affected the biosphere in some way, perhaps by changing the intensity of the radiation that reached the Earth's surface. Arguments have been leveled against this idea, but probably it is too soon to evaluate the pros and cons. Suffice it to say that the extinctions that accompanied the end of the Age of Dinosaurs are still one of the great mysteries connected with the history of life.

As this part of our story ends, we turn to a new world, different from the one that prevailed through much of Mesozoic time. Not only are the continents different in outline, but their inhabitants are different as well, and they continue to change rapidly. Let us begin the next chapter by setting the stage with a physical description of Cenozoic North America.

References

Augusta, Josef, and Burian, Zdenek, 1961, Prehistoric reptiles and birds: Paul Hamlyn, London.

Colbert, E. H., 1951, The dinosaur book: McGraw-Hill Book Co., Inc., New York.

Colbert, E. H., 1961, Dinosaurs. Their discovery and their world: E. P. Dutton & Co., Inc., New York.

Fenton, C. L., and Fenton, M. A., 1958, The fossil book: Doubleday & Co., New York, p. 329–374.

Kurtén, Björn, 1968, The age of the dinosaurs: Weidenfeld and Nicolson, London. (Paperback.)

Swinton, W. E., 1958, Fossil birds: British Museum (Natural History), London.

Swinton, W. E., 1970, The dinosaurs: Wiley–Interscience, New York.

15

Cenozoic North America

In broad view, Cenozoic North America is the North America of today, because it was during the Cenozoic that the continent we know took form. Through Cenozoic time the distinctive features—the mountain ranges, the hill lands, and the plains of today—came into existence one by one and in different ways, until gradually the landscape familiar to us was wholly in focus. For the first time the outline of the coasts appeared in recognizable form. This was made possible by withdrawal of the extensive Mesozoic seas such as those shown in Figure 10 - 7. It resulted from gradual uplift of the continent to its present average altitude of half a mile, by building of the Rocky Mountains and later the mountains farther west, by heightening of the Appalachian region, by the spreading-out of sediment in layers to form the broad plains of today, and by many other events.

All these things have happened within a time that in the geologic frame of reference was short, a mere 63 million years. While reading the earliest chapters of this book we should have called 63 million years a long time. But by now, with our experience of the long reach of Earth history as a whole, our point of view has broadened. We can accept the 63 million years of Cenozoic time for what it is: only about one and four-tenths percent of the time elapsed since the birth of the Earth and only one-third of the time during which dinosaurs inhabited the lands.

Throughout Cenozoic time North America was generally rising, now here and now there, in many different movements. The rise was unsteady; it amounted to much more in some areas than in others, and it was discontinuous in time. In central Colorado terrain that late in Cretaceous time lay beneath the floor of a shallow sea, today stands as high as 14,000 feet, more than 2½ miles above sea level. In contrast, since the Cretaceous ended, the Atlantic and Gulf Coasts of the continent have been bent gently downward again and again. Even now parts of those coasts seem to be sinking very slowly. But the Pacific Coast has been moving far more actively because it is the site of one of the Earth's major rifts, described in Chapter 6.

The Cenozoic rise of the North American and other continents and the erection of high mountain ranges, which tended to divide each continent into compartments, resulted in cooler climates generally. It resulted also in local extremes of climate, with much rainfall on the windward sides of some mountains and very dry conditions on the opposite sides. This cooling and diversification spurred on the great spread of angiosperms, the flowering plants that first appeared in Cretaceous time and that increased at the expense of less-evolved plants all through the Cenozoic. In the new, varied climates these angiosperms formed hardwood forests, savannas (grasslands with scattered trees), grassy plains, and desert brush, thus accentuating the diversity of the land environments that surround us today. The spread of angiosperms was accompanied by the diversification and dominance of mammals, whose natural populations are the ones most familiar to us. It will be a useful prelude, then, to the story of Cenozoic life if we set the stage with mountains, plains, and vegetation before we people it with a cast of animal characters.

As we interpret them from strata and from the forms of mountains, hills, and valleys, the major Cenozoic events in North America neatly illustrate the large-scale interplay of external and internal processes, especially the interplay of the forces of uplift with the forces of erosion. Repeated broad uplift of mountains or of plateau regions resulted in steepened slopes and quickened streams, and it activated erosion. Streams bit into their valleys, and mass-wasting swept loose rock material from all the highland slopes. The resulting great volumes of rock waste moved downward and outward in the rivers toward terrain that was lower and less steep. There the sediment, now too abundant to be carried down the gentler slopes, began to be deposited, first gravel, then sand, then in places even finer sediment. The deposits, stratified in layers, grew thick and spread over lower regions of vast extent, accumulating to thicknesses of hundreds, even thousands of feet in places. Although most such accumulations have since been elevated somewhat and deeply eroded, great bodies of them still remain and form distinctive natural regions of the continent.

As might be expected from the principle illustrated in Figure 2-8, such regions tend to occur in pairs: an uplifted, eroded area accompanied by an adjacent lower area that receives the sedimentary waste shed from the slopes of its higher neighbor. One such pair consists of the Appalachian Region and the Coastal Plain that flanks it on the east and south. Another pair consists of the Rocky Mountains and the Great Plains in the broad, western-interior part of the continent. Although they differ from each other in superficial ways, these two pairs are fundamentally similar, as we shall find later in the present chapter.

The "Old Continent" (1) *

A thorough description of the physical features of Cenozoic North America would demand a thick book fortified with many diagrams. (Such books are cited among the references at the end of this chapter.) But even within the scope of one chapter we can explain the highlights, the major physical elements of the structure and surface of the land, the elements that determine the chief environments in which we live.

It is logical to begin with the "Old Continent," the continent depicted in Figure 5 - 6, the continent that was built before the beginning of Phanerozoic time 585 million years ago. That continent is the basement of North America, the platform on which all the Phanerozoic sedimentary strata were deposited. Although much of it is buried beneath these younger strata, a vast area is exposed at the surface today in a wide region from which, at one time or another, erosion has stripped the sedimentary layers that once covered it. This great region of exposed basement consists almost entirely of metamorphic and igneous rock. It lies mainly in eastern Canada and extends a little into Minnesota, northern Michigan, and the Adirondack region of New York (Fig. 15 - 1). Except for parts of the far north and northeast that are mountainous, the region is characterized by irregular, low, rolling hills with a thin blanket of regolith that is discontinuous, exposing the igneous and metamorphic rock beneath. Glancing at the vegetation map in Figure 10 - 2, we note that a broad southern belt of this old region is covered with a subarctic forest of close-growing conifers and birch trees. A northern belt, too cold for such vegetation, is the broad, treeless tundra Canadians call the Barren Lands. The rocks there are mostly too rich in quartz to make fertile soil, but in any case soil has hardly yet begun to develop because of the very short time that has elapsed since northern North America was freed from the glacial cover it received in the last ice age (Chapter 17). Because of cold climate and poor soil the human population is sparse, and there is little agriculture. Because human enemies are few, and despite systematic trapping of fur bearers, the animal populations have survived better here than in most regions with more people.

* Numbers are keyed to the map, Figure 15–1.

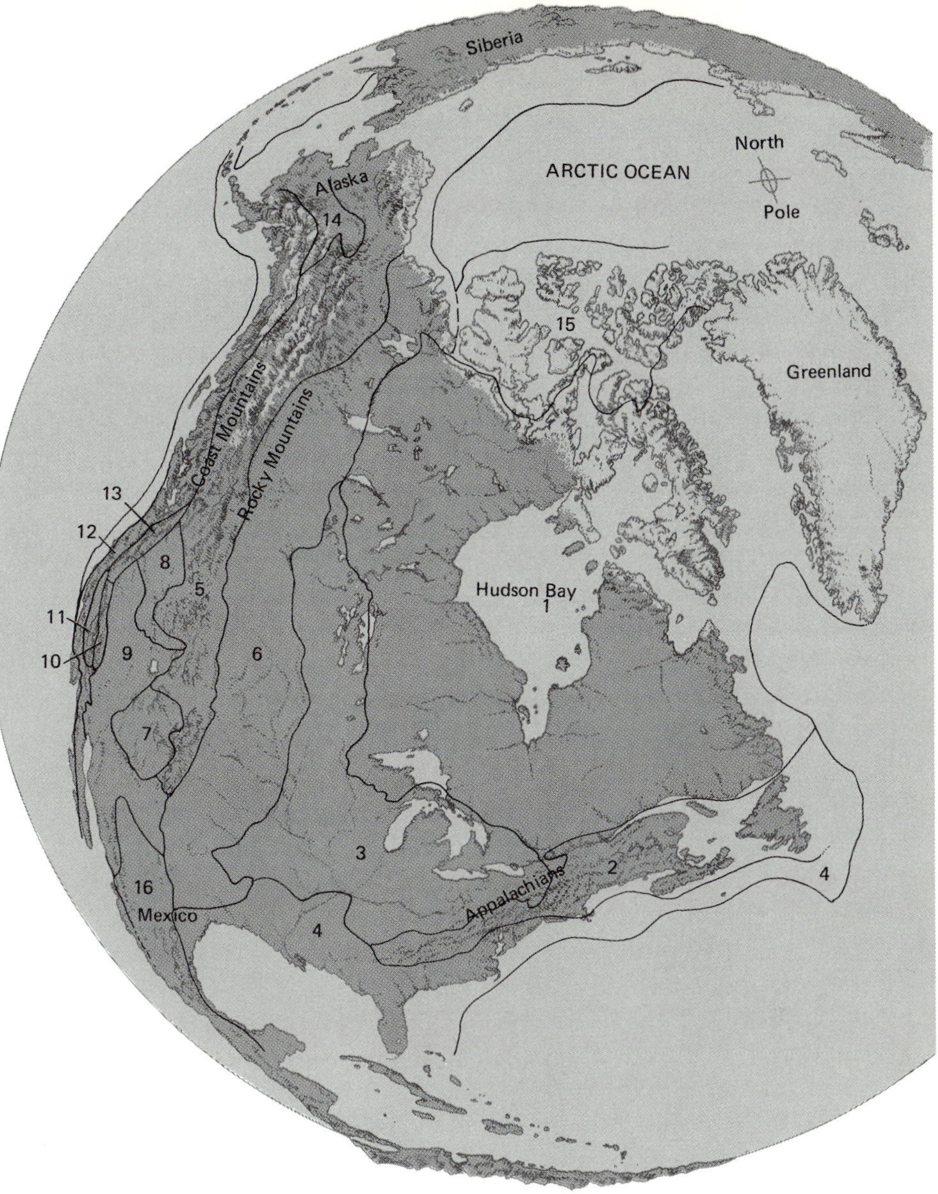

Explanation

1. "Old Continent" (Canadian Shield)
2. Appalachian Region
3. Central Lowland
4. Coastal Plain including Continental Shelf
5. Rocky Mountains (and mountains west of the Rockies in Canada)
6. Great Plains (and various hill regions in Canada)
7. Colorado Plateau
8. Columbia Plateau
9. Basin-and-Range Region
10. Sierra Nevada
11. Central Basin in California
12. Coast Ranges
13. Cascade Mountains
14. Yukon River Basin
15. Arctic Border Region
16. Mexican Plateau

Figure 15-1. Principal natural regions of North America (generalized). Not all the numbered regions are mentioned in the text.

Appalachian Region and Atlantic Coastal Plain (2, 3)

Along its entire eastern side the "Old Continent" is bordered by the Appalachian Region, a belt of country in which the trend or "grain" of the underlying rocks is northeast-southwest. This Appalachian belt continues beyond the "Old Continent" southwestward as far as northern Alabama. East of the Appalachians is an even longer belt, the Atlantic Coastal Plain, consisting of Mesozoic and Cenozoic sedimentary strata that slope very gently seaward. This plain includes the entire surface of the continental shelf as well as the coast proper. From Georgia northward to Cape Cod it consists of a landward part and a submerged part (Fig. 15-1). North of Cape Cod the coast has been bent downward sufficiently to submerge the entire Coastal Plain, so that the ocean laps directly against Appalachian rocks.

Earlier in this chapter we mentioned regions that occur in pairs, in which one part of each pair has been eroded while the other part has been built up by deposition of the eroded waste. The eroded Appalachian Region and the built-up Coastal Plain (including the surface of the Continental Shelf) are such a pair (Fig. 15-2). As was brought out in Chapter 10, most of the strata in the Appalachian

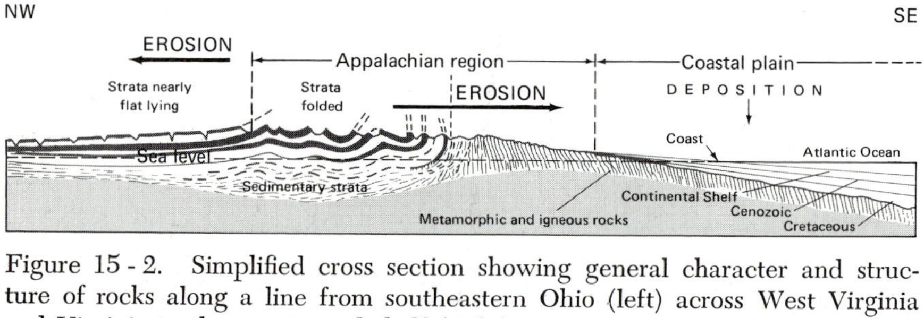

Figure 15-2. Simplified cross section showing general character and structure of rocks along a line from southeastern Ohio (left) across West Virginia and Virginia to the continental shelf (right).

Region were deposited on the floors of Paleozoic seas and were repeatedly squeezed and wrinkled into long narrow folds. The last and greatest of the squeezes occurred in Permian time. In the eastern part of the region squeezing was so intense that most of the rock there was converted into metamorphic rock. Farther west the strata, although bent into folds, still retain their sedimentary character, at least through the region between Alabama and the Hudson

River. Northeast of the Hudson, through the New England states and Maritime Canada, most of the rocks are metamorphic.

Throughout Mesozoic time the Appalachian Region was being eroded. The great volume of sediment derived from erosion of the eastern flank of the region was carried to the Atlantic by the rivers of the Atlantic slope, from Georgia and the Carolinas to New England and Canada, and was deposited on the Continental Shelf. The Cretaceous strata depicted in Figure 15-2 are among the deposits of sand, silt, and clay that rivers poured into the ocean as finely chopped-up Appalachian wreckage, adding still further to the Shelf. Later, of course, the general uplift that characterized Ceno-

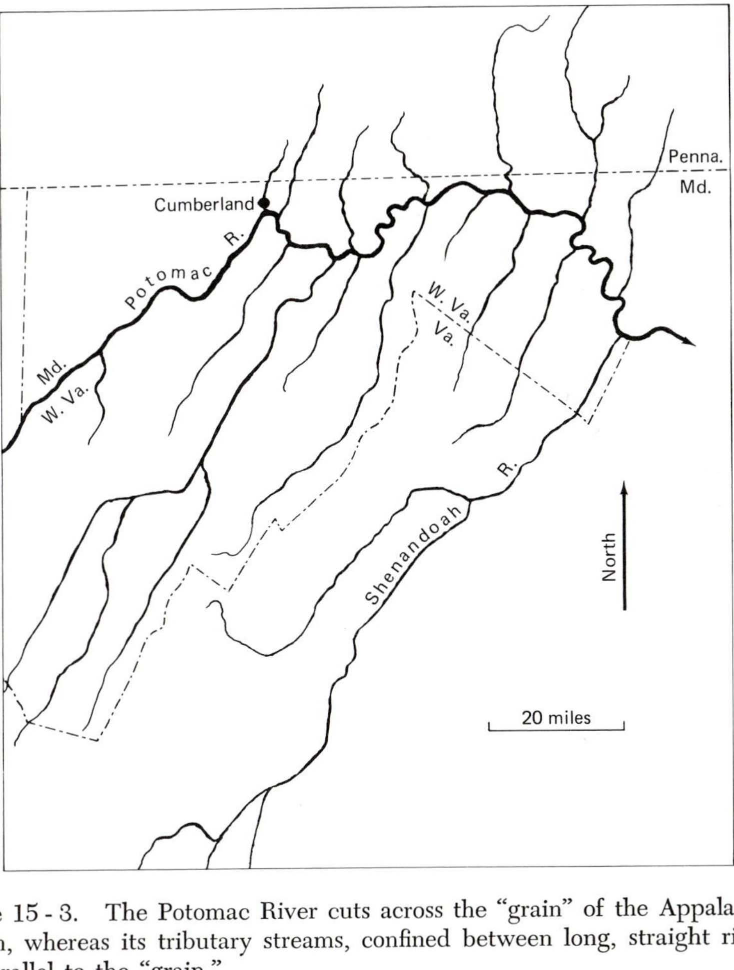

Figure 15-3. The Potomac River cuts across the "grain" of the Appalachian Region, whereas its tributary streams, confined between long, straight ridges, are parallel to the "grain."

zoic time affected the Appalachian Region. By steepening the slopes, uplift rejuvenated the rivers, which responded by bringing a new supply of somewhat coarser sediment to the ocean, where it was spread out over the Shelf.

Most parts of the Appalachians south of the Hudson River are characterized today, as they have been throughout most of the last 250 million years, by long narrow ridges separated by broader belts of lower land (Fig. 10-8). Because of this arrangement the terrain has an almost geometric appearance, emphasized by streams that flow between the ridges (Fig. 15-3) and by a distinctive pattern of vegetation. The steep, narrow ridges are forested, and so stand out from the lower lands, which are patchworks of cultivated fields.

The Appalachian Region and the Coastal Plain, then, are a complementary pair. Through a quarter of a billion years one part has risen gently several times, while the other part has at times gently subsided by tilting seaward. The eroded form of the mountains and the layers of sediment spread on the Coastal Plain are the response of external processes to movement caused by internal processes. They reflect the basic conflict that is always occurring at the Earth's solid surface.

Central Lowland (4)

Flanking the "Old Continent" on the south and southwest is a Central Lowland underlain, like the Appalachians, by marine strata of Paleozoic age. But the strata beneath the lowland, unlike those beneath the Appalachians, are comparatively thin, and apart from local exceptions such as the Ozark region in Missouri and the Ouachita region in Arkansas and Oklahoma, they have never been much deformed. So these strata have remained nearly flat-lying. Although they shared uplifts that affected the Appalachians in Mesozoic and Cenozoic time, they were not lifted to any great height. Streams in the Mississippi and other drainage systems have cut them into a myriad of low hills. Because its strata are little deformed, the Central Lowland lacks the long, straight ridges that characterize much of the Appalachian Region. Unconfined by straight ridges, the streams and valleys form gently branching patterns like the pattern of veins in a leaf (Fig. 15-4).

The long-continued erosion that cut the lowland into hills produced a great deal of sediment, which was carried to the Gulf of Mexico by the Mississippi and other rivers. At times the tidewater of the Gulf extended northward as far as the mouth of the present Ohio River at the southern tip of Illinois, and at others it receded seaward beyond the present coastline. But at all times the Continental Shelf was being built up by the spreading of sediment over its upper surface. The result is the Gulf Coastal Plain as it exists today.

The Gulf Coastal Plain and the Atlantic Coastal Plain are merely two parts of a single province that has experienced a common history. Streams that drained the western side of the uplifted Appalachian Region mingled their sediment with that derived from erosion of the Central Lowland across which they flowed. The mixed sediment then reached the Mississippi River, was combined again, and so was carried southward to the Gulf.

In this connection the sediments of the Gulf Coastal Plain throw a sidelight on the history of the Mississippi River. Adding to the text description that explains Figure 4 - 7, we can say that the Mississippi, in its main stem at least, is not only the Father of Waters

Figure 15 - 4. The Green River and its tributaries, in Kentucky, typify the branching pattern of streams in the Central Lowland. Dams have been built across some of these streams, creating lakes for recreation and the generation of electric power.

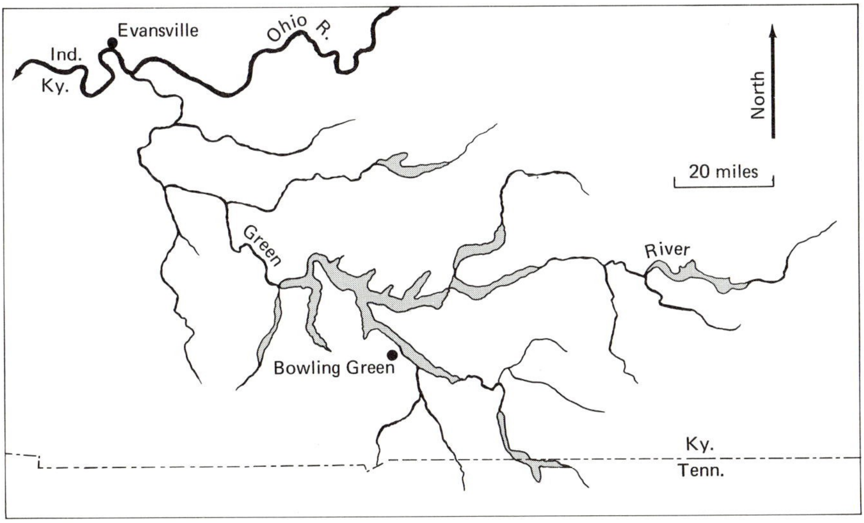

but also a father of venerable age. The river has been flowing in or near its present position at least since later Mesozoic time—during the last 150 million years or so—and possibly even longer. We know this because it seems to have been delivering its load of sediment to the same part of the Gulf of Mexico since before the Cretaceous Period began.

Rocky Mountains and Great Plains (5, 6)

The Rocky Mountains and the Great Plains east of them are another complementary pair of regions: erosion of the Rockies supplied the huge volume of sediment with which the Great Plains region was built up. A glance at Figure 15 - 1 reveals that the Rockies and other mountains related to them are together more than 3000 miles long. They extend from central New Mexico northward to the Bering Strait. Throughout that long distance the mountain structures vary a good deal, but the mountains as a whole are alike in having been created by crust movements that began in Cretaceous time and continued spasmodically far into the Cenozoic. The Rockies therefore are much younger than the Appalachians, which they far surpass in height. The altitudes of the lower ranges are 7000 to 8000 feet, and the higher ones are more than 11,000 feet. Mt. Robson in the Canadian Rockies is nearly 13,000 feet high, while Mt. Elbert, Mt. Evans, and Pikes Peak in Colorado all exceed 14,000 feet.

West of Denver, Colorado, the Front Range consists of a gigantic upward fold in which a whole sequence, 10,000 to 20,000 feet thick, of strata of Paleozoic and Mesozoic age was bent up (Fig. 15 - 5). The fold was eroded so as to expose basement granite and metamorphic rocks of the "Old Continent" in the high parts of the range, while the eroded stumps of the sedimentary strata form foothills at altitudes 5000 feet lower. The more prominent stumps form *hogbacks,* conspicuous ridges that run along the base of the Front Range. More such great folds form similar ranges farther west. The Continental Divide lies among them, its eastern slope draining toward the Mississippi River and the Gulf of Mexico while its western slope drains toward the Colorado River and the Pacific Ocean.

No sooner had the upward folds started to take form than streams flowing down the slopes of the folds began to carry away rock waste

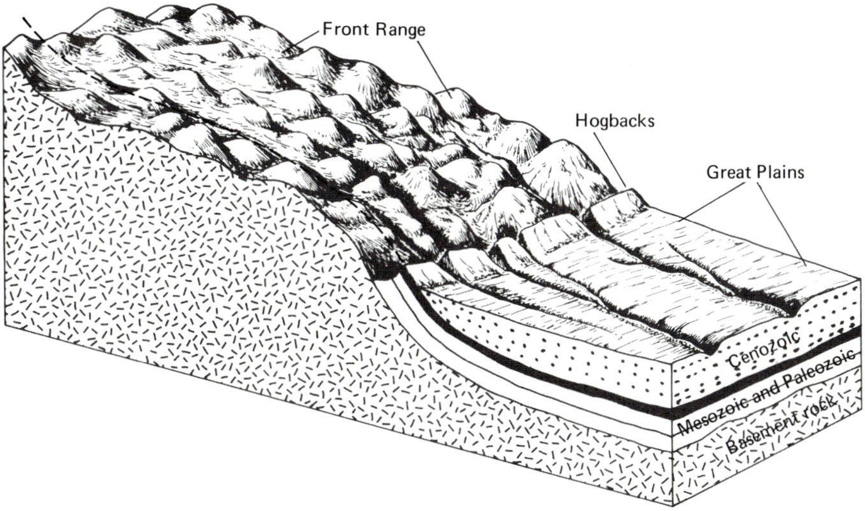

Figure 15 - 5. Chief features of the Front Range of the Rocky Mountains in Colorado.

A. Diagram showing arched-up sedimentary strata. Broken line at left represents the top of the deformed Mesozoic strata before erosion. (Adapted from a sketch by W. M. Davis.)

B. View from the air, looking north along hogbacks. Cenozoic strata appear at right. (T. S. Lovering, U.S. Geol. Survey.)

and spread it out as alluvium in the areas between the folds and also in the country that is now the Great Plains, extending eastward from the Rockies toward the Mississippi River. By filling up the valleys and finally burying the hills between them, the deposited alluvium made the Plains country almost completely flat.

At first the Plains Region enjoyed good rainfall and was well furnished with trees on which browsed many kinds of mammals. Buried in the alluvium, now sedimentary rock, have been found skeletons of these early Plains animals, which very likely drowned at times of really big floods. But by slow degrees the rainfall diminished, the climate became drier, and south of about latitude 50° the trees disappeared except along the immediate banks of the rivers. Even some of the streams themselves began to go dry during the hot summer season. The explanation of this change is found in the rise of the Rockies and other mountains farther west. These highlands interposed a barrier as much as a mile high to the rainfall that formerly had been brought by west winds from the Pacific Ocean. Today, after 63 million years of Cenozoic mountain making, the Plains country is drier than it has been at any time since the Mesozoic. Before European man settled the region, the native animals were grazers, grass eaters such as antelope and buffalo, a very different collection from the browsers, the leaf eaters, of Eocene time, fifty million years earlier.

In the later Cenozoic the Plains region was tilted up toward the west. The steepened streams trenched down into the alluvium they had built up earlier and carried off the waste to the Mississippi and the Gulf of Mexico. This is why, today, principal Plains rivers such as the Missouri, the Platte, and the Arkansas flow in trenches sunk 200 feet or more into the Plains surface.

In this history of the Rocky Mountains and Great Plains we have a grand example of the faithful response of an external process, river activity, to an internal process, mountain making. The response created a vast new plain, and the mountain making itself caused a change in the climate. Together they altered the whole environment so that changes in plants and animals followed. The whole story is a long chain of cause and effect. And at the end of the chain is the additional effect of industrial man, who has cut down mountain forests, plowed up the grass cover of the Plains, substituted cattle for buffalo, dammed or dredged rivers, and covered an appreciable part of the surface with concrete. The effects of these activities are being inscribed in the geologic record of sediments and in the record of fossils. They will be there, to be read in the future by anyone able to read.

Two broad plateaus (7, 8)

Figure 15-1 shows the Columbia Plateau and the Colorado Plateau, two regions named for the two rivers that traverse them. The pair are alike in that they consist of high and generally rather flat land, but they differ in the character and history of their strata. One is volcanic while the other is sedimentary.

The Columbia Plateau forms an area four times greater than that of the six New England states combined. It occupies more than a quarter of Washington, about half of Oregon, and much of the southern part of Idaho, extending all the way to Yellowstone Park in Wyoming. It is dry for the same reason that the Great Plains country is dry. A high mountain range, the Cascade Mountains, stands between it and the Pacific Ocean, forming an effective barrier to rainfall. This is a country of basalt, of ancient lava flows spread out widely, one on top of the other (Fig. 15-6). Along the boundary between Idaho and Oregon the Snake River has cut a huge canyon as deep as the Grand Canyon. In its west wall is exposed a pile of lava flows, some of them no more than ten feet thick but together adding up to an aggregate thickness of 4000 feet. In places where the surface of the basalt laps against adjacent mountains, the plateau resembles a frozen sea lapping against a mountainous coast.

Fossil leaves collected from layers of alluvium between some of the flows near the plateau margin, and evidence of other kinds, suggest that the lava poured out intermittently during Miocene and early Pliocene time. No comparable field of basalt exists elsewhere in North America, although similar fields do occur elsewhere in the world. The reason why there is so much basalt in the northwest part of the United States is a puzzle not yet solved. It is somehow related to the system of oceanic rifts, one of which is not far away (Fig. 6-6). Presumably the Columbia Plateau was built up by basaltic lava from underneath the American crust plate. The lava welled upward through the plate and the continent floating in it,

Figure 15 - 6. The strata that form the Columbia Plateau consist of basalt and are a record of long-continued outpouring of lava flows. In this view of one wall of the Grand Coulee in eastern Washington at least twelve such flows have been exposed by stream erosion. Many flows are cut by closely spaced vertical cracks caused by shrinkage of the basalt as it cooled. The cliff is more than 300 feet high. In the foreground is rubble created by weathering of the basalt. (U.S. Bureau of Reclamation.)

to emerge at the surface of the continent. There it created an enormous, shallow lake of lava, and then another and another. It would have been something like what is happening on the floors of the oceans today, on both sides of the oceanic rift.

The Colorado Plateau is very different in origin and appearance. Instead of basalt of perhaps oceanic origin, the material of which it is constructed consists of sedimentary strata, mostly of the kinds that were deposited in shallow seas that flooded over parts of the continent at times during the Paleozoic and Mesozoic Eras. Most of it therefore is much older than the rock of the Columbia Plateau.

In area the Colorado Plateau is roughly similar to its sister farther north. It includes large parts of Colorado, Utah, New Mexico, and Arizona. Near its center the boundaries of the four states meet at a point (the only such point in the nation). Also the plateau is transected from northeast to southwest by the Colorado River, which in

Figure 15 - 7. Many Colorado Plateau landscapes are combinations of horizontal and vertical elements. A scene in Monument Valley, Utah. (Alan Pitcairn from Grant Heilman.)

the western part flows through its famous Grand Canyon (Fig. 3 - 1). The canyon owes its great depth to the unusual height of the plateau, much of which stands a mile and a half above sea level. The outstanding characteristic of the plateau landscapes is that they consist mainly of horizontal and vertical elements, flat plateau surfaces separated by vertical cliffs (Fig. 15 - 7). These step-like landscapes in turn are controlled by two features: one is that the underlying strata are nearly horizontal; the other, that there are marked differences in the resistance of the strata to erosion. The way in which these controls operate is shown in Figure 15 - 8. The picturesque quality of the landscapes is enhanced by yellow and red colors in some of the strata, caused by the oxidation of iron minerals before or after the sediments were deposited.

A noteworthy feature of the Colorado Plateau is that its strata are little disturbed. Although most of the strata are marine Paleozoic and Mesozoic sedimentary rocks that were once below sea level, they have been gently elevated to great height without undergoing the drastic deformation that characterizes Paleozoic and Mesozoic strata in the Rocky Mountains.

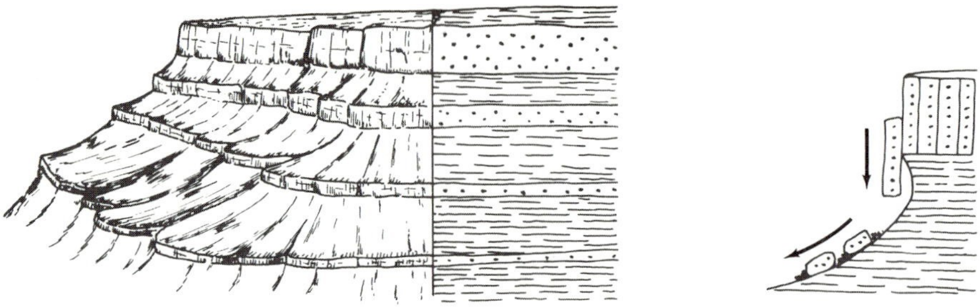

Figure 15 - 8. Cause of the step-like landscapes in the Colorado Plateau: alternating flat-lying strata that react differently to weathering and mass-wasting. Layers of sandstone and other resistant rocks contain many vertical cracks and form cliffs parallel with the cracks. Claystone and other weak rock are weathered more rapidly and form gentler slopes. The more rapid weathering undermines the cliffs above, allowing blocks of cliff rock to fall, break up, and creep downslope.

The few specially high parts of the plateau receive enough rain to promote the growth of pine forests, but most of the region is semiarid grassland and brush, and some of it is desert. Because of scanty rainfall, rivers are few and far apart and generally flow through steep-walled canyons. Their waters are brown with silt and clay washed in, mostly during floods, from the rocks exposed in canyon walls and in the step-like plateaus above.

Basin-and-Range Region (9)

Separating the Colorado Plateau from the Columbia Plateau and reaching southeastward far into Mexico is the Basin-and-Range Region. Touching Oregon in the north and including nearly all of Nevada, it reaches as far east as western Texas. With high mountains between it and the Pacific Ocean, it receives little rainfall (less than 10 inches per year; a few weather stations record less than 3 inches per year) and so is mostly desert. In the 19th Century it was traversed from east to west by the Overland Trail, the route followed by the wagon trains of emigrants and the "Forty-niners." One of several serious hazards of the route was scarcity of water, because of which some pioneers lost their lives.

The name of the region describes exactly its surface form: a series of mountain ranges separated by basins. Although numerous, the ranges are discontinuous and mostly short. With their generally north-south trend, they look, on a map, rather like a crowd of caterpillars crawling in a single direction. Some of the basins are less than 10 miles wide; others are as much as 50. Bare or dotted with desert brush, the slopes of the mountains are sculptured into rows of steep, dry valleys through which water flows only after rare rainstorms. At the mouth of each valley is an alluvial fan. The fans built by the streams on one slope of one range are aligned in a row along the base of the range (Fig. 15 - 9). They spread out sideways, touch their neighbors, and merge to form a smooth plain of gravelly and sandy alluvium that slopes down toward the center of the basin. Where the plain of alluvium that is built out from the range on the west meets the plain built from the range on the east, a shallow lake forms after each big rainstorm. The lake, which may contain a few inches or a few feet of water, gradually evaporates, exposing its

Figure 15 - 9. An idealized desert basin lying between two mountain ranges, typical of much of the Basin-and-Range Region but not representing any specific basin. Most basins are far wider than the one shown here.

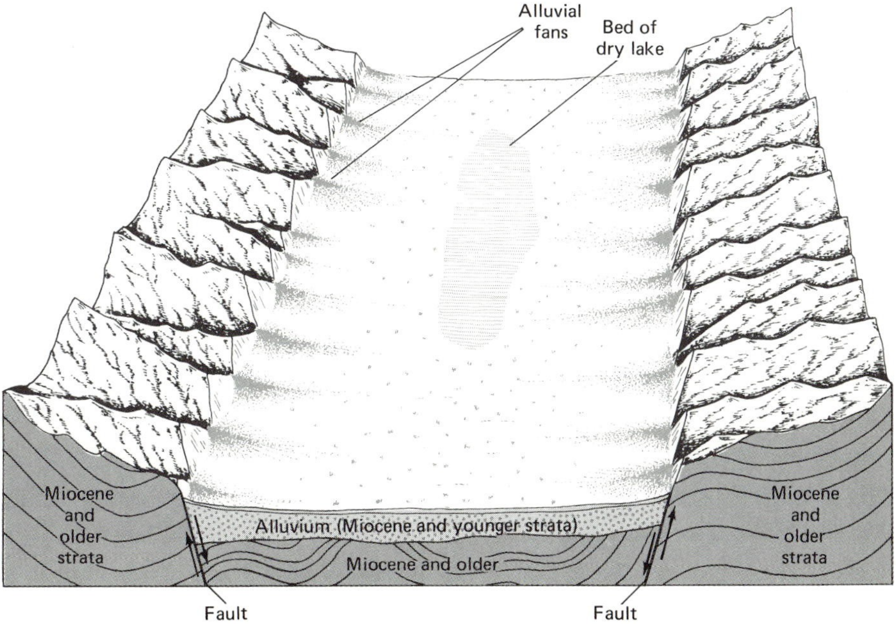

bed as a broad area of dry mud and white precipitated salts. Some of the biggest lakes, such as Great Salt Lake and Pyramid Lake, are fed by rivers that originate in high mountains at the very edges of the Basin-and-Range Region and do not run dry. Accordingly such lakes, although shallow and incapable of overflowing, always contain water.

Because of the dry climate, there are but two major rivers, both of which exist only because they are fed with water from outside the region. The Colorado River, which is fed from the Rocky Mountains and the Colorado Plateau, has only a single large tributary that gets its water from within the region. That is the Gila River in Arizona. The Rio Grande in New Mexico likewise traverses the region, but it too carries in most of its water from the Rocky Mountains. Throughout the rest of the Basin-and-Range country the flowing water never reaches the sea. Some of it sinks into the ground and the rest evaporates.

Nearly all the ranges and basins are the result of faulting. Blocks of the crust, miles or tens of miles in diameter, have slipped downward slowly, forming basins and leaving other blocks standing high to form mountain ranges. Some blocks have tilted like a trap door with hinges along one of its edges. The whole region, 1300 miles long even without its Mexican part, is like a pavement of tiles, but many of the tiles have come loose and sunk a little, making the pavement very uneven.

Figure 15 - 9 shows the way in which basins and ranges are related to faults. Because of the thick alluvial fans that cover the mountain bases, the faults are more often hidden than exposed to view. But detailed study of the exposed rocks, supplemented by the records of many water wells (and wells have to be very deep in that country) has revealed much of the history of the region. The blocks of crust that form both ranges and basins generally consist of rocks not younger than Miocene. In contrast, the alluvium that partly fills the basins is not older than Miocene. (Of course the top part of it is modern and is being added to at present.) So we conclude that the fault movements began generally in Miocene time. Although the details of their origin are not yet well known, it is likely that the movements were a result of interaction between the American and Pacific crust plates.

Whatever the cause, the tile pavement began to loosen and shift. From time to time slipping still occurs today. Every few years new breaks appear, some of them accompanied by earthquakes and some of them creating long, straight little cliffs right in the alluvial fans themselves. In these we see again the Principle of Uniformity at work.

Pacific Region (10-13)

California. West of the Basin-and-Range Region in California and west of the Columbia Plateau in Oregon and Washington lies the Pacific Region. In California it consists, in the simplest terms, of two high masses and a lowland (Fig. 15 - 10). On the east is a great mountain block, the Sierra Nevada, at its highest point reaching an altitude of 15,000 feet. On the west, along the coast itself, is a series of lower mountain ranges that trend northwest, oblique to the coast. Between the two high masses and parallel with them is a low basin some fifty miles wide, known as the Great Valley. The streams that drain the basin merge and, as a single stream, the Sacramento River, pass into the Pacific through the Golden Gate, a valley that is now an arm of the sea.

The Sierra Nevada consists mainly of big granitic batholiths (Fig. 2 - 4) together with metamorphic rocks into which the batholith magma was injected. Along the east base of the range runs a great, complex fault that marks the western boundary of the Basin-and-Range Region. In Cenozoic time the Sierra Nevada tilted upward along the fault while its western part tilted downward, underneath what is now the Great Valley. During the uplift, erosion moved great volumes of rock waste westward from the rising block into a great trench, in which the sediment accumulated in vast thickness. Very late in Cenozoic time the trench and the sediment it contained were compressed and deformed, creating the coastal mountains, and long faults began to form in it and along its eastern side.

These events have been explained as the work of moving crust plates. North America has collided with the Pacific plate (Fig. 6 - 6) in such a way that the edge of the plate has been slightly overridden by the continent, somewhat as, in Figure 6 - 8, plate D is overriding plate C. The extreme coastal part of California has become attached to the Pacific plate and is moving with it as the two plates slide past each other, along a related series of faults called the San Andreas Fault system (Fig. 15 - 10). This bundle of deep fractures enters

Figure 15 - 10. Sketch map of California and adjacent territory showing major features. Shaded belt is San Andreas Fault zone (generalized); wide line is the San Andreas Fault itself. (Compiled.)

California from Mexico and passes out to sea in the general region of San Francisco. It is the site of movement that has been in progress for at least 25 million years and is continuing today.

Since they collided, the two plates have been sliding past each other. The strip of California that lies west of the fault system is moving northwest, along with the Pacific plate to which it is attached, relative to the country east of the faults, country that is part of the North American plate. The movement, which can be measured by various means, amounts to more than 2 inches per year, and is the indirect cause of the major California earthquakes. Part of the energy involved in the movement is stored in the rock material along the faults, like the energy stored in a coil spring as the spring is compressed. With continued movement the built-up energy becomes so great that it exceeds the strength of the rock material to resist it. So the stored energy is dissipated by sudden movement along a fault, creating earthquake waves, a series of tiny vibrations analogous to the vibrations of a suddenly struck tuning fork. This is what happened, for example, in the San Francisco earthquake of 1906. In a few seconds the release of huge pent-up energy caused offsets, amounting to as much as 21 feet, along faults. The released energy broke roads, railroad lines, and other structures. The cumulative displacement along the San Andreas Fault, caused through the years by many repetitions of this process of alternate slipping and snapping, runs into hundreds of miles.

Although, as we have already seen, intense activity characterized the Appalachian Region at times in the Paleozoic Era, and the Rocky Mountains belt at times in the earlier part of the Cenozoic, the site of spectacular activity today is in California. Intermittent earthquakes have occurred along the San Andreas Fault system since the time of the earliest Spanish colonists, and we can confidently predict that as long as the two crust plates continue to slide past each other as they are doing now, there will be more earthquakes in California.

This explanation of the forces at work in coastal California is no more than a greatly oversimplified sketch. Part of it is still theory, and knowledge of most aspects of the geology will be improved in time. But the basic picture of two crust plates interacting with each other is probably here to stay. It makes the West Coast, especially California, much more exciting geologically than the East Coast, whose days of glorious activity were Paleozoic days in the distant past.

Pacific Northwest. The Pacific Northwest does not have a San Andreas Fault cutting along its coastal region, and in consequence its recent history has been more sedate. Like California, however, the region has two mountainous highlands with a lowland between. The Cascade Mountains, a long and mostly high chain, are separated from parallel but lower mountains along the coast by the Puget Lowland that extends through Washington and northern Oregon. Part of the low belt has been submerged to form Puget Sound. The Cascades are a long, high mass that extends from northern California into Canada. The mass is marked by scores of volcanic cones of many sizes, from very small pimples to the mighty Mt. Rainier (Fig. 15 - 11), 14,408 feet high, southeast of Tacoma, Washington. Among them they have poured out so much lava and have blown into the air so great a volume of solid particles that they have managed to bury the pre-existing metamorphic and igneous rocks al-

Figure 15 - 11. The top of Mt. Rainier (14,408 ft.) with several of its glaciers, dominating the Cascade Mountains. At left is Mt. Adams (12,307 ft.), 55 miles south, and at right Mt. St. Helens (9671 ft.), 60 miles southwest. These three great cones are typical of the volcanoes of the Cascades.

In the foreground the head of a glacier occupies a great indentation shaped like half a bowl, hollowed out by frost wedging. In its western side are visible the sloping layers of volcanic material with which the cone was gradually built up. (Miller Cowling, Washington National Guard.)

most completely. In northern Washington, however, such rocks are exposed in a broad complex of mountains 6000 to 8000 feet high.

A similar cone in Oregon, named "Mt. Mazama" although it has not existed in complete condition for thousands of years, literally blew its top in a huge eruption some 6600 years ago. The event, which must have been truly spectacular, is dated by the radiocarbon age of the wood of a tree buried in fallout from the eruption. The entire upper part of the cone was shattered. Its wreckage collapsed into the volcanic vent beneath it and left the base of the cone, a mere saucer-shaped stump, to cool off, fill with rainwater, and form the lake we call Crater Lake, 5 miles in diameter and 2000 feet deep.

The volcanic rocks and volcanic cones of the Cascades region are of Cenozoic age, and some are very recent indeed. Mt. St. Helens, in its present form at least (Fig. 15 - 11), is thought to be little more than 2000 years old. It erupted last in the 19th Century. Lassen Peak, a cone in northern California, erupted as recently as 1914 and 1917. Since then steam has continued to escape from it.

Conclusion

With this brief characterization of the Pacific Northwest our overview of Cenozoic North America ends. Through the action of various geologic processes over a span of 63 million years the low, mild-climate, maritime Cretaceous continent was converted into the higher and far more diversified continent we see around us today. Extremes of temperature and rainfall are greater than they were in the Cretaceous continent, and in consequence the range and variety of environments have increased. Both plants and animals have responded by adapting to the new opportunities. In the succeeding chapter we shall follow the responses of mammals as they penetrated far and wide and diversified in a dramatic way.

References

Anderson, D. L., 1971, The San Andreas Fault: Scientific American, v. 225, p. 53–68.

Bostock, H. S., 1970, Physiographic subdivisions of Canada: p. 10–30 *in* Geology and economic minerals of Canada, Econ. Geol. Report No. 1, 5th ed., Dept. of Energy, Mines, & Resources, Ottawa.

Fenneman, N. M., 1931, Physiography of western United States: McGraw-Hill Book Co., New York.

Fenneman, N. M., 1938, Physiography of eastern United States: McGraw-Hill Book Co., New York.

Hunt, C. B., 1967, Physiography of the United States: W. H. Freeman and Co., San Francisco.

Shelton, J. S., 1966, Geology illustrated: W. H. Freeman and Co., San Francisco.

Thornbury, W. D., 1965, Regional geomorphology of the United States: John Wiley & Sons, Inc., New York.

16

Mammals

As Chapter 3 tells us, the geologic column was built on the basis of the fossil record. The Phanerozoic, the part of the column represented by abundant, easily visible fossils, logically embraces three big units, Paleozoic ("ancient life"), Mesozoic ("medieval life"), and Cenozoic ("recent life"). The word *life* in each of these names emphasizes the fact that the subdivisions exist because of the fossils they contain. Cenozoic history was dominated by mammals, Mesozoic history by reptiles. Yet back of this difference lie the physical changes in continents and seas, brought about mainly by the spreading of ocean floors and the movements of continents. As was indicated in Chapter 15, Cenozoic North America differed greatly from the Mesozoic continent, and the new environments shaped new forms of life by evolution. We are faced now with the story of mammals. In telling the story we had best begin by looking about us at the mammals of today, so that we can better appreciate the evolution by which they reached their present state.

As we said in Chapter 13, mammal characteristics include mammary glands for suckling young, warm blood (which really means constant body temperature day and night), hair to provide thermal insulation, a four-chambered heart, teeth differentiated into biting and chewing functions, and a well-developed brain with a large memory center (the *cerebrum*). Sharing these characteristics in common, mammals are divided into three groups, based essentially on differences in reproduction.

1. Most primitive are *monotremes,* because they lay eggs (even though they suckle their young) and because of anatomical characteristics not visible in fossil remains. Today only two genera survive, the duck-billed platypus and the spiny anteater. Both are highly specialized to particular ways of life and both are confined to the Australian region, where competition seems to have been less fierce than in other parts of the world, offering a better chance of survival to very primitive creatures. Because the only known fossil monotremes are likewise Australian, perhaps monotremes never existed anywhere else.
2. An intermediate group are the *marsupials,* pouched mammals, of which the kangaroo is perhaps the most familiar. Although the young are born alive, they are very premature; a newborn kangaroo may be only 3 inches long. They climb into the mother's pouch and are suckled there for a long period until they have grown to several times their birth size and can emerge and fend for themselves. Marsupials are now principally Australian mammals, including such well-known kinds as wallabies, koala "bears," and wombats. The only marsupial remaining in North America is the opossum, which seems to have minimized competition by living in trees. Marsupial birth is safer than egg laying because the embryo is continuously protected by the mother. But it has been improved upon by the third mammal group.

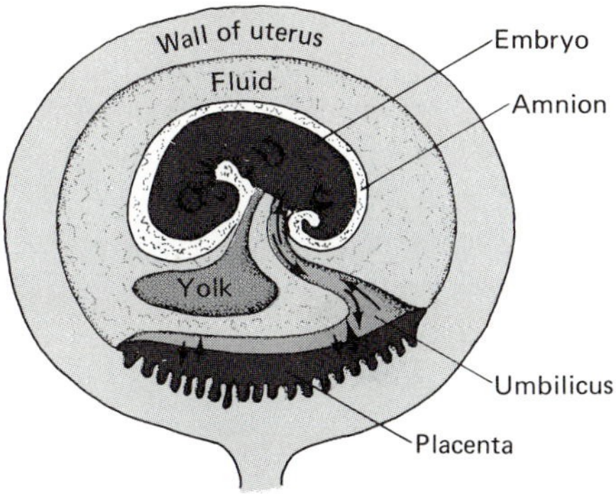

Figure 16 - 1. Essential features of embryo and uterus in placental mammals. (Compare Figure 13 - 3). Yolk functions during early phases and then disappears. Embryo receives oxygen and food and discharges carbon dioxide and excretory products via the placenta. These exchanges are shown by arrows.

3. Most advanced are *placentals*, so named because they possess a *placenta*. This is a membrane-like organ, within the uterus, filled with blood vessels that supply the embryo with food and oxygen and remove carbon dioxide and excretory products. The whole apparatus (Fig. 16 - 1) resembles the amniote egg (Fig. 13 - 3) from which it was obviously derived. But with the placenta, mammals evolved not only beyond the egg but beyond the marsupial system as well. With the placenta to look after it, the embryo could remain in the uterus until it had developed far beyond the size and complexity of the marsupial embryo. A young individual could begin to look after itself (in most genera at any rate) very soon after its birth. The placenta was a very real improvement in the efficiency of the reproductive process.

Mesozoic mammals

From our own particular late-Cenozoic point of view, today's mammals are the near end of the evolutionary line. The mammal-like reptiles of Permian and Triassic times are the far end. We must not forget that the evolutionary path from the most primitive mammals toward man was not straight, nor was it followed at a uniform speed. During nearly 250 million Mesozoic years mammals evolved in no spectacular way. We could almost say they marked time. Yet during those same years true reptiles spread far and wide and adapted to many varieties of mild-climate environment. Mesozoic lowlands and mild Mesozoic climates favored reptiles but did not particularly favor mammals. The newcomers had to wait. They waited until dinosaurs thinned out and disappeared and Cenozoic environments offered something new before they in their turn could begin an explosive expansion and move into the position of dominant creatures. This story illustrates clearly how the path of evolution turns this way and that as it responds to the opportunities that pull it in various directions. As long as crust plates move and continents are carried into higher or lower latitudes and sometimes collide, environments will change and give new opportunities to living things.

Very likely it was *Cynognathus* (Fig. 13 - 7) or one of his relatives that was ancestral to true mammals. Or perhaps the honored ancestor was a reptile not yet discovered as a fossil. It does not matter greatly, because mammals and mammal-like reptiles are very closely related. Whatever the ancestral reptile, it gave rise to mammals of very small size. Fossil mammals have been found in Triassic, Jurassic, and Cretaceous strata, but most are the size of rats and mice, the largest no bigger than a cat. No complete skull, let alone a complete skeleton, has yet been found, but there are a good many tiny jaws, with mammal-style differentiated teeth, and scattered bones. The bones look more like those of marsupials or monotremes than like placentals. One fossil, from Cretaceous strata in western Canada, is clearly an opossum.

Their small size suggests that Mesozoic mammals were unable to compete with dinosaurs and other reptiles, despite their comparatively large brains. They seem to have been second-class citizens, keeping as far out of the way as possible and perhaps feeding on insects and seeds. Lacking a complete skeleton, we cannot form an accurate idea of what they looked like, but from fossil skulls it is possible to suggest the look of a head (Fig. 16 - 2).

Figure 16 - 2. Head of *Ptilodus*, a late Mesozoic mammal. The rest of the body cannot be restored because only the skull has so far been found. (Adapted from W. B. Scott.)

Adaptations to Cenozoic environments

An overall view. This was the existing state of things as the Mesozoic Era drew to a close. Mammals began to emerge from obscurity. With changing landscapes and cooling climates, dinosaurs slowly disappeared group by group, leaving empty spaces that could be filled. No longer threatened by reptiles, mammals became bold. Their hair and their warm blood enabled them to withstand the cooler climates. Their larger brains enabled them to cope with the new and more varied environments in ways impossible for dinosaurs. So, as the reptiles faded away, the mammals advanced into what was almost a vacuum, a series of unoccupied spaces in which there was little competition. The reptiles were gone. The only thing a pioneering mammal had to fear was other mammals. Of course among the "other mammals" predatory meat eaters soon developed and fought for their place in the food chain. Today, in many countries, that place has been usurped by man.

In this story we find a parallel with the Mesozoic history of reptiles. A new race expands into new territory and diversifies into several kinds as it encounters a variety of available environments. Some kinds prey on others. A sort of running balance is reached in the competition among various kinds. The balance is "running" because it is continually changing, with some kinds dying out and others developing. It is an endless series of battles, with the tide going now one way and now another way.

Looked at in perspective, the Cenozoic history of mammals seems to consist of two phases: an early phase of experimentation and a later phase of perfection of detail, in which the directions taken by evolution were more clearly defined. The two phases are of about equal length. The early one lasted well into Oligocene time, the later phase through the rest of the Cenozoic. Of course it is still going on, but the rise of man to a dominant position has given the evolution of mammals since the Pliocene a very sharp twist.

The fossil record of the early phase makes it *seem* as though evolution were experimenting (often rather clumsily), developing and checking over different models to find those which could withstand the competition and make a go of it in this environment or that. Many of the early models were found wanting and were rejected one by one. The survivors still in the race got through the early phase

Figure 16 - 3. *Uintatherium,* a heavy-built plant eater that stood about 5 feet high, restored in an Eocene environment. Its grotesquely ornamented skull could hardly have been useful in attack or defense. Possibly it was a source of sexual attraction; if so it would have been favored in natural selection. (Courtesy of the American Museum of Natural History.)

and in the later phase were shaped and improved in well-defined directions. Many of them are still here as the dominant mammals of modern time. That, in a nutshell, is the Cenozoic history of mammals.

Early, unspecialized groups. The rather primitive character of early Cenozoic mammals (Figs. 16 - 3, 16 - 4) sprang from two chief causes. First, these mammals descended directly from late Cretaceous stocks that were themselves primitive. Second, the widespread forest environments of the time compelled no great or sudden change in form or structure. Probably most of the early Cenozoic mammals belonged to the placental group. As a rule they were of small or medium size (although one attained the bulk of a modern rhinoceros). Their feet were flat or nearly so (Fig. 16 - 4, A), some of them stumpy like those of an elephant, and nearly all equipped with five toes. They included both herbivores (mostly browsers) and carnivores. The latter are distinguished mainly by their teeth and primitive claws rather than by any striking differences of form. Some, such as *Uintatherium,* seen in Figure 16 - 3, possessed grotesque bony bumps on the skull and, in the males, long upper canine teeth.

Environment and diet. One by one these primitive kinds died out or evolved into more advanced kinds under the guiding influence of two principal factors, environment and diet. Let us examine how these factors operated. In the earliest part of Cenozoic time forest, woodland, and savanna were widespread. In such environments the

Figure 16 - 4. Four representative early and primitive Cenozoic mammals. A. *Patriofelis,* a carnivore. B. *Phenacodus,* a small herbivore. A relative of *Phenacodus* may have been an ancestor of the horse line. C. *Coryphodon,* a heavy-built herbivore that may have frequented swamps. D. *Barylambda,* an herbivore some eight feet long.

leaves of trees and bushes are the chief potential food for plant-eating animals, which move slowly from tree to tree. To escape their enemies such animals depend on concealment; most of them do not run far. Carnivorous enemies are present, and they too make use of concealment in stalking their prey. A long chase is usually unnecessary.

In contrast, grasslands, which began to appear in Eocene time and increased in extent throughout the rest of the Cenozoic, offered little beyond grass and herbs as food for plant eaters. Such mammals took their food from close to the ground. They had to keep moving in order to reach new grass and to find water, and so traveled long distances. Likewise they had to escape the inevitable predators by running, for there was nowhere to hide. They could improve the chances of survival of their species by reproducing in quantity and traveling in large groups or herds. The carnivores themselves were faced with the necessity for greater speed just to follow their prey, for they too had no place to hide. To keep themselves on a diet of meat, they had to run for their food.

Some mammals are omnivorous. Modern bears, for example, eat fruits, berries, ants, and fish. They can survive in a variety of environments, usually those in which there are streams and at least some trees. But they have few enemies and to them speed is not important.

We can expect that such differences of environment and diet would have influenced strongly the forms and structures of mammals. Indeed they did, as becomes evident when we examine a series of fossil skeletons taken from a long succession of strata.

Evolution of limbs and teeth. Fossil skeletons collected from various Cenozoic strata reveal progressive changes in structure that are obviously related to environment and to the kinds of feeding the environments demanded. Mammals that browsed in forest and woodland retained the primitive, rather flat feet we suppose were typical of Mesozoic mammals. They also retained rather unspecialized molar teeth for chewing up their leafy food. Rather slowly their legs lengthened, in part so as to improve the chances of escaping from predators that were gradually becoming more efficient.

The carnivores that stalked and preyed upon the browsers were likewise conservatively flat-footed at first. But little by little their limbs became adapted to running fast through short distances.

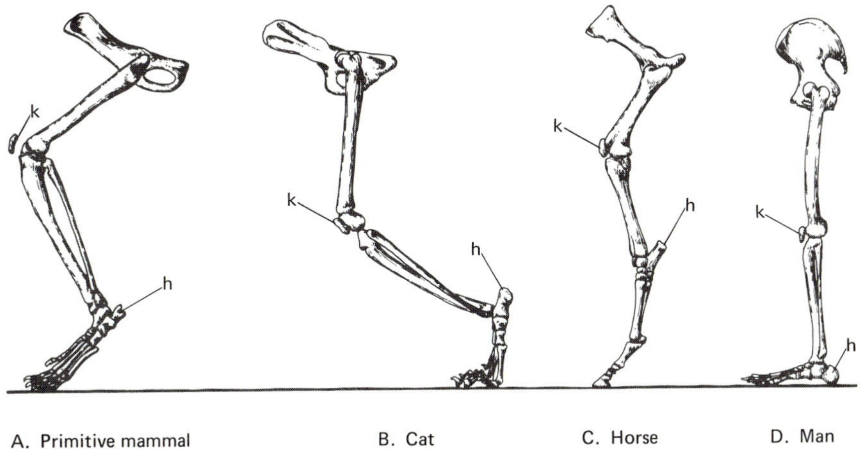

Figure 16 - 5. Leg and foot in four mammals.

Heels came up off the ground (Fig. 16 - 5, B), limbs became longer and lighter, and muscles bunched at the hips to provide a quick thrust at the start of a chase. Claws developed for seizing and pulling down prey. Front teeth became sharper for biting and jaw muscles strengthened for gnawing.

The limbs and teeth of grassland mammals were changing too, but in a different way. Because the grazers were obliged to run, both legs and feet lengthened greatly and muscles became bunched more effectively. Heels were lifted so far up off the ground that feet were nearly vertical (Fig. 16 - 5, C). Toenails enlarged and thickened, protecting the toes. While this was happening, certain toes lengthened, leaving others out of contact with the ground, to shrink in size and disappear (Fig. 16 - 6). This change reduced the weight of the foot and so increased both the speed and the range of grassland mammals. Because it took place more or less simultaneously in many different kinds of mammals, the change is strong evidence of the compelling influence of environment. But as environments subdivided, the change was carried further in some kinds of mammals than in others. Pigs ended up with four functional toes, rhinoceroses with three, camels, deer, and cattle with two, and horses with only one.

The process was a long one. From their first known appearance in the Eocene, horses succeeded in attaining a one-toe condition by early Pliocene time, a lapse of nearly 60 million years. If the average time needed for creating each new generation of horses was 4 years,

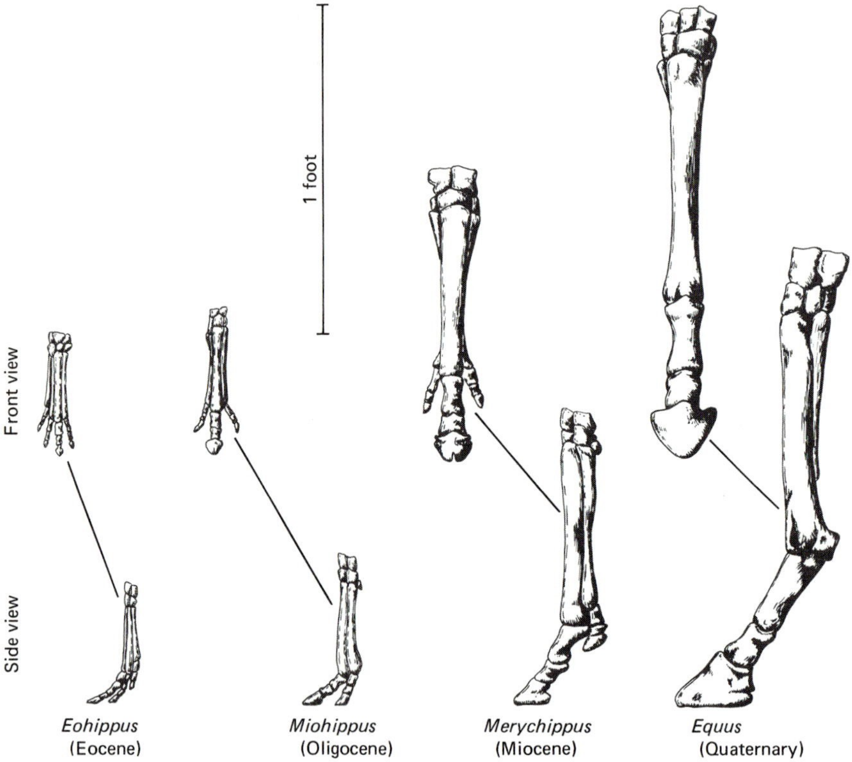

Figure 16 - 6. Forelimb bones of four genera of fossil horses. In fifty million years of evolutionary progress, toes decreased from 4 to 1 and overall length increased by nearly 12 times.

the process would have spread over 15,000,000 generations—15,000,000 opportunities for change—with the environment, in the background, pressing for change as each new generation appeared. If we took 15,000,000 horses, one from each generation, and formed a parade in single file, the little Eocene ones first and the big modern ones last, all walking past the reviewing stand at 4 mph, the parade would last more than four months. Watching the procession during any one hour of that time, we would see no apparent change in the animals that passed before us. But by the end the passing horses would have increased in size from little four-toed browsers some 12 or 15 inches high to one-toed horses walking 5 feet high at the shoulders (Fig. 16 - 7).

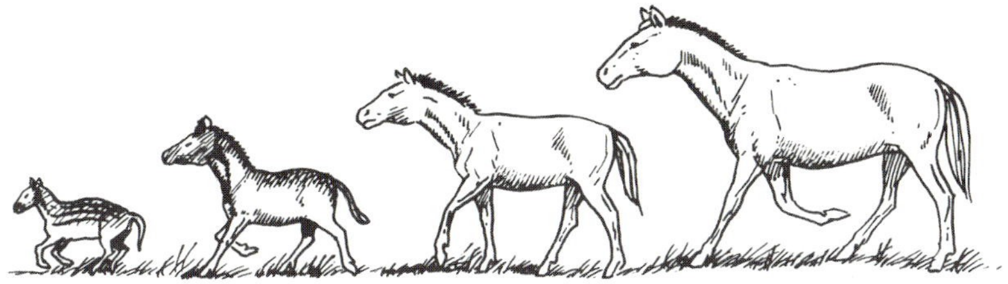

Figure 16 - 7. Four genera of horses from the parade. *Eohippus* and *Mesohippus* were much smaller in proportion to the two later horses than the sketch shows.

The fossil skeletons of grassland mammals show us other evolutionary changes. Jaws lengthened, with cropping teeth in front well separated from chewing teeth behind. But because grass leaves contain a good deal of silica within their structure, and because most grassland is dusty as well, the erosion of molar teeth by tiny, hard particles of silica was a serious matter. Evolution met the difficulty by modifying the molars. In horses these teeth developed a continuously growing folded structure (Fig. 16 - 8), like a miniature model of folded strata in the Appalachian Mountains (Fig. 15 - 2).

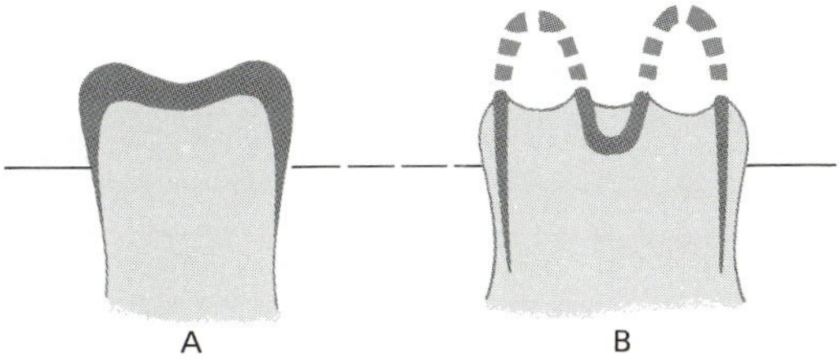

Figure 16 - 8. Molar teeth of mammals. A. Low-built molar of non-grazing mammal. B. Molar of grazing mammal, high-built and with fold structure. The softer materials are worn down most readily, leaving the hard enamel (black) standing as tiny stumps or ridges.

The folds included both a hard layer of enamel and softer materials adjacent to it. As chewing dusty grass eroded the teeth, the enamel was eroded least, and ridges formed, making the surface rough and well adapted for grinding up green food.

In contrast with the extreme specializations we have described, some mammals specialized very little. These were neither herbivores nor carnivores, but mammals that could and did eat almost anything that was handy. Their favorite environment was neither specifically forest nor specifically grassland, but any region that offered good eating in the form of insects, berries and other fruits, fish, and small mammals. They were searchers and wanderers. Typical living examples include bears, raccoons, some kinds of monkeys, and man. Mammals that lived this kind of life were not pressed by environment or diet into changing limbs or teeth. They have always retained the primitive flat foot (Fig. 16 - 5, D) and low-built molar teeth (Fig. 16 - 8).

Evolution of mammals—land, sea, and air

When fossil mammals are assembled in evolutionary lines, as in our parade of horses, the lines branch out into a tree much like our chart of the evolving dinosaurs. The mammal chart is a good deal more complicated, but Figure 16 - 9 is a stripped-down version that includes only the principal branches. Even this simplest version, however, makes it clear that most of the chief evolutionary lines branched off from their main stem in Eocene time. The story of evolutionary changes along the principal branches is far too detailed

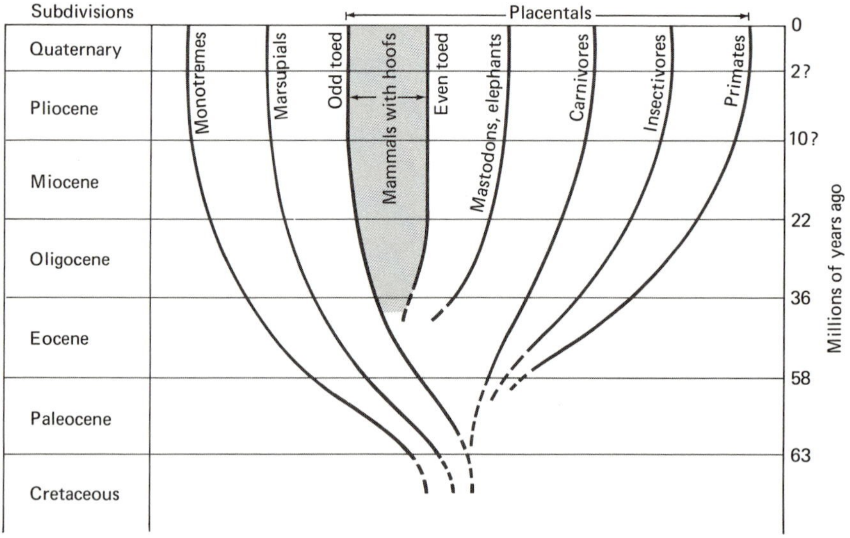

Figure 16 - 9. Evolutionary tree of mammals.

B. *Dinotherium*

A. *Platybelodon*

Figure 16 - 10. A few of the mammals with odd structures that proved unsuccessful.

A. *Platybelodon,* an elephant-like mammal with a mouth like a power shovel, for scooping up swamp vegetation. Miocene. B. *Dinotherium,* of elephant size, had a lower jaw and tusks that curved down. Miocene. C. *Moropus,* somewhat horse-like but larger than any horse. An herbivore with short hind legs. Its three-toed feet were armed with claws, possibly used for grubbing up roots. Miocene. D. *Brontotherium.* A browser more than 12 feet long. Oligocene. E. *Syndyoceras,* a deer relative, with more horns than seem necessary. Miocene. F. *Epigaulus,* a gopher with horns and extraordinarily long claws. Pliocene.

C. *Moropus*

F. Horned gopher

E. *Syndyoceras*

D. *Brontotherium*

to be recounted here. It is interesting, though, to see a few of the many curiosities, models that became extinct along the way, occupying dead-end branches not shown on our chart. Some of these are assembled in Figure 16 - 10. However, just because they are extinct and may look a bit strange to us, it would not be fair to them to say they were unsuccessful. The Model T Ford automobile is extinct, but in its day it was successful to a unique degree.

In contrast with the strange models shown in the figure and with the extreme evolutionary changes that took place in horses is the comparative stability of large carnivores. Early in Cenozoic time we find cats branching into two kinds, biting cats such as the lions, tigers, and domestic cats of today, and the more heavily built saber-tooth cats. In the latter (Fig. 16 - 11) the upper canine teeth are lengthened and flattened and the jaws articulated so that they can open wide. Leaping on its prey, the cat stabs deeply with its sabers and then pulls, ripping two long gashes to let out the blood.

Figure 16 - 11. In this scene from Oligocene days in western United States a saber-tooth cat (*Hoplophoneus*) is making a kill among primitive horses (*Mesohippus*). (Courtesy of the American Museum of Natural History.)

By Oligocene time both biters and stabbers had become so well adapted to their bloody ways of life that they changed little thereafter. The saber-tooth finally died out little more than ten thousand years ago, but simply because he is extinct we are not justified in classifying him as a failure. In the form of two or more genera he lived with little modification through about 30 million years. That is no dead end; it is evolutionary success. Indeed it is possible that the saber-tooth became extinct only because Stone-Age man with sophisticated weapons competed with him for food too successfully. More of this in Chapter 18.

As we have seen, Cenozoic mammals repeated the multiplication of kinds and numbers to a position of dominance, just as Mesozoic reptiles had done. Like Mesozoic reptiles they also succeeded in entering the air and the sea. A small tree-living mammal developed wings built of skinny membrane and started an evolutionary line of bats. The bat's wing is better engineered than a flying reptile's (Fig. 14 - 11) but it falls short of a bird's wing both in structure and in its lack of feathers. Perhaps this is the chief reason why bats have never achieved the success that birds have attained.

In the realm of the ocean, too, mammals repeated the story of the pushing, expanding horde of Mesozoic reptiles. No sooner had the Eocene Epoch got under way than land carnivores appeared in the sea, converted into primitive whales with sharp teeth. Probably they had passed through an intermediate phase of wading into shallow water in pursuit of fish. One of them later evolved into true whales, which diversified into the several kinds known today, as well as into walruses. The largest marine mammal, the living blue whale, is as long as the longest dinosaur ever was, and is more than twice as bulky.

Exactly as their reptile predecessors had done 200 million years before them, all the marine mammals evolved streamlined bodies and converted their limbs to paddles or flippers. Here again we see with startling distinctness the ability of environment to mold and remold the form and structure of living things. The lure of the Cenozoic seas did not end with whales. In Eocene time a plant-eating mammal, possibly an early and distant relative of elephants, waded in and joined the marine contingent, founding the line that led to modern sea cows and manatees. And to cap the story, in Miocene days a carnivore possibly related to cats joined the earlier sea mammals and evolved into the modern seals.

Long-distance migrations

The fossil record tells us clearly that mammals have migrated repeatedly from one continent to another. We find the same genus, or closely related genera, in two separate lands. How could this distribution have come about? Pollen is blown thousands of miles through the air, but pollen is microscopic in size. Insects are blown hundreds of miles over water. Seeds can drift across the Atlantic. Logs of trees have drifted long distances. A snake and a crocodile (both used to water) are known to have drifted hundreds of miles to islands. Polar bears walk hundreds of miles over the frozen surface of the sea. But these are all special cases, which cannot explain a fossil record that shows that dozens of kinds of mammals, ranging in size from mice to elephants, traveled from one continent to another.

The movement of crust plates, breaking up a continent and carrying its parts in different directions, does not solve this problem, because a wide Atlantic Ocean has existed since long before Cenozoic time. As for floating across a sea, possibly very small mammals could float on logs, but the only way to get most mammals from one continent to another (without the use of a fleet of arks much bigger than Noah's famous craft) is to walk them across a land connection, a land "bridge." At this point Figure 5 - 3, our map of the continents with their shelves, comes in handy. The map shows two things: (1) narrow land connections that exist today between the two Americas and between Europe and Africa, and (2) shallow places where, with a rather small change of sea level relative to land, two or more lands could have been connected (or almost connected, with islands between). Examples are Eurasia→North America via Siberia and Alaska, and Eurasia→East Indian Islands→Australia. Thus if we allow for a little movement of lands or change of sea level, we could provide either dry bridges or narrow bodies of shallow water to connect all the continents except Antarctica. (Remember we are still talking about only the last 63 million years.)

Alaska-Siberia bridge. A land connection between Alaska and Siberia would be re-created if the land were to rise or the sea to fall by only 150 feet. There is reason to believe that a broad, dry bridge existed there throughout the greater part of the Cenozoic. It carried a varied traffic of mammals that moved in both directions. This is why so many American and Eurasiatic fossil mammals are alike, and why a few indeed are nearly identical. We cannot credit those mammals with intelligence to plan and execute a long, one-way trip. They just drifted along in the direction of good, satisfying food and not too much competition in the eating of it. Insect pests made the trip too, on and off the sweaty bodies of the herbivores, while large cats and other carnivores played the part of assassins and plunderers along the highway.

The two Americas. The history of the Isthmus of Panama, which connects the two Americas today, was different, yet it profoundly affected the development of Cenozoic mammals. In Paleocene and early Eocene time the Isthmus was land. It permitted a number of the primitive, unspecialized kinds to filter into the southern continent. Then, later in the Eocene, it cut the intercontinental link by sinking into the sea and remaining submerged for almost fifty million years, until the end of the Pliocene. During that time the primitive mammals (Fig. 16 - 12), some of them marsupials, that had drifted into South America before the link was cut evolved slowly,

Figure 16 - 12. Exotic South American mammals shown in Argentina in Pleistocene time. Left: Ground sloths; Right: Glyptodons with tortoise-like protective shells. (Courtesy of the Field Museum of Natural History.)

protected against the invasion of the more advanced kinds that were developing elsewhere. But when the bridge was restored a lot of advanced, modern mammals (including the big cat called the jaguar) crowded in from the north and began competing with the more primitive occupants. One by one the local kinds became extinct, and our evidence suggests that the death toll was much increased by Stone-Age men.

Like the Alaska-Siberia bridge, the Isthmus of Panama carried two-way traffic, but its northbound traffic was not heavy. Armadillos, ground sloths, and glyptodons trickled into North America, and the first-named still live in Texas today.

The marsupials of Australia. Everyone knows the kangaroo as a national symbol of Australia. Kangaroos are marsupials, as indeed are almost all the rest of Australia's mammals. Their primitive character makes marsupials no match for placental mammals, which, had they been able to enter Australia in force, would probably have caused the extinction of most marsupials long ago. In North America, where marsupials and placentals once competed, all the marsupials that flourished earlier became extinct, except for the line that led to the opossum, the only marsupial that remains today.

In Australia marsupials obviously have fared much better, and they have fared better because competition from other land animals has amounted to very little. Apart from marsupials, Australian land animals today consist only of two kinds of monotremes, some placental rat-like rodents, bats, and small reptiles and amphibians. The almost complete absence of native placentals is best explained by this hypothesis, which although not proved is a very likely picture of what happened: The map (Fig. 5 - 3) shows a great deal of shallow water between Australia and nearby Asia, and islands are so numerous that the straits between them are not wide. It has been suggested that in very late Cretaceous or early Paleocene time, primitive marsupials arrived in Australia by "island hopping." That is, individuals were accidentally rafted across narrow straits between islands. They colonized an island, and in time another rafting accident occurred, and so on until marsupials reached Australia. In twenty or thirty million years a series of such accidents would be statistically almost certain to occur.

If we accept the suggestion that Australian marsupials are descended from island hoppers, we must admit that placental mammals should have been able to hop just as well if not better. Possibly therefore, when Australia became detached from the supposed former great continent shown in Figure 6 - 10, at some time before the Cretaceous, and drifted northward on a crust plate, early marsupials there were carried along too. Placental mammals had not yet evolved.

Regardless of which idea we may prefer, the result would have been the same: marsupials in Australia were isolated. So they were free to evolve along various paths without ruinous competition from more efficient placentals. Their evolution consisted mainly of gradual adaptation to the many different environments within the broad Australian land. Each environment molded marsupials into bodily forms that were adapted for particular life styles—forms that resemble those of placental mammals in similar environments. Thus the modern marsupials include the wombat (comparable with a North American placental groundhog), the bandicoot (placental rabbit), the phalanger (placental flying squirrel), the Tasmanian wolf (placental wolf), and the marsupial mouse (placental mouse). In these parallels we see once more the strong influence of environment on evolution, molding bodily forms and capabilities to fit ways of life in various surroundings regardless of the hereditary material with which such evolution started.

At times later in the Cenozoic the placental rat-like animal arrived, followed by bats and snakes. The bats could easily have been blown across the water in storms; probably the others arrived floating on trees or logs. They were not a serious competitive threat, and do not seem to have affected the marsupials in any radical way. Not until man, a hunter, arrived in comparatively recent time, accompanied by another hunter, the dog, was marsupial life affected. The arrival of European man in Australia has had catastrophic results. One by one, marsupials are disappearing.

One of European man's imports, the (placental) rabbit, throws an interesting light on the expansion of new races into empty or little-defended territory. In 1859 two dozen rabbits were brought to Australia from England. During the next 69 years the progeny of these rabbits spread over almost the entire area of Australia. An area of about one-third of the country, in which counts were made in 1928, was estimated to contain half a billion rabbits. Such a figure makes the expansion of reptiles in the Mesozoic Era more understandable.

Conclusion. The distribution of land mammals around the world is, then, very nearly what one would expect if there had been land or island bridges between northern continents, and no bridges, during the Cenozoic at least, extending southward from either Australia or South America. Those two continents are dead-end streets because they contain mostly dead-end faunas. Mammals drifted down them at times through a period of more than fifty million years, but could go no farther. No land bridges to the Antarctic Continent existed. Owing to its remoteness and its low temperatures, that continent received only a few stray visitors. Even today its population of higher animals consists only of seals and penguins, both of which could have arrived, like modern human explorers, without any bridge at all.

Meaning of mammal history

Although the Cenozoic history of mammals has not been nearly as long (thus far) as the Mesozoic history of reptiles, it possesses greater meaning for us. It is based on many more fossils and represents more different kinds of animals. The geography and environments of the Cenozoic world are recognizable in our terms, the Mesozoic ones much less so. Therefore the stage on which the action takes place is more familiar. Finally, mammals are our own kind, and as such they are inherently of more interest to us. In their development we see more clearly the structures and functions that exist in our own bodies.

In this chapter we have not emphasized structures and functions, nor have we even mentioned primates, the particular group of mammals to which we belong, except incidentally in Figure 16 - 9. Primates are so important for us that they deserve separate discussion. After we have discussed late-Cenozoic environments, and especially the climatic background of those environments, we shall turn to primates in Chapter 19, and see where they will lead us.

References

Fenton, C. L., and Fenton, M. A., The fossil book: Doubleday & Co., Inc., Garden City, N.Y. p. 375–435.

Romer, A. S., 1959, The vertebrate story: University of Chicago Press, p. 219–308.

Scott, W. B., 1937, History of the land mammals in the western hemisphere: The Macmillan Co., New York.

17

The Glacial Ages

Cenozoic change of climate

As the Cenozoic Era developed, the mammals described in the preceding chapter began to participate in an experience very different from anything known to have happened in Cretaceous time. That experience was the onset of colder climates. So to the several changes that affected the continents during the Cenozoic Era we must add another, a change in the prevailing climates. The Earth's lands became colder. The cooling was greatest in the polar regions and least in the equatorial regions, but it was worldwide. In broad terms the change was important, not only for mammals but for other living things as well. It will be easier to understand what happened if we begin by reviewing the chief kinds of evidence upon which we base our belief that temperatures changed, starting in the earlier part of Cenozoic time.

Evidence of the change. In particular we need to note three occurrences.

1. Fossil shells of microscopic invertebrate animals in layers of Cenozoic fine sediment brought up as cores (Chapter 5) from beneath the deep-sea floor. The shells are of kinds known to live in cold water, in contrast to the kinds found in layers below and above, which are characteristic of warmer water.
2. Grains of quartz sand having on their surfaces microscopic features caused by rough handling by glaciers. Such sand grains occur sprinkled liberally through certain layers in the fine sediment brought up in cores from the deep-sea floor around the Antarctic Continent. They must have been floated seaward in icebergs (large broken-off chunks of glacier ice) which as they melted dropped their content of sand grains down onto the floor of the ocean. Sand grains of this kind have been found in sea-floor layers as old as Eocene, and indicate that there were glaciers on the Antarctic Continent as early as that epoch. The layers in which they occur are the same as those that contain the cold-water kinds of fossil-invertebrate shells.
3. Fossil leaves embedded in certain Cenozoic strata exposed on land, leaves of plants that live in cold climates. Fossil plants characteristic of warmer climates are found in older and younger strata.

Here then are three kinds of evidence, all different, yet all pointing to one conclusion: lowered temperature in Cenozoic time, most conspicuously in the high latitudes of the Southern Hemisphere. From these and other kinds of evidence has been put together a curve (Fig. 17 - 1) that depicts the broad downs and ups of temperature through the Cenozoic Era. Except for the part near its right-hand end, the curve has been constructed entirely from the kinds of information we have just mentioned. It shows that the change was slow and gradual but not continuous.

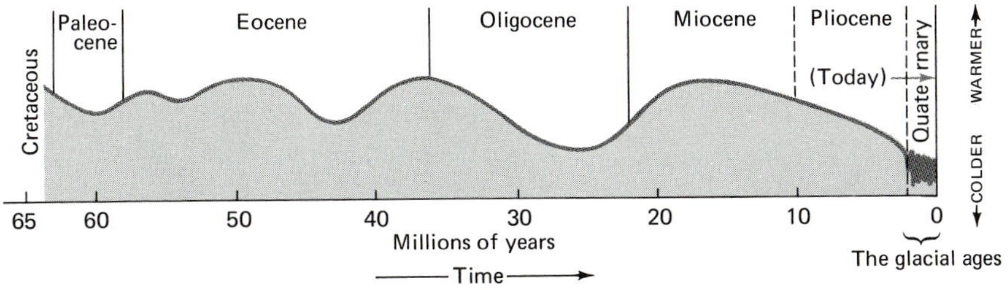

Figure 17 - 1. An attempt to suggest the fluctuation of temperature at the Earth's surface through the Cenozoic Era up to today. The curve cannot be accurate because it has been generalized for the whole Earth. It shows when some of the chief lows and highs occurred. If we possessed full information the curve would probably be seen to have many smaller undulations superposed on the large ones shown here.

Fluctuation of climate: glacial ages. The fact that change of climate was not continuous is striking. Temperatures fluctuated again and again, from warmer to cooler and back to warmer. The cooling was felt first in the Antarctic and then also in Alaska and other far-northern regions. But not until two million years ago, more or less, did the colder climate begin to affect middle latitudes greatly. When it did, the effect was obvious and decisive. Snow accumulated in those latitudes and built itself up into huge, thick glaciers that covered much of North America and northern Europe. Those comparatively recent times, when great sheets of ice derived from built-up snow flowed over middle-latitude regions, are what we are accustomed to call the *glacial ages,* and they are so labeled in Figure 17 - 1. Yet strictly speaking there were similar glacial ages in such regions as Antarctica and Alaska many millions of years earlier, as the figure shows. Those earlier glacial ages are less well known, not having been discovered until the 1960s, and it has not yet been decided just how to redefine the term *glacial age* so as to include them.

However, the question of definition is comparatively unimportant. What is important is that within Quaternary time alone there were several glacial ages, probably more than the few indicated roughly by the zigzag curve in our figure.

The latest glacial age. The latest of these glacial ages was surprisingly recent. It reached its peak barely 20,000 years ago when a thick sheet of ice, a great glacier, covered nearly all of Canada and much of the United States, reaching southward over the sites of

New York City, Chicago, and Seattle, while another glacier spread over Europe as far south as Copenhagen, Berlin, and Leningrad. The ice that lay over North America and Europe at that time covered a combined area of more than nine million square miles and may have been more than a mile thick, burying the tops of most or all of the mountains it overran. So its volume may possibly have amounted to 9,000,000 cubic miles of ice. Yet today the aggregate volume of glacier ice in the United States exclusive of Alaska is probably less than 20 cubic miles. This modern ice exists in the form of a thousand small glaciers in high mountains, most of them in Washington and Oregon. In Canada, however, the volume of ice existing today is much greater, perhaps 10,000 cubic miles, because Canada extends into the cold Arctic region and so offers better conditions for the survival of ice. But even 10,000 cubic miles is only a tiny fraction of the volume that existed in Canada 20,000 years ago.

When we think of the staggering volume of ice that formerly lay upon the Earth's lands—and so very recently—two basic questions come to mind. First, is a glacial age an event unique and peculiar to the Cenozoic Era? And second, what made such a thing happen? Before going further let us try to answer these questions as far as it is possible to do so.

Early glacial ages. First, then, have there been glacial ages in earlier geologic times, times long before the Cenozoic Era began? The answer is certainly *yes*. The evidence consists of bits and pieces—we will detail it presently—but it is unmistakable, and some of the bits and pieces are very big, being spread over wide areas. Evidence of a Permian glacial age is present on several continents (not impossibly these were all part of a single continent in those days), and we find traces of glaciers on the lands at various other times during the Paleozoic Era right back to its beginning, early in Cambrian time. Even in much older rocks, rocks dating from long before the Phanerozoic, we find marks made by glaciers and sediments deposited by glaciers. Some of these features are more than two billion years old, perhaps half the age of the Earth as a whole (Chapter 3). And who is to say that there were not still older glacial ages as yet undiscovered?

At any rate, considering only the glacial ages now known to us, events that span a length of more than two billion years, we must admit that they fit well into the Principle of Uniformity, the principle that says, as to geologic activity, that there is nothing new under the Sun. So apparently the icy events of 20,000 years ago—or of today in the Antarctic—are mere replays of similar events that have happened, somewhere or other, again and again during the Earth's span of existence.

So the first of our two questions is answered. A glacial age is not unique any more than the raising of a great mountain range is unique: it will happen whenever and wherever the right combination of conditions occurs. This answer makes it easier to deal with the second question: what makes an ice age happen? What we have to do is define "the right combination of conditions" and then understand what happens when that mix occurs.

What makes a glacial age happen?

The basic conditions. The answer to this question can be understood only in the light of a little knowledge of glaciers in general. In many regions in middle latitudes, such as the United States and Europe, some of the annual precipitation falls not as rain but as snow. Even in high mountains the snowfalls generally occur in the winter season. If winter temperatures are low enough, snow remains on the ground, but the warmth of spring and summer melts it and it disappears. However, in very high mountains such as the northern Rocky Mountains, temperatures are so low even in summer that patches of snow remain through the summer season, and subsequent winter snowfalls add to the snow already there. The snow on a mountain slope, added to year after year, becomes compacted and is pulled upon by the force of gravity. Under the pull it spreads downslope with a creeping or flowing motion, because snow consists of ice, a weak substance that yields easily to the pull of gravity. By the act of flowing, the compact snow has become a glacier. Now if the snowfall is sufficient, and if the temperature is low enough to keep it from melting away, a glacier can take on a tongue-like shape and can continue to grow longer, flowing down a mountain valley somewhat as a stream of water flows down a valley, although of course much more slowly.

The snow that lies on a steeply sloping house roof after a winter storm, forming lobes and wrinkles as it creeps down the slope of the roof a little each day, is essentially a small, thin glacier. Similar but much bigger lobe-like tongues of ice are seen by the hundreds at close range in mountains such as the Alps. Glaciers in adjacent valleys merge where the valleys join each other. And at the foot of a mountain range all this ice creeping slowly down through the valleys can merge at the base of the mountains and spread out as a single, continuous sheet of ice. What is to prevent its creeping on indefinitely? Just one thing, and an effective one: melting. The lower we go in altitude or in latitude, the higher the temperature rises. Sooner or later the temperature at the outer edge of our creeping glacier will become high enough—just high enough—to melt all the ice brought to the edge by creeping flow. There the outer margin can advance no farther. True, the creeping flow continues, but the ice brought to the margin melts just as fast as it arrives there. The melted ice flows away as streams of water.

This is the condition of the tongue-like glaciers commonly seen by tourists in the Alps, the Canadian Rockies, and other mountain regions. Such glaciers occupy mountain valleys, and the positions of their downstream ends are determined by the balance between rate of creeping flow and rate of melting. Under the climates that now prevail, the glaciers don't change much. But lower the Earth's surface temperature only a little, and they would all begin to lengthen. Lower the temperature more, enough more, and we would have another glacial age with half of North America made uninhabitable for man and most other animals.

The point of all this is that a glacial age is nothing more than the logical result of the Earth's temperature falling by a very few degrees. The mystery about a glacial age is not the snow and ice; it is the cause of the low temperature. As long as the Principle of Uniformity is at work and as long as the water cycle operates, there will be snow and ice in the coldest places. It is only when the Earth's temperature falls to a low enough point so that rainfall becomes snowfall over wide regions, summers become chilly, and melting diminishes, that we get a glacial age.

The balance is a very sensitive one. We are not as free today from the danger of a glacial age as many people think. A considered estimate based on year-to-year weather conditions in the mountains of southern Norway, the ski-resort country between Oslo and Bergen, indicates that if average annual temperatures there were lowered by only as much as 5.5°F over a long period, the resulting change in the glaciers would be sufficient to start another glacial age in Europe. Indeed much of the ice that spread over northwestern Europe to a maximum about 20,000 years ago came mainly from snowfall in those very mountains of southern Norway. It was augmented of course by snow that fell over a much larger area of ice, once the glacial age had got well started, had "caught on," like a snowball rolling downhill.

Clearly, then, the state of health of a glacier depends mainly on climate. Where temperatures are high there are no glaciers. Where temperatures are low, glaciers form but spread only as far as the line where melting balances inflow of ice. It follows that a glacial age, in which glaciers are very big and very abundant, is a time of low temperature, a time therefore when precipitation falls in the form of snow. The natural result is that the line along which melting balances inflow moves far into lower latitudes, permitting glaciers to cover vast regions. After the peak of a glacial age, as temperatures grow warmer, the critical line moves back into higher latitudes, glaciers shrink, and the glacial age ends.

Today we are far past—20,000 years past—the peak of the latest glacial age. Most of the ice, some 9 million cubic miles in volume, that was right here 20,000 years ago has melted away, and the meltwater has run off through streams into the sea. But even today, 20,000 years past the peak of that cold time, ice still remains in places high enough and cold enough to prevent its melting. Even today there are still more than a thousand glaciers in the United States exclusive of Alaska, and there are more than twelve hundred in the Alps alone. In Greenland there is still one big glacier that covers most of that land, 1500 miles long and 500 miles wide. The volume of ice in Greenland alone adds up to nearly 800,000 cubic miles, by far the biggest chunk of ice in the Northern Hemisphere; and all of it was created by ancient snowfalls that have not yet melted away.

When we turn to the Southern Hemisphere, we see the Antarctic Continent right in the center of it, squarely astride the South Pole. The ice on that continent dwarfs the big chunk of Greenland ice. Its volume is nearly five million cubic miles—more than 90 percent of all the ice in the world and more than 75 percent of all the world's fresh water, both liquid and frozen. It covers an entire continent to an extent nearly one-third greater than the area of the whole United

States including Alaska. So it is true that Antarctica has not emerged from a glacial age in any visible way, as North America has done. Ice still nearly covers it, although 20,000 years ago ice may have covered it even more extensively. Ice has overwhelmed North America again and again. It has come and gone, but as far as we can tell, an ice sheet has blanketed the Antarctic Continent continuously through at least the last ten million years. It has swelled and shrunk as the climate fluctuated, but it seems never to have disappeared like the ice sheets over North America and Europe. The reason for this different behavior is obvious. The Antarctic is the world's highest continent (that is, its average altitude is highest). Moreover it is centered on the South Pole, where temperatures are always very low. All the precipitation falls as snow and melting does not occur. So the ice, once formed, is maintained not only from year to year but throughout millions of years. It creeps downward and outward across the continent it covers, like a huge stiff mass of pancake batter on a griddle. Reaching the coast and entering the sea, it breaks up into huge pieces, forming broad, flat-topped icebergs. A few of the largest bergs have been measured and have been found to be enormous. One of them proved to possess more than twice the area of the state of Connecticut. Once reduced to bergs floating in the sea, the ice gradually melts away, but there is always more ice creeping seaward across the surface of the Antarctic Continent.

Pulsation. Summarizing the basic conditions necessary for the development of glaciers, we need only have high enough lands, or lands in high enough latitudes, so that temperatures will be low enough to maintain snow from year to year. As we have seen, high lands result from movements of crust plates and from collisions between continents. High lands form from time to time as the movements continue. But movements of this kind are very slow; the movements of crust plates are measured in terms of centimeters per year. If the movement of crust plates and the making of new mountains were the *only* events that cause glacial ages, it would be impossible for a glacial age to disappear in less than twenty thousand years, as we know has happened. If the movement of crust plates were the whole story, once ice had formed and spread over a large part of a continent there would be nothing to prevent its staying there through millions of years, until the mountains were slowly worn away by erosion or until the continent, afloat in its crust plate, was very slowly carried into a warmer latitude where the blanket of ice could melt.

Glacial ages, at least those in middle latitudes, begin and end far too quickly to be explained by such a slow and clumsy process. The changes happen not in millions of years but in mere thousands. Thanks to a wealth of radiocarbon dates it has been possible to construct a crude but fairly good calendar of the melting away of the great mass of ice that covered so much of North America only 20,000 years ago. The process of destruction began roughly 15,000 years ago and was complete by about 6000 years ago. In other words, the complete melting of that huge blanket of ice took only about 9000 years (Fig. 17 - 2). Perhaps more than nine million cubic miles of ice was converted into water, which flowed into the nearest rivers and so into the ocean.

Not only did the process take only 9000 years, but also the early part of the process was marked by a few reversals during which the ice thickened and spread outward again before resuming its wasting away. Such reversals occurred in Europe, in North America, and in New Zealand at approximately the same times. The obvious conclusion is that there exists a *second cause* of climatic change, one that operates quickly and simultaneously throughout the world and that is independent of the making of mountains and the movement of crust plates.

Many attempts have been made to identify this cause. Several theories have been suggested, but no single one is accepted by all the scientists who have studied the problem. We shall have to content ourselves with a single theory, one that explains the facts but that is itself unproved. That theory proposes that the heat energy received at the Earth's surface from the Sun varies in a slowly pulsating way, with the result that tempertures continually fluctuate through a rather small range. The idea is simple enough, but the means of proving whether it is right or wrong still escapes us. If we accepted it for want of a better idea, we could argue that in a time (say Cretaceous) of low lands and widespread seas, few if any glaciers could exist in the world and so the supposed slow pulsations of heat at the Earth's surface could have only a small effect on climate. But in a time (say Cenozoic) of high lands and many mountains, and with a good share of the continents lying in rather high latitudes, many glaciers could exist in high places. In such a situation, a pulsation sufficient to lower temperatures only a little might in-

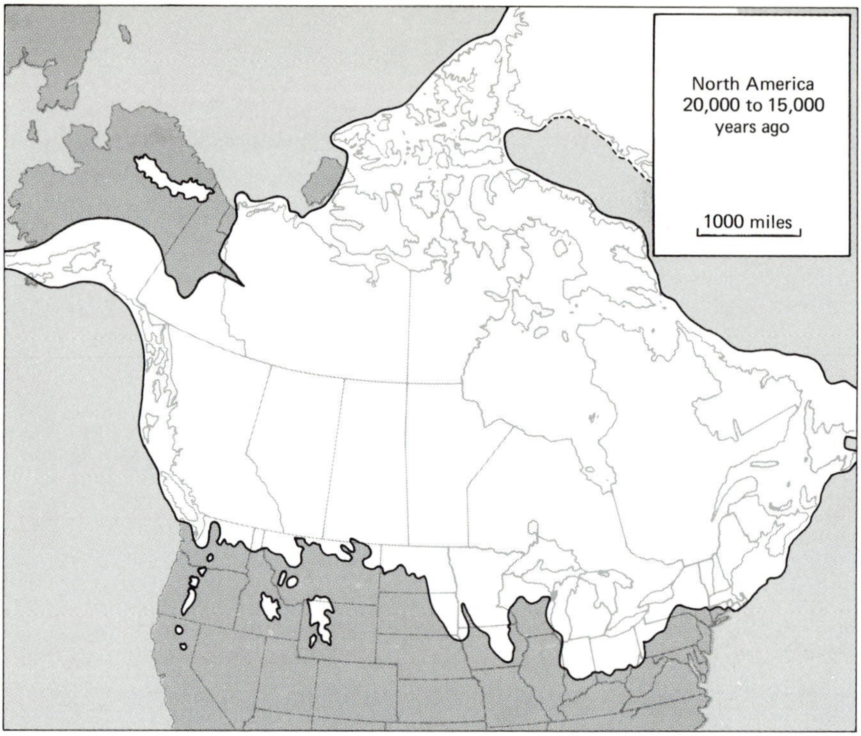

A

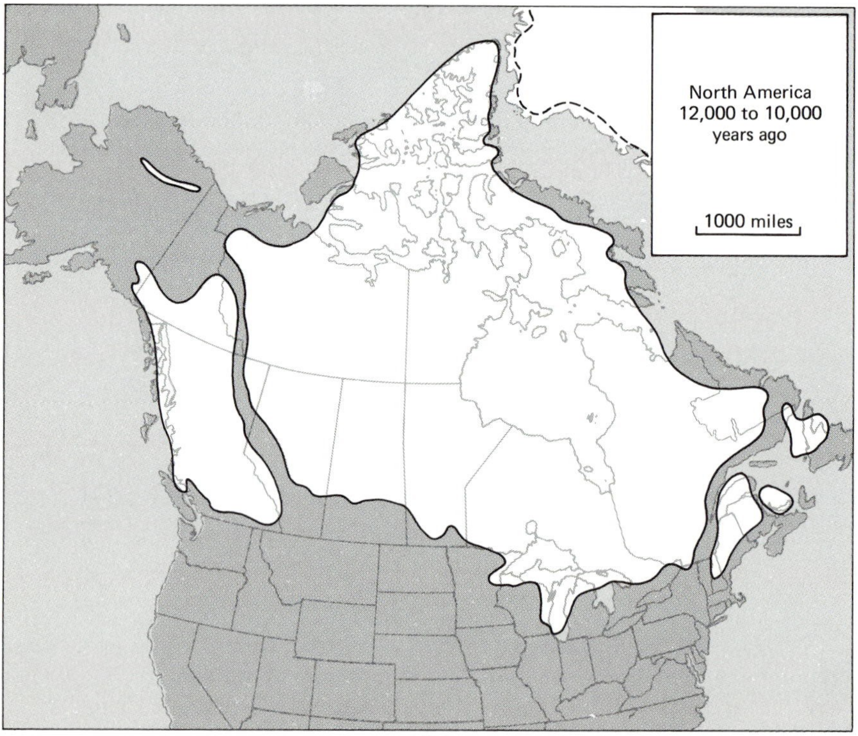

B

Figure 17 - 2. Motion picture of the melting of glaciers from North America as the latest glacial age ended. Small details are omitted. (Data mainly from Canada Geological Survey.) A. North America 20,000 to 15,000 years ago.

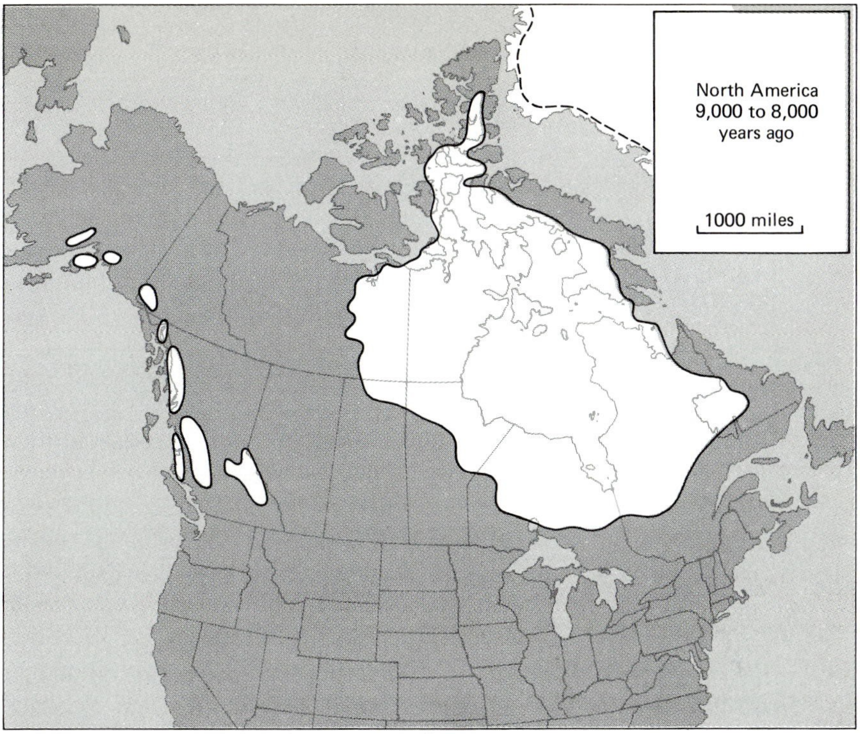

C

D

B. About 12,000 to 10,000 years ago. C. About 9,000 years ago. D. About 7,000 years ago.

crease the size of glaciers in a catastrophic way. Conversely, a slight general rise in temperature might have the opposite, but equally catastrophic, effect. Here, for the present, the matter rests.

Effects of glaciers on the land

Erosion. One of the principal reasons why it is possible to make a map showing the extent of former glaciers is that creeping ice makes noticeable changes on the land over which it flows. Ice rasps, files, and otherwise erodes the surface. Then it deposits the rock waste it has acquired by erosion. In consequence, what is frequently seen is an eroded surface, abruptly overlain by a deposit of rock waste. Both the surface and the deposit are distinctive and in most cases are easily recognizable as marking the former presence of a glacier.

Bits of rock of all sizes picked up by the creeping ice become frozen into the under surface of the glacier. These act like the sand on a sheet of sandpaper, gouging and scraping the rock floor beneath, and leaving many discontinuous parallel scratches and grooves on its rocky bed (Fig. 17 - 3) that are unlike any marks made by a

Figure 17 - 3. Glacial scratches and grooves on surface of sandstone bedrock. The surrounding rubble was left by the glacier, which was flowing directly away from the position of the camera. Southern Connecticut. (R. F. Flint.)

stream of water. In other places chunks of bedrock are pulled out of place along cracks and carried away, adding to the load of debris in the base of the glacier.

Deposition. The rock debris in the ice is strung out and deposited along the way, forming sediment that in places near the outer edge of the glacier can be very thick. Being a solid, ice does not deposit its load as a river does. A river drops its sediment in the order of decreasing diameters of the particles (Chapter 4). But the rock waste in the base of a glacier becomes stranded on the rocky floor beneath, just as it was carried in the ice, with coarse and fine pieces side by side, boulders next to silt particles, without sorting or stratification (Fig. 17 - 4). The result often looks like rubble pushed together by a

Figure 17 - 4. A deposit of rock debris made during the latest glacial age. The individual fragments are of many different sizes and are not sorted, not stratified, and not rounded. These characteristics distinguish the deposit from one made by water. Pick handle is 18 inches long. North slope of Mt. Rainier, Washington (the volcanic cone shown in Figure 15 - 11). (D. R. Crandell, U.S. Geol. Survey.)

Figure 17 - 5. Six pebbles taken at random from a glacial deposit in New York State. Each has one or more flat sides made by grinding.

bulldozer. Also, instead of being rounded as a stream would round them by turning them over and over, the pieces of rock in a glacial deposit have irregular shapes and include flat facets acquired by grinding as they scraped, firmly held in the base of the glacier, across a rock surface (Fig. 17 - 5).

In some places along and near the outer margin of a glacier the deposited rubble is moved by water that runs along the base of the ice as the glacier melts. In such places the material loses its distinctive glacial character and becomes sorted and stratified by the running water. In this way layers of stratified sediment become sandwiched between bodies of rubble in a mixed-up, confused arrangement.

Regardless of the proportion of stratified sediment it may contain, the glacial deposit as a whole tends to form a ridge-like heap, large or small, along the glacier margin. Such a ridge is an *end moraine,* a distinctive feature of glaciation. In some regions many moraines lie one behind the other, each marking the position of the glacier margin at the time of the deposition of the moraine.

Streams of meltwater that flow outward from an end moraine at the margin of a glacier deposit gravel and sand that has been sorted and stratified in typical stream fashion, in the valleys in which they flow. Some of these deposits are 100 feet or more in thickness and are as wide as the valleys that contain them. Much gravel and sand seen along rivers such as the Ohio and the Mississippi has this glacial origin, and is distinctive even as far downstream as the Mississippi Delta. Yet despite the great volume of such deposits plus the volume of glacial debris spread over the glaciated country farther north, the aggregate thickness of regolith and bedrock ground up and moved away by the great ice sheets that formerly overran North America and Europe is surprisingly small. Although we cannot be exact about it, we might guess that the average thickness of rock material removed is perhaps no more than 25 feet.

Lake basins. A more obvious effect of glaciers, and in particular great ice sheets, on the landscape is the creation of a great number of small and large basins, many of which fill with water and become lakes. Any good large-scale map of Canada and the United States or of northern Europe shows at a glance that most of the lakes are clustered in the glaciated regions. In North America alone the number of lakes runs into hundreds of thousands.

The basins are formed by glaciers in several ways. Some are gouged out of cracked and fractured bedrock as the glacier flows across it. Others are nothing more than the low parts of the surface of an irregular deposit of glacial sediment. Still others are river valleys that have been dammed by glacial sediments. (The Great Lakes are at least partly of this origin.) And many small basins are made by blocks of glacier ice, from a hundred yards or so to more than ten miles in diameter, that were buried in sediment deposited by or on the glacier. When such a block of ice melts away, it creates a basin into which any sediment that had covered the ice block collapses. Many of the thousands of lakes in Minnesota occupy the sites of former blocks of ice.

Smaller pulsations of climate

Climate since 1800. The thermometers maintained by government agencies in most countries provide us with a record of what world temperature has been doing since the beginning of the 19th Century. The story is condensed in the curve seen in Figure 17-6. It tells us that within the last hundred years the world's average annual temperature increased irregularly by more than one degree Fahrenheit.

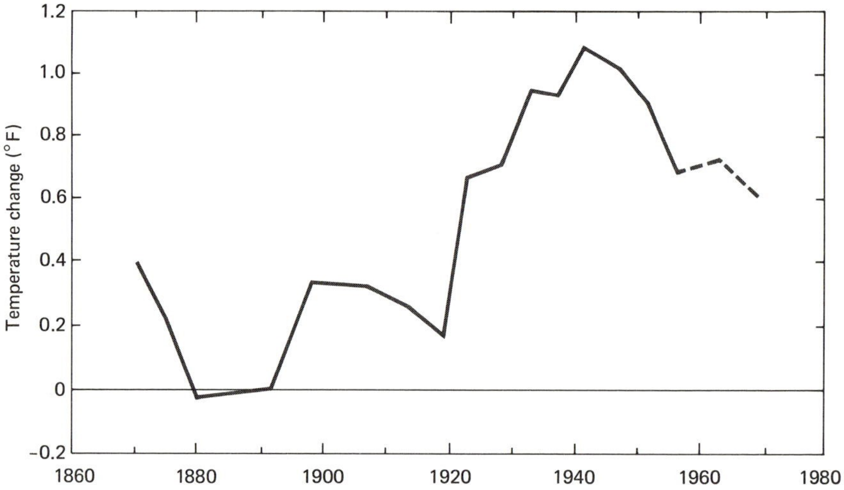

Figure 17 - 6. Curve showing change of temperature (5-year means) in the Northern Hemisphere from 1870 to 1970. (After J. M. Mitchell and C. H. Reitan.)

The change affected most parts of the world, including the tropics and high latitudes and northern and southern hemispheres. Then, just after 1940, the trend reversed and a cooling-off period began. Temperatures became lower, so that by 1970 they were back at the point they had crossed at about 1920. Here we have proof that the Earth's climates are not fixed and unvarying, but are changing significantly. The warm winters and hot summers of the "Dust Bowl" years in the 1930s in North America are seen as part of a generally warming climate of wide extent and, in terms of human affairs, long duration.

Not surprisingly, the record of the sizes of small glaciers in mountains in North America and in the Alps is rather similar to the temperature curve in Figure 17 - 6. Measurements made on the same glaciers year after year show that between the later part of the 19th Century and the middle of the 20th, many glaciers generally lost more by melting than they gained by snowfall, and so generally shrank to smaller size. But since about 1950 some glaciers have again begun to increase in size. They seem to be reflecting the reversal seen in the temperature curve, but it is too soon to say whether they are on a new trend.

Climate through the last 1000 years. Weather records measured by thermometer do not exist for the time much before the 18th Century, but in a number of indirect ways it has been possible to form a good general idea of broad fluctuations of temperature in Europe and also in Japan during the last thousand years. Evidence of several kinds shows that from about the 11th to about the 13th Century the climate was warmer than it has been at any time since. This was the Viking period, when summers were so warm and dry and when northern seas were so free of floating ice that Norsemen were able to sail far and wide in small open boats. They even established colonies of 3000 people or more in southern Greenland, from which agricultural products were traded to Europe. But after about 1300 the climate began to get colder and floating ice returned and blocked shipping. After about the year 1500 trading ceased and there was little or no communication with Europe. The colonies were cut off. In the 18th Century a ship did arrive, but found no survivors of the once-flourishing colonists.

A 20th-Century archeological study of 100 churchyard burials in one of the colonies revealed part of the later history of the colonial venture. The soil in the burial ground was solidly frozen, as it is today in much of the Arctic region, although obviously it had not been frozen at the times of burial. The remains were of people who were very young, suggesting short life expectancy, and of small stature, suggesting poor diet implied also by malformation of the skeletons and by most unusual wearing down of the teeth. All in all, it seems likely that the people died of disease, hunger, and other causes brought about by a long period of gradually worsening climate.

The general decline of temperature from the Viking period to the 17th Century was widely felt in Europe. In Norway and in the Alps, mountain farms were abandoned before advancing glaciers. The upper limit of trees gradually retreated down mountainsides in the Alps, and high vineyards in Germany ceased to produce and had to be abandoned. Winters were longer and colder. Anyone who has looked closely at 17th Century Dutch landscape paintings will recall that many of them are winter scenes showing people skating on frozen canals. Such scenes have not been witnessed very often in the present century.

In summary, the climatic record of the last thousand years includes an early Viking period, warmer than today, and a later cold period, colder than today. The warming that occurred through the first half of the present century was a recovery from that very cold time. As a whole, the record clearly shows that climate is not fixed. It fluctuates.

The last 10,000 years. In Sweden and Finland, as in other high-latitude countries, the vegetation grows in distinct zones (recall the map, Fig. 10 - 2) that are defined mainly by temperature. Those countries are dotted with lake basins created by the former great glaciers in the ways we have described. Nearly all the basins are less than 15,000 and many are less than 10,000 years old (Fig. 17 - 2). Some of the lakes have become entirely filled with sediment, mostly plant remains in the form of peat, and so have become converted into bogs. Others are still lakes but are being slowly filled with peat. The sediment includes not only stems and leaves but also great quantities of pollen discharged each season, like thin greenish smoke, by the surrounding plants.

Scientists thought that by drilling a core out of the full thickness of peat in a bog or lake and identifying the kinds of plants present in each layer or stratum, they could reconstruct in detail the generations of vegetation that had lived around the lake (Fig. 17 - 7). The change in the vegetation from one layer to the next should reflect the change of climate that had begun during the melting of the glacier. They expected that the kinds of plants would change from tundra at the bottom (representing the arctic shrubs and grasses that grew while the glacier was still nearby) through a smooth transition to today's kinds of trees at the top.

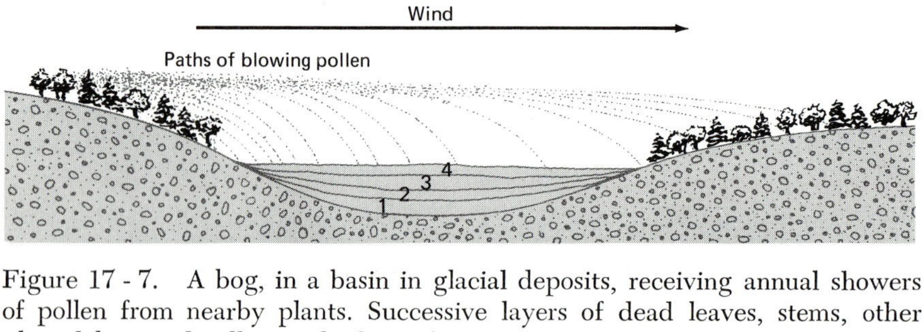

Figure 17 - 7. A bog, in a basin in glacial deposits, receiving annual showers of pollen from nearby plants. Successive layers of dead leaves, stems, other plant debris, and pollen are built up, forming peat.

They found and identified the fossil plants (chiefly in the form of pollen), but the change from bottom to top was a surprise. It started with tundra and went through fir and spruce forest into birch and pine forest, and on to oak, beech, alder, and hazel, all together showing a gradual warming. But higher up, in the upper strata, the plants changed back to birch and pine, which are the chief trees growing in the neighborhood today. The oak, beech, and hazel are no longer there; they now grow farther south. But radiocarbon dating of the oak-beech-hazel layer showed that that layer was formed about 5000 years ago.

Evidently, then, the climate had warmed to a maximum by about 5000 years ago (about 3000 B.C.). At that time average temperatures were higher by about 2°F than they are in the same places today. Thereafter the trend reversed, the climate became colder and wetter, and the oak trees around our bog died away and were replaced by birch and pine. Here once more we have sound evidence that climates fluctuate, and that far from warming steadily since the great ice sheets started to melt, the climate was warmer and drier 5000 years ago than it is today. Glaciers in the Alps and Rocky Mountains were then much fewer and smaller than they are now. Many of those we see today did not begin to form until less than 5000 years ago, and so are "modern" glaciers rather than relicts of the ice bodies of the latest great glacial age.

The future

Scientists whose work involves the history of climate are often asked two questions. The first is, "Will there be another glacial age?" and the second is, "If so, when will it come?" The answer to the first question is the easier one. Most scientists would agree that that answer is "Yes, probably," because several glacial ages have occurred within the last couple of million years, and the basic conditions that favor them—high lands, many mountains, and the presence of a great ice sheet at the South Pole—are still with us.

The answer to the second question is much more elusive. Our climatic information is not yet accurate enough to tell us whether glacial ages occur in some recognizable pattern of time intervals. *If* we knew that they did, and *if* we could then measure the time intervals between glacial ages of the past with real accuracy, it should then become possible to predict what the climatic future holds for us. Perhaps prediction will become possible, but it is not possible at present.

References

Flint, R. F., 1971, Glacial and Quaternary geology: John Wiley & Sons, New York.

Hovgaard, William, 1925, The Norsemen in Greenland: Geogr. Rev., v. 15, p. 605–616.

Lamb, H. H., 1965, The early medieval warm epoch and its sequel: Palaeogeography, Palaeoclimatology, Palaeoecology, v. 1, p. 13–37.

Post, Austin, and LaChapelle, E. R., 1971, Glacier ice: The Mountaineers: University of Washington Press, Seattle.

Schwarzbach, Martin, 1963, Climates of the past: D. Van Nostrand Company, Princeton, N.J.

18

Glacial-Age Environments

Shifting populations

Of course glacial ages were times of stress and change for living things. Climates were growing colder, glaciers were forming and spreading, and plants and animals were on the move, drifting toward better environments and competing actively for living space. It was like a slow-motion streaming away of refugees from a battle zone toward safer places, but instead of one individual making the entire journey in days or weeks, it was more likely his distant descendant who, many generations and hundreds or even thousands of years later, reached a place where he and his kind could settle without having to move still farther on.

Plants and animals that lived high in mountains moved little by little down the mountainsides toward environments that were more like what they had been accustomed to before the glacial age began. Those which lived in the far north had to shift their living places southward through many hundreds of miles to avoid being overwhelmed by ice. The margins of the huge ice sheets covered up the land at rates that sometimes greatly exceeded 100 feet per year. A few mammals, as we shall see, compromised with the glacial conditions and evolved woolly coats and protective blubber that enabled them to live in the coldest places, but this process was slow and needed the stimulus of two or three glacial ages before it could be fully accomplished.

If they moved far enough most organisms could find territory at least something like what they had been accustomed to, but it was never quite the same, if only because of the competition between natives who had not moved away and newcomers, refugees who were really interlopers. If evolution were not a slow process compared with the rather short duration of a glacial age, many new species might have resulted from the displacements and the competition, instead of the rather modest evolutionary changes we find in the fossil record. Of course, what we know about glacial-age environments we have learned from fossils, just as we have learned about the earlier history of life from them.

North American environments

In building up a map of environments during the latest of the glacial ages we define former life zones on a basis of the plant and animal fossils we have available. Although they are fairly abundant, those fossils have been collected from a comparatively small number of localities—bogs and swamps, caverns, alluvium, and other places in which remains of land organisms have accumulated and have been preserved. In between these localities we must base our map of the environments on logical guesses, which must be gradually improved and refined.

The region south of the ice sheet. The most dramatic element on the map of 15,000 to 20,000 years ago (Fig. 18-1) is the southern margin of the great ice sheet, stretching across the continent for more than two thousand miles between the base of the Rocky Mountains and the Atlantic Ocean. The border of the ice sheet was like a huge broad ramp, in wintertime covered with white snow but in summertime showing dirty, grayish ice. It sloped upward to a height of 5000 feet in perhaps the first 50 to 60 miles northward from its edge and onward, at a decreasing rate, to perhaps 10,000 feet. Small, shallow streams of meltwater flowed swiftly down the ramp in summer, in some places forming lakes at the edge of the ice and in other places flowing down the nearest valleys that led away from the ice sheet. Some of the biggest of these valleys are sketched on the map. These streams of glacial meltwater deposited such great quantities of gravel and sand on the floors of their valleys that the sediment covered up new vegetation as fast as it could become established. So, near the ice sheet at least, the valley floors were wastes of bare gravel and sand traversed by networks of interlacing stream channels. The valleys themselves were longer than the same valleys are today, because, as we shall see, the Atlantic coast then lay far seaward of its present position.

Apart from the bare valley floors, the country adjacent to the edge of the ice sheet was carpeted with tundra and sparse clumps of spruce trees. The tundra soil was frozen all winter and was extremely wet in summer. In early summer it was probably the home of clouds of mosquitoes. In the winter season the margin of the ice sheet enjoyed a good deal of bright, cold weather between the rather frequent heavy snowfalls. In summer the climate was hardly good because of persistent cloudy weather and frequent falls of rain or snow. This rather unattractive climate resulted from the presence of the ice sheet, which we might almost say brought its own bad weather with it. The ice sheet, nearly as high as the Rocky Mountains and much colder to boot, was a great topographic barrier to the comparatively warm and moist masses of air that flowed northward and eastward from the Gulf of Mexico and to a lesser extent eastward from the Pacific. The edge of the ice sheet was a boundary, like the boundaries we mentioned in Chapter 1, and as such it created conflict. Encountering the ice barrier, the warm air was forced to rise over it, and in doing so became chilled; its content of moisture

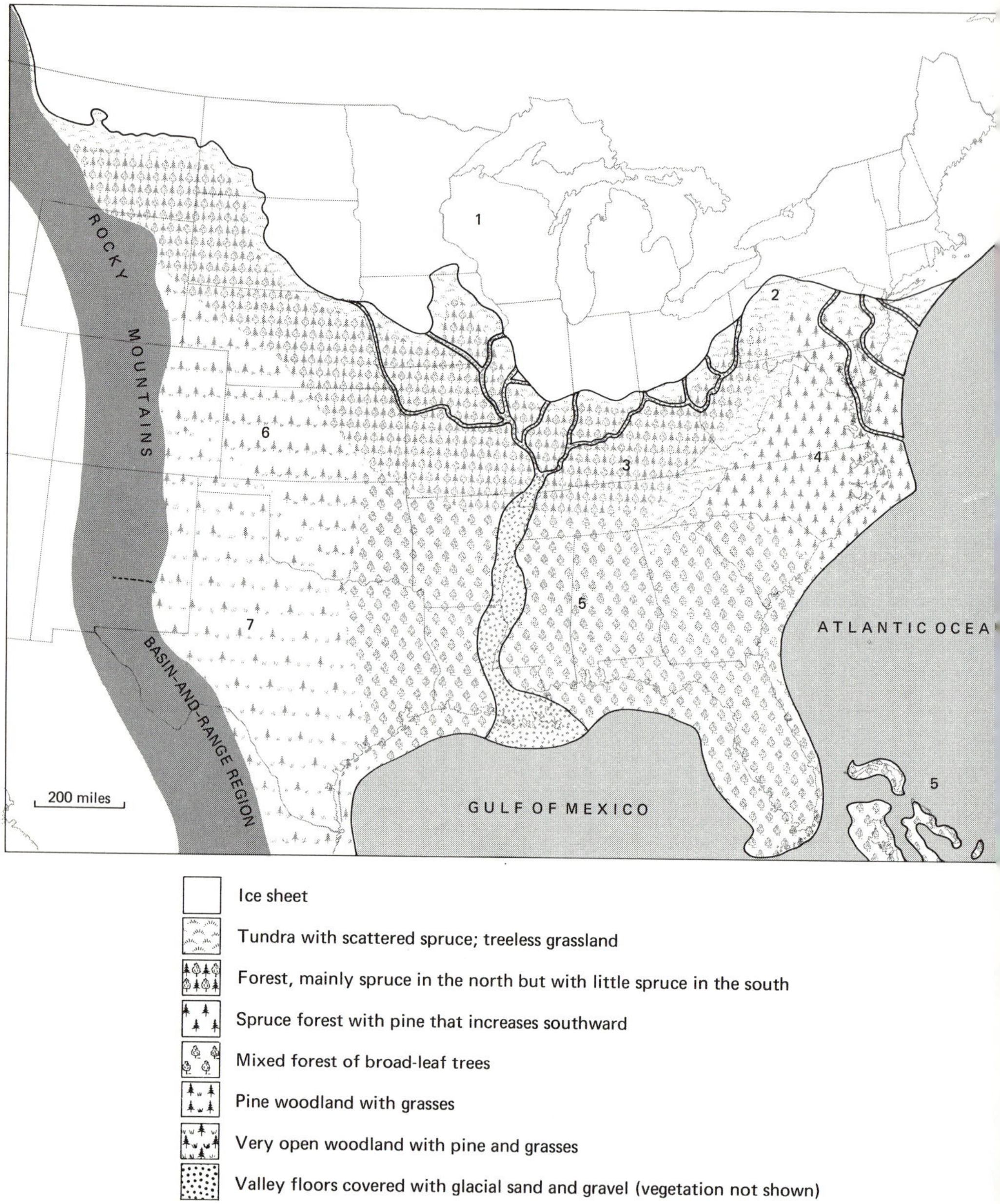

Figure 18 - 1. United States and southern Canada east of the Rocky Mountains, as the terrain may have looked between 20,000 and 15,000 years ago. The boundaries between the various belts of vegetation are not accurate, particularly toward the south and southwest.

condensed, created clouds, and fell as rain or snow on the ice sheet and on the belt of country adjacent to it. Some of the storms first hit the ice sheet in areas west of the Mississippi River, and then, deflected by the thick mass of ice, moved eastward along the ice, dumping rain or snow continuously along the way, even as far east as Newfoundland.

Returning to the narrow belt of tundra, we can see that it forms a long narrow prong extending down the length of the Appalachians. The comparatively high altitude of the mountains and ridge tops induced cold climate along them even in summer.

South of the general belt of tundra the scattered clumps of spruce merged to form a forest of northern type, stretching away to the south with spruce gradually decreasing in favor of broad-leaf trees. East of the Appalachians the principal trees other than spruce were pine. Still farther south these northern forests merged imperceptibly into forests of more temperate type, consisting mostly of broad-leaf trees. Such forests extended from Georgia and Florida westward, merging somewhere west of the Mississippi River into woodland. Woodland differs from forest in that the trees grow far apart, with grasses and other low-growing plants between them. It covers considerable areas in southeastern Montana today, and develops generally where the climate is somewhat too dry to support close-growing trees.

If we look again at Figure 18-1 and view the overall pattern of glacial-age vegetation we have just described, we can compare it with today's pattern, seen in Figure 10-2. The difference clearly reflects the lower temperatures of glacial time, when the various belts of vegetation were pushed southward and changed in width. Where today the Great Plains are natural grassland with few trees other than those planted artificially as windbreaks, the glacial-age vegetation was probably forest in the north and pine woodland farther south, with the trees becoming sparser southward into Mexico. Along the immediate banks of rivers were "gallery forests," strings of trees whose roots could reach water at all times of the year. Such string-like forests grow along riverbanks on the dry Plains even today because the ever-present water makes them possible.

The western region; pluvial lakes. Our map of glacial-age vegetation stops at the Rocky Mountains, partly because information on the terrain farther west is particularly scanty and partly because the

Figure 18 - 2. Part of Basin-and-Range Region and surrounding territory, showing (A) existing lakes, mostly salty and shallow, and (B) lakes that existed 20,000 to 15,000 years ago. The greater number of lakes in the northern part of the region results from the less-hot climate there.

many mountain ranges make it impossible to show what is known on a small-scale map. In general the belts of vegetation, very narrow because the slopes are steep, simply moved down the mountainsides through a vertical distance of at least 2000 feet, and in some areas perhaps considerably more.

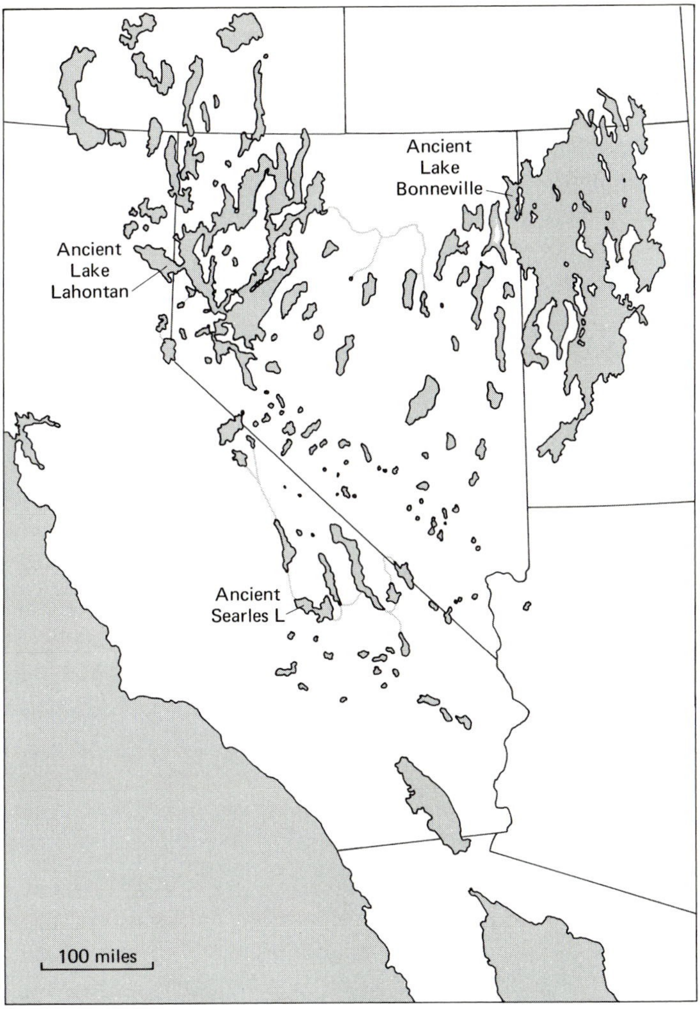

B

A striking change, however, did occur in the Basin-and-Range Region. A large proportion of the basins (which, as we remember, lack outlets) became filled or partly filled with water and formed lakes (Fig. 18-2). The largest, Ancient Lake Bonneville, was more than 1000 feet deep and compared in volume of water with modern Lake Michigan. While the lakes existed, beaches and deltas were formed and thick bottom sediments containing fossil fishes and other organisms were deposited. These features were exposed as the water

Figure 18 - 3. Searles Dry Lake, California. View from the air looking southeast across the white salty flat that is today's dry lake. The dotted line is at the position of the highest shoreline, 650 feet above the salty flat, of Ancient Searles Lake. When the lake surface stood at that shoreline, the water spilled out through an outlet channel, now dry, into the basin seen in the left distance, where it formed another, even deeper lake. (American Potash & Chemical Corporation.)

evaporated and the lakes dried up; in today's desert conditions they are an astonishing sight.

Figure 18 - 3 shows a typical basin in the Mojave Desert region in southeastern California. The central area, dry, white with salt, and almost perfectly flat, embraces about 12 square miles. Very little rain (6 to 10 inches annually) falls there, and summer temperatures are extremely high. Obviously, when such basins were filled with water the climate was very different. It was once supposed that the difference was mainly a matter of rainfall, and because of that supposition the ancient lakes were called *pluvial lakes* (meaning lakes formed by the agency of rain). The name has stuck to them even though today it is realized that low temperature, which greatly reduced the rates of evaporation, especially in summer, was more responsible for creating the lakes than increased rainfall was.

The low temperatures (very likely 12° to 13°F lower than today's) that prevailed in the Basin-and-Range Region while pluvial lakes existed were the low temperatures of the glacial ages. Radiocarbon dates of the lake sediments show that the basins became filled with water shortly before the great glaciers reached their maximum, and that as the glaciers melted away the lakes dwindled, until by about 6000 years ago most of them had become dry. This does not mean that the glaciers caused the lakes. It means rather that both lakes and glaciers were different responses in different regions to the lowering of temperature that characterized the glacial ages.

Of course the lowered temperature and the increased proportion of water that did not evaporate permitted desert vegetation to remain in only the hotter and drier places. Juniper and pine woodland occupied wide areas around the basins, with forests of pine, spruce, and fir higher up on mountain slopes. These changes attracted many animals to the region, as we will point out a bit later.

Sea level in a glacial age

One of the most remarkable of the many changes that characterized glacial ages was that the level of the sea became lower, not because lands were pushed up out of the sea but because the actual amount of water in the sea decreased. Looking again at the water cycle (Fig. 2 - 1) we can see why this had to happen. The amount of water on the Earth is fixed. Most of it is in the sea, some is on its way to the land, some stays on land as lakes and glaciers or beneath it as ground water, and some is en route back to the sea. The level of the sea's surface at any time reflects the balance among the quantities of water in these four positions.

When temperatures decrease in a glacial age, more precipitation falls as snow and less as rain. Glaciers are created. Less water returns to the sea because a greater part of the water substance in the water cycle stays locked up on land in the form of ice. So the depth of the sea must decrease.

Today's sea level (Fig. 18 - 4, A) is determined by the amount of ice that stays locked up on the Antarctic Continent and on Greenland, simply because there is very little ice anywhere else. But 20,000 years ago, glaciers had formed and spread so far over middle latitudes that a huge amount of additional ice occupied the land.

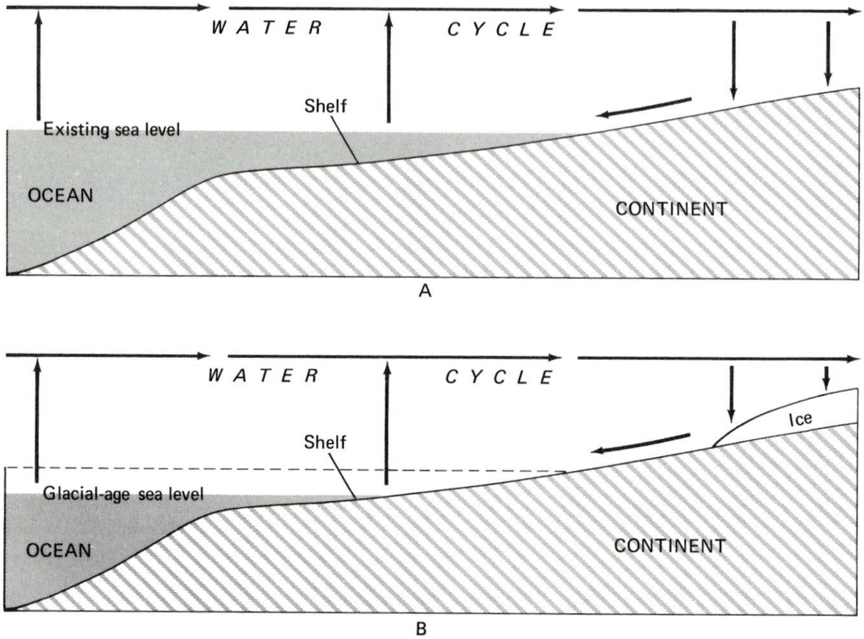

Figure 18 - 4. Under today's conditions, (A), with middle-latitude lands almost free of ice, all the water evaporated from the ocean returns to the ocean. But in a glacial age, (B), a large quantity of the water is held on the lands in a frozen condition. So less water returns to the ocean and sea level must be lowered in proportion to the additional quantity of ice on land.

Conversely, if there were no ice on the Antarctic Continent or on Greenland, as was the case in Cretaceous time, the sea level would be much higher than it is today, perhaps as much as 140 feet.

Calculating the amount of that additional ice and knowing the area of the ocean, we can determine how much lower the level of the sea should have been at that time. Sea level should have been nearly 300 feet lower than it is today (Fig. 18 - 4, B). This is why the map in Figure 18 - 1 shows the east coast of the United States lying far east of its present position, along the line where water depth today is 300 feet. When the coast stood there, much of what is now continental shelf was land in the form of a wide coastal plain covered with forest and probably teeming with animals. At that time the Hudson River, flowing past the site of New York City, continued onward for nearly 100 miles before emptying into the Atlantic Ocean. The shallow valley it occupied is still there, cutting across the continental shelf. Long Island and, farther east, Cape Cod and the islands near it did not exist separately. They formed part of a monotonous plain, at first covered by the ice sheet and later exposed as the ice melted away.

Some of the forests and swamps that grew along the coast while the sea stood lower have been encountered in drill holes made in the search for water or petroleum. Others are openly exposed on the sea floor; in some places these consist of the stumps and even trunks of standing trees, rooted in place. Teeth and bones of glacial-age animals have been dredged up from the shelf in the nets of fishing boats. But many more were covered up by sediment deposited while the sea was rising to its present level, and now lie buried beneath the sea floor.

The Hudson is not the only river with a deep, drowned valley. Many other rivers along the Atlantic coast occupy deep valleys that were cut into rock when the sea level was much lower and were later filled or partly filled with sediment. Similar filled and drowned valleys have been charted in many places off other coasts.

Along parts of coasts where water is shallow today, the glacial-age lowering of sea level converted islands into parts of the mainland. A well-established example is Britain, which was connected with the European mainland by a wide plain that extended to France on the south and to Denmark on the east. Large mammals such as elephant, rhinoceros, and hippopotamus passed freely across it at various times, and smaller mammals such as man did likewise. The site of the present English Channel was the valley of a west-flowing river, while the Rhine River extended northward through a valley in what is now the floor of the North Sea. It emptied into the ocean well to the east of northern England.

Similarly Alaska was connected with Siberia, Siberia with Japan, the Malay Peninsula with Sumatra, Java, and Borneo, and Australia with New Guinea and Tasmania. Although first suggested by present-day water depths, most of these connections have been confirmed by finds, at both ends of a "bridge," of identical fossil mammals which, it is thought, could not have got across except over dry land.

Glacial-age mammals

North America. Earlier in this chapter we described the vegetation zones that existed in North America during the latest glacial age.

Still earlier we noted that much of our knowledge of them is based on fossil pollen. We can now say something about the mammals that inhabited those zones. As the fossil animals are studied, it becomes clear that their distribution throughout the continent is consistent with the fossil vegetation. The animals we find in the glacial-age spruce forest are approximately the *sorts* (although not always the identical species) that live in spruce forest today, and so on through the various zones. This consistency confirms our general reconstruction of the vegetation zones.

In the belt of tundra along the edge of the ice sheet the largest and most conspicuous mammal was the woolly mammoth (Fig. 18 - 5), a short-tailed elephant two or three feet less tall than a modern African elephant. Although its ancestors reach far back into early Cenozoic time, the woolly mammoth had specialized very recently to withstand a severe climate. Its guard hairs, 10 to 15 inches

Figure 18 - 5. Woolly mammoths crossing the tundra amid clumps of spruce in glacial-age central Europe, where they were at least as numerous as in North America. At the right are a pair of woolly rhinoceroses, much at home in Europe but never present in North America. (Courtesy of the Field Museum of Natural History.)

long and shaped like the tails of enormous rats, were interspersed through a coat of thick wool. Just beneath the skin was a layer of blubber two or three inches thick, and a fatty pad covered the top of the skull. The four-toed feet were very broad, a useful characteristic in wet pasture. This elephant grazed on tundra plants, as we know from one specimen, a body complete with hair, skin, and flesh recovered from frozen ground. In its mouth was a mass of food, not yet swallowed at the moment of death, consisting of tundra vegetation. The molar teeth, with folded structure like that of the later horses (Chapter 16), were adapted for grazing and so were suited to tundra conditions.

Although it is frequently stated that the cadavers (skin, flesh, and all) of mammoths have been found frozen in glacier ice, and that the whole race of mammoths met with some sudden climatic catastrophe, this is not true. It *is* true that four nearly complete and about 35 partial cadavers have been found in Arctic Siberia, in the zone in which the ground is still frozen today. There is no evidence that Siberian mammoths as a population met with a catastrophe. On the contrary, the cadavers indicate nothing more than that individual animals occasionally met with fatal accidents, either by drowning in a lake or in mud, or by burial alive in the slump of a river bank. The Principle of Uniformity is not contradicted by frozen mammoths.

These animals ranged very widely—right across America from Atlantic to Pacific, and onward across the land bridge into Eurasia, where they reached westward to the Atlantic again. Their numbers were large. We noted in Chapter 9 that during the 18th and 19th Centuries between 40,000 and 60,000 fossil mammoth tusks were collected in Siberia alone. They entered the market and were sold for the manufacture of billiard balls, piano keys, chessmen, and other articles long before the introduction of plastics destroyed the demand. Besides mammoths, the population of the tundra belt included caribou and musk-oxen, both still living today in Arctic North America or Greenland. In addition there were many rodents, especially Arctic lemmings.

In the spruce-forest zone into which the tundra merged, the most conspicuous mammal was the American mastodon (Fig. 18 - 6), with molar teeth appropriate to a browsing life. Although it would not have been much at home on the tundra, it could feed wherever there were forest trees. Consequently it ranged through nearly all parts of the United States that were then forested, as far south as Florida and even to central Mexico. Besides its teeth, the mastodon had two other features that fitted it for a life in forests: it was smaller than the mammoth and its tusks were much smaller in proportion to body length. Thus it could move between close-growing trees with less difficulty.

Along with mastodons in the spruce-forest zone were moose (especially in the swampy places), a curious-looking extinct "elk moose"

Figure 18 - 6. The American mastodon, one of the commonest glacial-age large mammals in North America, shown between a swamp and a spruce forest. (Courtesy of the Field Museum of Natural History.)

Figure 18 - 7. An elk moose, a glacial-age browser in northern forests of eastern North America.

Figure 18 - 8. Two giant beavers. These creatures were nearly ten feet long and probably weighed 500 pounds—about the size of a modern black bear. Compare the squirrel (top center). (Courtesy of the American Museum of Natural History.)

(Fig. 18 - 7), an extinct browsing musk-ox-like animal adapted to forest and woodland, and a really extraordinary giant beaver (Fig. 18 - 8), other sorts of rodents, deer, bears, and large cats.

In the broad-leaved forest still farther away from the ice sheet were horses, camels, bison, large cats both biting and saber-tooth, bears, wolves, at least two kinds of elephants (Fig. 18 - 9) without arctic adaptations, and mastodons. In addition there were tapirs, peccaries, and armadillos from South America, and toward the west great box-turtles several feet long.

Out in the grassy woodland that covered much of the Great Plains were elephants, horses, camels, bison, elk, antelopes, wolves, ground sloths (Fig. 16 - 12) from South America, and a host of small mammals, mostly rodents.

One of the richest localities for glacial-age fossils, where collecting has been going on for many decades, is Rancho La Brea, now swallowed up by the city of Los Angeles, but preserved as a small park and museum. It consists of a series of oil seeps, now hardened

Figure 18 - 9. The Columbian elephant (*Mammuthus columbi*), a large American elephant that lived in central and southern United States and Mexico. (Courtesy of the American Museum of Natural History.)

to the firmness of asphalt, that formed pools at low places in the ground. Shallow pools of water covered the seeps during the winter season, and these attracted animals in large numbers. Mammals and even birds became trapped in the oil (Fig. 18 - 10), as were carnivores that came to feed on their carcasses. Over a long period between about 40,000 years ago and about 15,000 years ago the bones of

Figure 18 - 10. Glacial-age scene at the Rancho La Brea oil seeps. Two sabertooth cats feed on carrion while a third cat fends off hungry vultures. At left, wolves await their turn. In right background are horses. (Courtesy of the Field Museum of Natural History.)

hundreds of thousands of trapped animals accumulated in the seeps. The bones, black with asphalt, include those of elephants, camels, horses, bison, ground sloths, wolves, biting and saber-tooth cats, giant birds of prey, and a host of others. Bones of more than 2400 individual saber-tooth cats alone have been studied. And in addition thousands of fossil insects have been found, as well as fossil cypress, willow, and other kinds of trees.

Altogether, in many parts of the United States the teeming mammal life must have approached in numbers and variety what can still be seen in East Africa today, despite the pressures created there by man as a dweller, as a farmer, and as a hunter. In Nebraska alone more than 85 species of glacial-age mammals have been collected and studied.

The foregoing description of glacial-age mammals is based only on fossils from the latest glacial age, 15,000 to 20,000 years ago. To describe the animals represented by fossils from earlier glacial ages, some of them hundreds of thousands of years old, would take us into too much detail. All we need say here is that the older fossils show that similar movements of animals toward warmer places occurred in each glacial age, as did movements back again during the intervening times of warmer climate. Throughout the whole see-saw process, species and genera changed little by little with the slow progress of evolution.

Eurasia and Africa. It is time now to compare glacial-age environments in the Old World—Eurasia and Africa—with those in North America. Today one must go high up into the mountains of Norway and Sweden or to the coast of the Arctic Ocean in order to find tundra in Europe. But in the glacial age (Fig. 18-11), tundra and other treeless vegetation extended right across western and central Europe and southward almost to the Mediterranean Sea. Very few trees grew north of the latitude of the Alps. Northern Europe was covered by an ice sheet that in places was two miles thick, and the principal mountains farther south were likewise ice covered. Not only fossil vegetation but also the extent and depth of frozen ground at that time tell us that central Europe was colder than was North America south of its ice sheet. Little warm air could reach central Europe, a complicated result of a different pattern of land and sea, differences in the system of winds, and the presence of the Alps and other high mountains north of the Mediterranean.

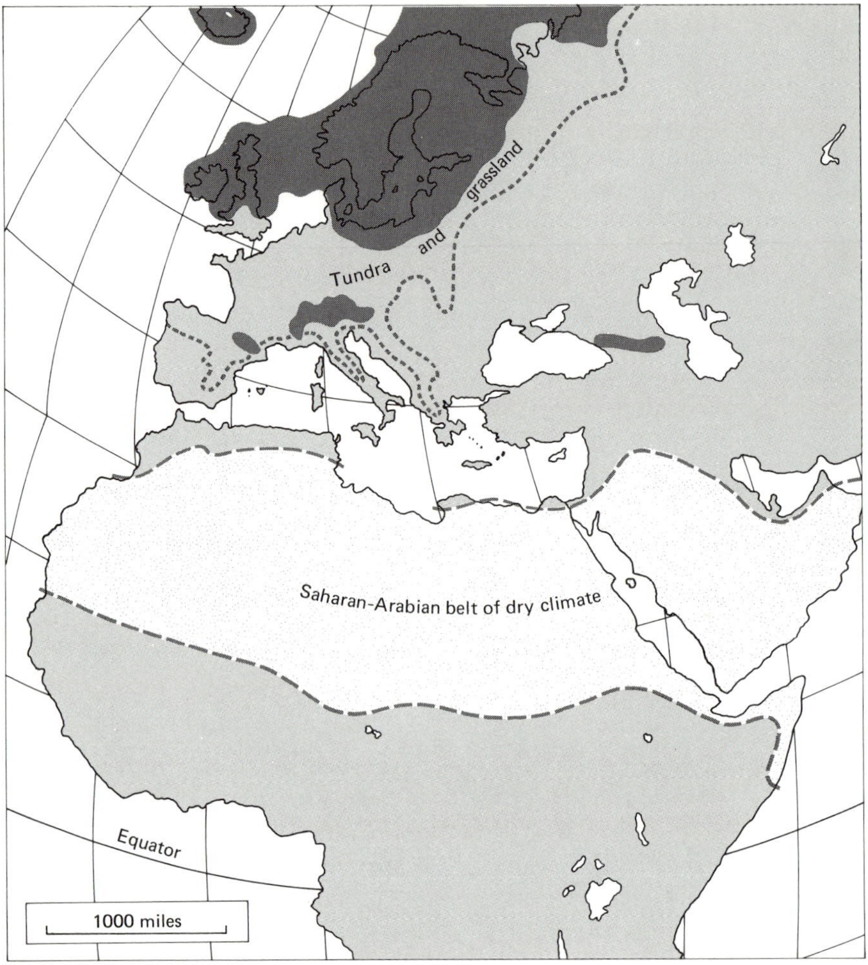

Figure 18 - 11. Map of Europe, western Asia, and Africa in the latest glacial age, showing the European ice sheet and the glaciers on Iceland, Pyrenees, Alps, and Caucasus Mountains. Farther south is a belt of dry climate.

The extreme cold in Europe, especially central Europe, had a striking effect on European trees. Before the glacial invasions Europe had a rich flora that included many kinds of trees. But during each glaciation some of the kinds that were sensitive to cold climate were destroyed. The Alps and other mountains on the south, backed by the Mediterranean Sea beyond, left these trees no refuge in which to take shelter during the cold times. So after each glaciation fewer trees returned to European territory. In northern Europe there are only six species of broad-leaved trees, the only survivors of the glacial ages.

The story in North America was quite different. Despite the glacial ages its tree flora has remained rich and varied. Because there were no mountains between the edge of the ice sheet and the Gulf of Mexico, plants could shift southward without difficulty and ride out the glacial time in a temperate climate, returning afterward to their former living places. Because of this pattern, North America today has many more species of trees of all kinds than there are in Europe.

In contrast to what happened to European trees, the wide, cold, treeless environment in central Europe was a splendid living place for woolly mammoths, reindeer (the European equivalent of the American caribou), musk-oxen, and a tundra animal, the woolly rhinoceros (Figs. 18 - 5, 9 - 11, A), that for some reason never got across to North America. He was a grazer equipped for extreme cold, with long hair and an inside coat of blubber much like those of the mammoth. Other tundra animals included Arctic hares, lemmings, and a variety of other rodents. In the forest and woodland areas were moose, a large elk-like creature, deer (Fig. 18 - 12), bears (Fig. 18 - 13), and carnivores such as large cats and hyena. Toward the drier east, far into central Asia, were several sorts of antelope, wild asses, and a host of rodents. Ranging through all these environments were horses, cattle, and the European bison, much as mastodons in North America ranged through forests from the fringes of the tundra southward all the way to Florida.

We noted in Chapter 16 that although there was some exchange between North and South American mammals along the Isthmus of Panama in Quaternary time, the faunas of the two continents remained quite distinct from each other. Although not a complete barrier to the movement of mammals, the tropical forests of the Isthmus were an effective screen or filter. Another effective screen stood between Europe and the southern half of Africa: the Sahara Desert region. The Saharan belt of dry climate (Fig. 18 - 11), in which annual rainfall is less than 10 inches, extends from near the Mediterranean coast southward for a thousand miles. East-west, it extends from the Atlantic Ocean to the Red Sea, a distance of 3200 miles, and the dry belt then continues eastward across Arabia. This screen must be effective today because the mammals of Europe include none from southern Africa. Indeed this is the chief reason

Figure 18 - 12. Giant deer (*Megaloceros giganteus*), the largest of the glacial-age European deer. About the size of a moose, it bore enormous, heavy antlers with a spread of as much as 13 feet. This giant became extinct about 10,000 years ago. (Courtesy of the Field Museum of Natural History.)

why European tourists fly across the Sahara to visit the great game parks in East Africa. There is nothing like those animals at home, except in zoos. Despite the presence of pluvial lakes and streams in the Sahara at least at times during and after the latest glacial age, the screen was almost as effective then. Few animals from southern Africa even reached the Mediterranean coast. The mammal faunas that lived south and north of the Sahara were distinctly different.

During the long intervals between the glacial ages some African animals that then lived north of the Sahara came into Europe. Among these were elephants, hyenas, and big cats. There were also hippopotamus (common in the lower Nile until only about 3000 years ago), which managed to reach as far north as southern England. That fact and other evidence indicate that in the times between the glacial ages climates were at least a little warmer than they are now. We see similar evidence in eastern North America,

Figure 18 - 13. The European cave bear (*Ursus spelaeus*), a huge glacial-age creature as big as the living Alaskan brown bear. The bones collected from caves throughout Europe represent hundreds of thousands of these bears. They hibernated in caves whose rock walls they polished with their shaggy coats as they passed in and out. (Courtesy of the Field Museum of Natural History.)

where the manatee, a subtropical marine mammal that today lives around Florida and in the Gulf of Mexico, ranged as far north as New Jersey between glacial ages. Also, plants and animals that today live no farther north than Maryland and southern Pennsylvania are found as fossils in the neighborhood of Toronto.

Extinctions. The rich glacial-age fauna of large mammals we have described, which made big-game country out of North America, would have been worth coming far to see. Even more remarkable than the fauna itself is the fact that during and after the melting away of the glaciers 70 percent of the genera of large mammals became extinct. The victims included the woolly mammoth and the two other North American elephants, the mastodons, all the camels, all the horses, all the large cats except two, all the kinds of bison except one, the elk moose, the giant beaver, the great wolf, the tapirs, and all the ground sloths. All that today remain from the great glacial-age fauna of large mammals are elk, deer, the small antelope, mountain sheep, mountain goat, buffalo, the small wolf, bears, cougar (mountain lion), jaguar, and in the north moose, caribou, and Arctic musk-ox. Marine mammals were entirely unaffected.

Examining the death list, we see that it consists of large mammals, nearly all of them dwellers on grassy plains or in open woodland. In general small mammals and most forest dwellers survived. This event was not like the gradual dying out of dinosaurs in Cretaceous time, involving tens of millions of years. It took place with astonishing speed. Radiocarbon dates tell us that it began about 12,000 years ago and for the most part ended about 10,000 years ago, with possibly a scattering of later survivors.

In North America the death toll seems to have begun in the northwest, progressing southward and southeastward and continuing down the Isthmus of Panama and over the length of South America, where the last ground sloths became extinct about 10,000 years ago.

How such an extraordinary extermination could have come about has aroused a scientific controversy. On one side are those who believe that the cause was either change of climate or disease, although few convincing facts have been brought up in support of these ideas. Nor has it been explained why there is no comparable evidence of extinctions at the ends of earlier glacial ages.

On the other side of the controversy are those who think such extermination could have been caused only by some new factor in the environment, one that had not been present in earlier times. They suggest that the factor, of whose presence there is evidence, was a predatory mammal that was entering the Americas for the first time—the mammal we call man. This predator possessed darts tipped with pointed, sharp-edged heads of quartz. He used a jointed throwing stick to launch his darts with high velocity. He had other killing methods such as the construction of pitfalls, the driving of an animal into a swamp or muddy lake, and the driving of herds of animals, sometimes through the use of fire, over cliffs to their deaths below (Fig. 18 - 14).

It seems to be true that the extinctions in any given region seem to follow closely behind the evidence of the first appearance of man there. According to the theory of human predators, the large native

Figure 18 - 14. A fire drive by Stone-Age hunters. Men with torches stampede a herd of wild horses, forcing them over a cliff to their death hundreds of feet below. Evidence of such drives has been found in various parts of Europe. Buried at the foot of a thousand-foot cliff in southeastern France has been found an 8-foot layer of bones, the bones of an estimated 100,000 horses. The date is between 20,000 and 15,000 years ago.

mammals did not recognize man as dangerous to them because he was new and unknown; so they were unwary and unsuspicious. Hence there would have been a strong element of surprise in the human attack. There was much mass killing, probably far beyond the immediate need for food and skins. The theory holds that as the large plant-eating mammals were killed off, the large cats and the great wolves were deprived of their food and became extinct through the effect of human competitors.

An extreme form of the theory proposes that Stone-Age men invaded from Alaska, making their way southward in small groups along the corridor at the east base of the Rocky Mountains, a pathway newly opened by melting of the ice sheet on the east and of the mountain glaciers on the west (Fig. 17 - 2, B). Encountering the rich big game, they systematically attacked it, and fanning out slowly in pursuit, moved their camps gradually southward and eastward at an estimated rate of 10 to 15 miles per year. At that rate they could have swept through both Americas in less than 2000 years. Many dart points and a few kill sites where the hunters butchered their meat have been found. Human bones are almost nonexistent, but the population was small and mobile, and unless their dead were buried, their remains are not likely to have survived the activities of scavengers and chemical weathering.

Opponents of this dramatic theory object that a population large enough to accomplish the destruction of so many big mammals could not have existed. Yet the other explanation, that the extinctions resulted from change of climate or from disease, meets just as much objection, if not more.

Similar events seem to have occurred in Australia somewhat earlier and in New Zealand considerably later, the victims of course being very different kinds of animals. Extinctions in Africa and Eurasia (Fig. 18 - 14) began much earlier and were much more gradual. But it is pointed out that Stone-Age man was already living on those continents long before the extinctions began and so cannot be ruled out as the prime killer.

It is not claimed by those who believe in the theory that Stone-Age hunters killed *every* individual of the species that were destroyed. When the number of individuals in a species becomes small, a combination of natural factors tends to decrease the number rapidly until extinction occurs. None of the theories is proved, and here the matter rests, awaiting the appearance of fresh evidence.

On the death list of North American mammals we mentioned camels and horses. Few people realize that both these animals are native American stock. They originated early in the Cenozoic and spread from North America via the land bridge to Asia; camels spread to South America as well. Yet when European men began to explore the Great Plains in the 18th and 19th Centuries they found large numbers of wild horses (mustangs). These were not survivors from the Pleistocene. They were descendants of horses brought from Spain by Cortés when he conquered Mexico early in the 16th Century. In barely 250 years the wild offspring of those Spanish horses had populated western North America.

The camels, although they became extinct in North America, survive in South America, most of them domesticated, in forms such as the llama, a camel especially common in Peru. They are small, and humpless like their North American ancestors. The large humped camels we see in zoos and circuses are specialized Asiatic types imported in recent times.

Our discussion of the possible relation of Stone-Age man to large mammals leads naturally to the absorbing story of the history of man himself, a story that is bound up with the glacial ages and their environments. So in the final chapters we will trace what is known of the ancestry and evolution of man.

References

Farrand, W. R., 1961, Frozen mammoths and modern geology: Science, v. 133, p. 729–735.

Flint, R. F., 1971, Glacial and Quaternary geology: John Wiley & Sons, Inc., New York, chaps. 18–20, 23, 28, 29.

Kurtén, Björn, 1968, Pleistocene mammals of Europe: Aldine Publishing Co., Chicago.

Martin, P. S., and Wright, H. E., editors, 1967, Pleistocene extinctions; the search for a cause: Yale University Press, New Haven, Conn.

West, R. G., 1968, Pleistocene geology and biology, with especial reference to the British Isles: Longmans Green and Co., Ltd., London; John Wiley & Sons, Inc., New York.

19

Physical Evolution of Man

Forerunners of the human body

The cooler, drier, much more diversified world of later Cenozoic time certainly influenced evolution toward man. But the basic structures were already there, having been gradually put together and added to throughout a very long period. Cenozoic environments merely acted to perfect them. At this point in our story we might well list the highlights of what evolution had accomplished towards developing the human body by, say, the beginning of the

Table 19 - A. Principal Features of the Human Body, and Their Origins.

Feature:	Inherited from:
1. Elongate body with distinct head and tail.	Ancestors of fishes
2. Bony skeleton.	Fishes
3. Walking limbs, five-fingered limbs.	Amphibians
4. Amniote egg.	Reptiles
5. Long legs, differentiated teeth, "warm" blood, hair, mammary glands.	Primitive mammals
6. Placenta, vivipary.	Early placental mammals

Cenozoic Era. We can compact them into six features or groups of features (Table 19 - A).

The developments that created these features are milestones along the zigzag path of evolution we are following toward man. Each resulted from an opportunity offered by environment. Although our story has now reached the beginning of the Cenozoic Era, there is more to come.

Origin of primates

We come now to the primates, the group of mammals to which man belongs. The place of the group among all mammals is charted in Figure 16 - 9. Primates are characterized by hands and feet that are flexible and that have five fingers or toes. Most primates live in trees. At or just before the beginning of the Cenozoic Era, the group branched off from insectivores, those small, insect-eating mammals that may have lived in trees and that were among the most primitive of placental mammals.

At first thought it seems surprising that man, the primate that has attained a commanding position among mammals, should look back to an ancestry so very primitive and so humble. But on further thought such ancestors turn out not to be surprising at all. In fact they are just what we would look for as the ideal raw material from which to build a man. The reason is that those humble tree dwellers were possessed of primitive, unspecialized characteristics that the process of evolution could mold in any direction. Reptiles and birds, of course, were available as raw material, but they had long before become too specialized. Each had already gone too far in some one evolutionary direction, and so was not good building material for evolution in another. The insect eaters, living inconspicuously in their forests, were potentially plastic, ready for the chance that gave them a starting push.

Among the early primates, already evolved from insectivores as a distinct group, was a small creature called *Notharctus* (Fig. 19 - 1), which we can take as typical of these probable early ancestors of man. It lived in a near-tropical forested region that included what is now the state of Wyoming. Its teeth suggest that its food consisted of young leaves and fruits. It ran and jumped from one branch to another. The picture shows clearly the flexible hands and feet of

Figure 19 - 1. One of the early primates, *Notharctus,* from Eocene strata in Wyoming. About the size of a half-grown house cat, it is typical of very early ancestors of man. (Courtesy of the American Museum of Natural History.)

which we have already spoken. Each had the primitive five digits that can be traced all the way back through geologic time to the first amphibians in the Devonian.

Once established as a group, primates of this general sort began to branch out, forming different kinds that evolved in divergent directions, as can be seen in Figure 19 - 2. However, all of them continued to dwell in forest trees. First to branch off were various rather primitive kinds, such as lemurs, that run along the upper surfaces of tree limbs as squirrels do. Next the line of monkeys diverged, gradually subdividing into many different kinds. After that came the important split between the ape line on the one hand and, on the other, the line of *hominids,* a word that loosely means *mankind.* The members of the ape line stayed in the trees and after a long evolu-

Figure 19 - 2. Tree of primate evolution.

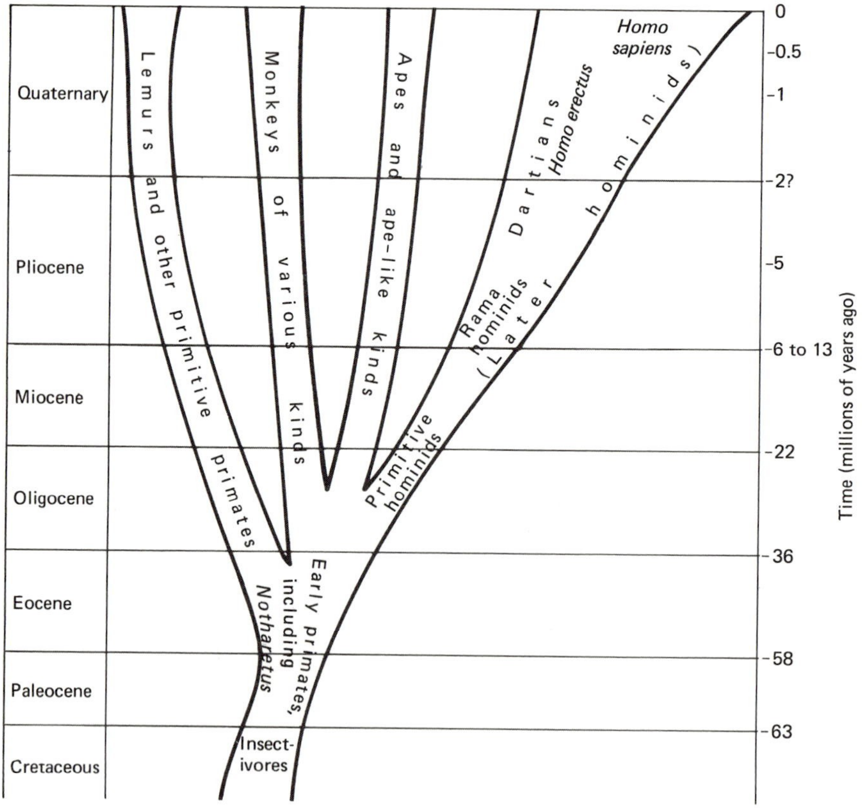

Tree of primate evolution

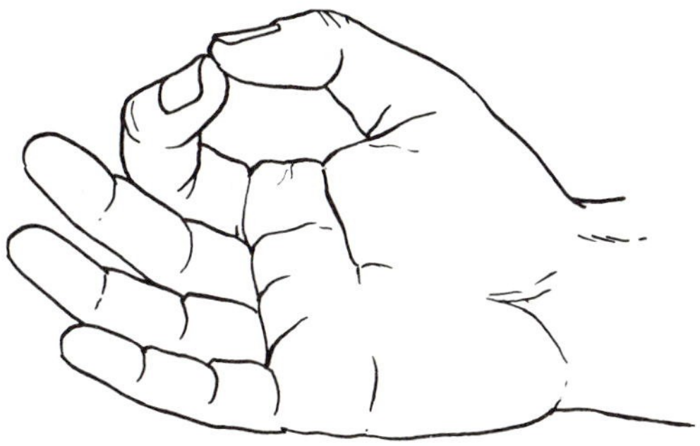

Figure 19 - 3. The human hand, with its opposable thumb, is the fundamental basis of man's industrial and artistic skill. Its ability to grasp goes back to tree-dwelling ancestors (Figure 19 - 1).

tion consist today of only four closely related genera: Chimpanzee, Gorilla, Gibbon, and Orang Utan. In contrast, the hominid branch includes not only human primates but also the pre-human primates that are distinct from the line of apes. As Figure 19 - 2 indicates, the split between apes and hominids occurred some 25 to 35 million years ago, in Oligocene time. For a long period thereafter both continued to live at least part of the time in trees.

Early in the present century, when little was known about man's ancestry, a popular controversy raged over the question, Is man descended from an ape or is he not? Sometimes a monkey was substituted for an ape. That controversy, based on both ignorance and prejudice and fortunately long since dead, was beside the point all along. Both apes and monkeys are evolutionary cousins of man, but to find a common ancestor of all three we must go a long way back, to the Oligocene Epoch.

The fact that the primitive primates lived in forest trees well past the time when the hominid line split off from the ape line was a great advantage to hominids in their later evolution. The advantage is based mainly on four factors in the anatomy of hominids:

1. Hands and feet, already *flexible* as we see in Figure 19 - 1, did not change but stayed flexible because of the constant need to climb. This helped perfect the ability to grasp boughs and branches—and, much later, tools.
2. *Grasping* was enormously aided by a thumb that could be curled in the direction opposite to the curled fingers (Fig. 19 - 3). This is seen in the right forefoot (we might almost say right hand) of the little animal in Figure 19 - 1. Grasp a ruler or any large stick

in your own hand and see how much more control the opposed thumb gives to you, a primate. A squirrel, although a five-fingered tree dweller, cannot match it.

3. A life of rapid movement in trees, following narrow paths and continually avoiding branches, was a lifelong obstacle race. It demanded very good *vision.* Most mammals that live on the ground have eyes at the sides of the head, but the eyes of primates gradually moved toward the front. Thus, as in a binocular telescope, the field of vision of one eye overlapped the field of the other, making it possible to perceive depth. This resulted in ability to judge distances and so to gain an improved knowledge of the immediate surroundings.
4. An *enlarged brain* accompanied these developments. Movement through a tree to the next tree demands agility and the ability to coordinate groups of muscles instantly. So the control apparatus, the brain, had to be improved.

These four advantages, flexibility, a grasping hand, better vision, and a bigger brain, were shared alike by primitive hominids and primitive apes. But the hominids possessed an additional advantage that forms a large part of the difference between them and their early ape cousins. They began to develop a short face without a muzzle. Since they were starting to use their arms and hands to defend themselves, they began to lose the need of a muzzle with long canine teeth for biting and snapping. Their canines grew smaller and shorter, and their molars moved closer together for more efficient chewing of a varied list of foods. The teeth of Modern Man have now degenerated because the need for chewing hard, rough food no longer exists. The change was brought about when the use of fire led to the practice of cooking. With cooking, food became more tasty and more tender as well.

Rama hominids

The new shorter face was a feature of the earliest certain hominid found as a fossil thus far, a Miocene-and-Pliocene creature which we had better call the Rama hominid. Its generic name is *Ramapithecus.* The first part of that word, *Rama,* is the name of a Hindu god; the second part, *-pithecus,* means ape. The first fossil specimen was found in India in 1934; at that time it was still believed that man had existed only since the beginning of the Quaternary. Hence it was natural that a fossil primate from Miocene strata

Figure 19 - 4. The Rama hominid restored, with one hand on a termite hill. This animal might well have hunted termites in a woodland environment, as chimpanzees do today. (Drawn by R. F. Zallinger under direction of E. L. Simons.)

should have been classified as an ancestral ape. But since then fossils of the same genus have been found in southern China, in Europe, and in East Africa. Detailed study shows all the fossils to be those of a hominid. The name therefore is misleading, but the rules of scientific naming do not allow it to be changed. So, to avoid confusion in our discussion, we will call it the Rama hominid.

Neither the original find nor the later finds are even nearly complete skeletons, but the collection of bones shows the short face that characterizes hominids. It also indicates a small-sized animal, three or four feet tall if it could have stood erect. (Just how nearly erect it could have stood we do not know.) The posture we see in Figure 19-4, a restoration from the available bones, is only an educated guess, as is the lack of a tail, which in *Notharctus* was very long. But later hominids and modern apes have lost the tails their ancestors possessed. The fossil bones do not in themselves tell much about Rama's environment. But the other fossil mammals collected along with the hominid bones are of kinds that lived in tropical woodland at rather low altitude. The bones include those of crocodiles and with them are fossil palm trees as well. So it is likely that our hominid lived in warm woodland too. However, the teeth of the Rama hominid suggest a diet of small fruits, buds, young leaves, grubs, and perhaps even ants and termites. Most of these foods would have been found not in the trees themselves but on the ground. So we might speculate that although a dweller in woodland, Rama foraged and fed on the ground. This was the beginning of a long slow change in the hominids' way of life. Instead of coming down to earth just for a meal, they came down and stayed.

A permanent life on the ground

Although, like their ape cousins, the earliest hominids continued to live in the forest, they must have made more and more frequent descents to the ground. Eventually they climbed down for good, leaving the trees to the apes. That movement was completed some time, perhaps a long time, before five million years ago. We know this because the fossils of later hominids are found along with the bones of mammals not of forest kinds but of kinds characteristic of woodland and dry grassland. During the middle part of the Cenozoic Era the environment in which hominids lived was changing, probably because of the activity of crust plates. As we remarked in Chapter 15, large parts of North America were then rising and mountains were forming. The cooler and drier climates that resulted caused grassland to spread, and this stimulated the rapid evolution of horses and other grazing mammals. The geologic record in southern Eurasia and in eastern and southern Africa, the regions where fossil hominids of the time have been found, tells a similar story. We can thus suppose that warm forests were shrinking and that woodland and grassland were taking their places.

With the more efficient jaws and teeth of their shortened faces, hominids could cope with a wider variety of food than had been available to their ancestors. So, as open country encroached more and more on their environment, it is likely that the temptation of the roots, tubers, and grain found in more open country is what persuaded hominids to desert both treetops and forest floor little by little and begin a life permanently on the ground and in the open. They spread gradually and timidly outward from forest borders, via woodland into grassland, in their daily search for food. At some point along the way, as we shall see presently, the hominids' menu began to include meat. These changes in environment and diet lead us to a more advanced group of fossil hominids, the Dartians.

Dartians

In the debris from a blast in a South African quarry in 1924, there was found, undamaged, a small skull. Luckily the skull was sent to Johannesburg, to the anatomist Professor R. A. Dart, who found that he had part of a six-year-old hominid child (Fig. 19 - 5) still with its baby teeth. Dart studied it closely and named it *Australopithecus africanus* (literally "African southern ape"). That name involves the same kind of confusion as does the name *Ramapithecus* because *Australopithecus,* along with other fossils of the same genus discovered later, is universally recognized as a hominid, not an ape. Therefore we shall follow the lead of the late Sir Arthur Keith, a British anthropologist, and use for *Australopithecus* and his close relatives the informal name *Dartians* in acknowledgment of the decades of study devoted to them by Professor Dart.

Of course Dartians are hominids as we ourselves are, but we are justified also in speaking of them as men, early men, even though some cautious scientists prefer to call them primitive subhumans.

Since that first discovery in 1924 the remains of more than a hundred individuals have been found, most of them in various parts of Africa but others in the Near East and in China. Although not one of the individuals is a complete skeleton, an important result of the

Figure 19 - 5. The skull of the six-year-old Dartian child, held in the hands of Professor Dart in his laboratory in Johannesburg. (R. F. Flint.)

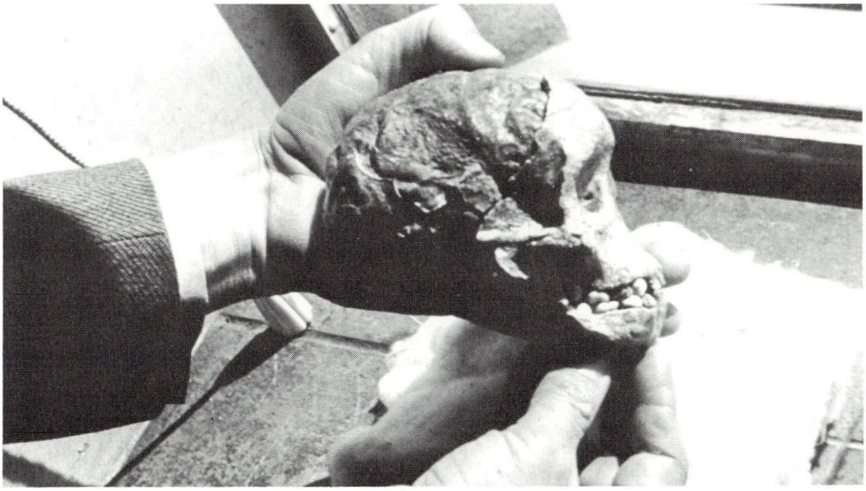

study of the fossils concerns the shape of the Dartians' pelvic bones. These bones are so shaped that their owners could have walked erect, as we do. Although tremendously interesting, this fact is not surprising. It was to be expected, because once hominids had descended from the trees for good and had got out into the open, an erect posture would have had a great advantage over a stooping one and would have been favored by natural selection. One of the chief reasons why we can think of Dartians as men is that their posture was erect (Fig. 19 - 6).

Figure 19 - 6. A Dartian, restored at the entrance to a hillside cave in southern Africa. Compare his more erect posture with that of the Rama hominid in Figure 19 - 4. The cave overlooks dry woodland and grassland rather than a forest. [By permission of the Trustees of the British Museum (Natural History).]

Another man-like characteristic of Dartians concerned their hands. No longer having to grasp one tree branch after another with their hands simply in order to move about, hominids could develop their arms and hands for more complex uses, including the making and handling of tools and weapons. But of course the skill of the hand has to be directed by a competent brain adequate to the task of directing. So these new uses of the hands by hominids demanded an ever-larger brain. And the fossils show that Dartians possessed brains larger in proportion to body weight than their Rama predecessors did. The volume of the hollow in the skull that contained the Dartians' brain measures, on the average, some 500 cubic centimeters (about that of a medium-sized orange). Contrast this volume with 400 cc (a small orange) in a Chimpanzee and with 1400 cc (a big grapefruit) in Modern Man.

If the fossils found thus far are representative, Dartians included various races. Some kinds were rather lightly built whereas others had heavier skeletons with thicker bones. They rarely exceeded 4 feet 6 inches in height, and so were smaller than most of the races of Modern Man. We can imagine them as covered with fur (really thick fine hair), probably somewhat thinner than Rama's. With the passage of time the hair probably thinned greatly. Running on the ground, a new occupation for hominids, builds up heat in the body and the heat must be made to dissipate as rapidly as possible. So a thick hairy coat (except in an extremely cold region) is a disadvantage to be got rid of through evolution.

The earliest known fossil bone that certainly belonged to a Dartian, part of a jawbone with a tooth, found in the northern part of East Africa, is dated radiometrically (Chapter 3) at 5.5 million years before the present. And because Dartians did not become extinct until about one million years ago, they had as a group quite a long life. But as individuals they did not fare so well. From the form and condition of various human bones it is possible to estimate the age of an individual when he died. Such estimates made from the available Dartian bones show that none of those particular Dartians lived to be more than forty years old and that the average age at death was only eighteen. This recalls the story told by skeletons of the lost colonists in Greenland, cut off by cold climate (Chapter 17). In both cases the short life expectancy resulted from the fact that life was hard. The Dartians' food was coarse and the supply uncertain, and there were predatory enemies to be reckoned with. Probably in part because of such enemies, Dartians took refuge in caves, at least at times if not continuously. Because they lived in country that was open and rather dry, vegetable food would have been limited and they would have had to depend on meat. Perhaps at first they drove carnivores away from their kills, tore off limb bones and heads and carried them to caves and shelters to eat raw. Later, they gradually became hunters instead of merely scavengers, but at all times they must have eaten what they could find. It is evident that the Dartians at one fossil site in East Africa ate frogs, mice, and other small mammals, and in addition the helpless babies of large mammals such as giraffes, which they could catch with their hands. At one cave site in South Africa there is strong evidence that Dartians killed baboons by a sharp blow on the head, perhaps with a leg bone of another animal.

In spite of the fact that Dartians occupied caves frequently if not continuously, we know of no evidence that they had learned how to use fire. They did, however, learn how to make crude tools, and this is part of the reason why we can call them men rather than just hominids.

The beginning of tool making represents a step of the greatest importance because it is the beginning of industry, the sort of activity represented today by complex electronic apparatus. Industry and other activities it led to are so much a part of the story of man and his environment that we shall save them for Chapter 20. Returning to the Dartians, we can summarize their contribution to the evolutionary progress of human anatomy in this list: (1) Steady progress in the size of the brain. (2) Evolution of a pelvis that made it possible to stand erect. (3) Evolution of tool-making hands. With those hands Dartians made a very crude sort of tools, which modern science has discovered and collected along with fossil Dartians themselves.

Homo erectus

The next succeeding group of hominids belongs to the genus *Homo*, our own genus. The group is named *Homo erectus* ("man standing erect") even though its members were probably not much more nearly erect than Dartians were. It includes a bundle of several varieties, differing somewhat from each other. All of these had earlier been given different names such as Java Man, Peking Man, and Heidelberg Man, but all belong to a single species.

Fossil remains of *Homo erectus* have been found in both Europe and East Asia (one of the fossils from Java was the skull of a year-old infant) and in both northern and southeastern Africa. It looks, therefore, as though this kind of man had gradually spread over most of the Old World. One of the finds in southern Africa consisted of *H. erectus* bones right among the bones of Dartians. This means that the two groups were living in that region at the same time, a million to a million and a half years ago. *Homo erectus* continued to exist until 400,000 years ago or even less. He lived his racial life entirely within the series of glacial ages depicted on the curve in Figure 17 - 1.

These people were a little taller than Dartians; they averaged about five feet in height. But their skulls indicate an average brain volume of around 900 cc, nearly twice as large as the Dartians'. Human brains had been growing faster than human stature. The skeleton of *Homo erectus* was rather similar to that of modern man. But the skull was smaller, thicker, and lower, with brows that jutted out conspicuously over the eye sockets (Fig. 19 - 7).

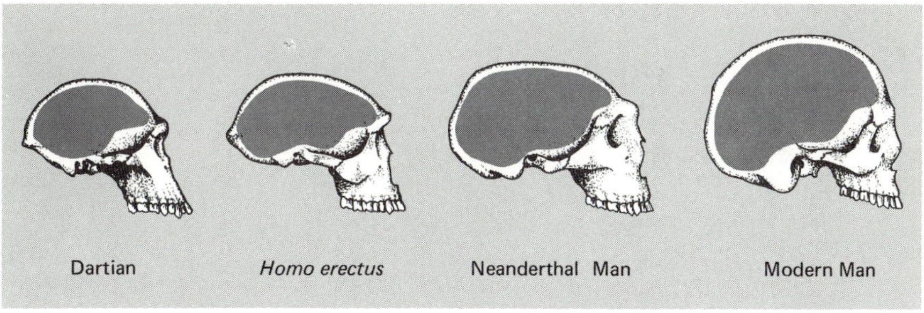

Figure 19 - 7. Evolution of the human skull through five million years. The skull became larger and rounder, the face shorter and more nearly vertical, and the brow ridges less prominent. Brain is shown in gray. (Not to exact scale.)

Much of what we know about the way of life of *H. erectus* comes from an ancient cave near Peking that is packed full of bones and rubbish accumulated through an enormous length of time. The bones include parts of more than fifty human skeletons. People were living in the cave at least as long ago as 500,000 years. The refuse they threw aside after their meals included bones of deer, rodents, horses, camels, bears, and other mammals up to and including elephants in size, as well as traces of vegetable food. The fact that adult elephants formed part of their diet shows that these people were more sophisticated hunters than Dartians had been. Dartians could pounce upon the babies of large mammals, but it takes careful organization and skilled group effort to bring off a successful elephant hunt. In this we see one of the results of the bigger brain of *H. erectus.*

Charcoal and charred bones among the refuse, and even layers of charcoal suggesting hearths that were used again and again as kitchen fires, show that meals were cooked. This is our earliest indisputable evidence of the human use of fire (Fig. 19 - 8). Probably fires were first used for warmth, and later their value for cooking meat was discovered accidentally. Obviously the menus were varied

Figure 19 - 8. Reconstruction of the appearance of the group of *Homo erectus* known as Peking Man, who lived perhaps half a million years ago. A man sits in the mouth of a cave making crude tools from stones. Another man drags in a deer he has killed, while a woman prepares to cook some of the meat. [By permission of the Trustees of the British Museum (Natural History).]

(roasted rats on Tuesday, elephant steak on Friday). The *H. erectus* people also used fire in hunting, and very likely in ceremonial games as well. Stone tools for cutting up meat had been in use since the time of the Dartians, but were now more skillfully made. We shall say more about them in Chapter 20.

Although precise evidence is impossible to obtain from fossil skulls, it is thought likely that *H. erectus* people possessed some capability for speech, at least of a primitive kind. It might not have been easy to organize large, complicated hunts without it, and there is evidence that not only the people near Peking but others far away in western Europe hunted in large groups, following plans carefully worked out in advance.

The evidence of human fossils suggests that in the Peking region, at least, people were killed and their brains extracted and probably eaten. Because there seems to have been abundant animal food, this rather unpleasant practice may have been less for nourishment than for some ritual purpose.

Homo sapiens

Early races. Before *Homo erectus* became extinct a new species of *Homo* had appeared. This was *Homo sapiens* ("man possessing wisdom"), our own species. The date of the earliest representative yet known may have been 400,000 years ago or a little earlier. This representative was not Modern Man, of course, but an early variety of our species. The names given to the many known varieties are too numerous to be listed in our discussion, but among those most widely known are Neanderthal Man and Cro-Magnon Man, and these two we will describe. It is not known with certainty just how the varieties were related—whether they were descendants in a single line or merely cousins. For example, we are not sure whether Neanderthal Man was a grandfather of the Cro-Magnons or only a great uncle. All, however, were characterized by skulls that were more globular in shape, with a considerable range in brain volume that averaged about 1400 cc.

Very little is known about the earliest races of *Homo sapiens* because fossils (all of them European) are few and incomplete. The restoration depicted in Figure 19-9 is therefore based mainly on indirect evidence, but it portrays a lively incident in their way of

Figure 19 - 9. Reconstruction of a deer hunt by a band of people who are among the earliest known members of the *Homo sapiens* species. In the distance elephants are feeding. The scene is near what is now Swanscombe, on the bank of the River Thames 20 miles or so downstream from the City of London. The time is more than 200,000 years ago. [By permission of the Trustees of the British Museum (Natural History).]

life as skilled hunters. Here we need to remind ourselves about the limitations of any restoration of an extinct mammal based on fossil material. As we noted in Chapter 9, careful measurements on fossil bones give us reliable information about the dimensions and form of the individual to which they belonged. But the contours of flesh and such characteristics as eyes, skin, and hair usually have to be recreated by analogy with individuals that are living today. For this reason restorations are never precisely accurate. We should think of those that are reproduced in this book as the best than can be done with the basic information available. Most people would agree that in spite of their limitations, good restorations are a far better aid to understanding the past than are no restorations at all.

Probably it was people of this general kind, earliest *Homo sapiens*, who constructed the earliest known human habitation that was a *building* rather than a cave. Discovered in a hillside not far from Nice on the Mediterranean coast of France and excavated about

1965, the remains showed this structure to have been 60 feet long by 20 feet wide. The roof had been supported by tree trunks that acted as columns. Within the structure were found two hearths with charred wood, bones of rabbit, boar, deer, and a long-extinct elephant, as well as fossil human feces. No other human remains were found, but the elephant and indirect evidence suggest the building may have been constructed as much as a quarter of a million years ago. Man was beginning his career as an engineer.

Neanderthals. Even though information about the earliest *Homo sapiens* is scanty, we know a good deal about the Neanderthals, a race that spanned at least the time from about 110,000 to about 35,000 years ago. Complete or incomplete skeletons of well over 100 individuals have been discovered in Europe, the Near East, Africa, and China. Our name for those people, *Neanderthal*, means simply "Neander Valley," a valley near Düsseldorf in western Germany where a skeleton was found by a quarryman in 1857.

There were two races of Neanderthal people, of which the later one was distinctive. The height of these later people was barely five feet, their arms and legs were short, their heads large, beetle-browed, and set upon thick necks. Although their brains were large, the skull was somewhat flat and the bulk of the brain was concentrated toward the back of the head, the frontal part which includes the memory center being comparatively small (Fig. 19-7). Were they living today, Neanderthals would hardly be thought handsome by our standards. But obviously they were strong and tough.

In Figure 19-10 we see a family group on a hillside. Farther up the slope are the mouths of caves, such as Neanderthals generally occupied, with other human figures outside them. The father puts on a short skin cloak before he starts off on a hunt with his spear. The mother pegs out a fresh skin to dry in the sun, while two children do what most children do. Lying on the ground are a duck and another bird waiting to be plucked before cooking.

This picture, however, may be deceptive. The Neanderthals' lives, in all probability, were by no means easy. Data on Neanderthal skeletons show that more than half the people they represent died while still in their teens. This grim statistic recalls the high mortality rate among the colonists of medieval time in Greenland (Chapter 17), when living conditions were very bad. Life in a cave, especially during a glacial age, must have been both hard and uncertain. The

Figure 19 - 10. A Neanderthal family group, some 100,000 to 50,000 years ago. The short cape perhaps marks the start of the wearing of clothes by man, as body hair gradually decreased. Fur clothing, however simple, should have been welcome in glacial-age temperatures. [By permission of the Trustees of the British Museum (Natural History).]

Neanderthals did what they could to mitigate the discomfort of their cavern homes. At the mouths of some caves are the remains of stone walls, and across the mouth of at least one is a row of round holes with pointed bottoms, now filled up with soil, which evidently had held sharpened poles between which skins had been stretched to keep out the wind.

Cold winds were not the only danger faced by Neanderthals. In a cave in the Middle East was found a skeleton from which one arm had been amputated with a stone knife. We can speculate that a man had been wounded, very likely in a hunt, and his people knew enough about the results of infection to realize that amputation was necessary. The radiocarbon date was 44,000 years ago. In another position in the same cave a skeleton showed that its owner had been stabbed in the ribs with a dagger; the wound had been healing for a week or two when a mass of rocks from the cave ceiling fell in on him and killed him. This event happened 45,000 years ago.

The contents of other caves record events that are less grim. A cavern in northwestern Italy contained bones of Neanderthal people, stone weapons, and bones of cave bears. The floor of one of its chambers was covered with a thin continuous layer of limestone deposited by dripping water. Beneath the stone was a floor of moist clay. In the clay were many scratches made by bears as they stretched their claws much as a house cat does on a carpet. Also in the clay were the marks of human fingers as they scraped up blobs of clay. Nearby, at the base of the cavern wall, were small balls of clay that had been rolled between the hands of Neanderthals, had been thrown against the wall and flattened, and had dropped to the floor. On the walls were soot from torches and the sooty marks of human fingers. Finally, in the clay floor were the impressions of a human foot (Fig. 19-11) perfectly preserved, like the other marks on the floor, by the protective layer of stone. Although we cannot be sure, it looks as though Neanderthals were either playing a game or just idly fooling around.

Figure 19-11. Cast of a Neanderthal footprint in the plastic-clay floor of a cavern in Italy. The shape of the foot is broader and more irregular than that of Modern Man. (A. C. Blanc.)

Although it was popular long ago to think of Neanderthals as brutish, they were certainly not so. The results of excavations tell us that they lived in small tribes of a few dozen people, improving cleverly on the stone tools of their predecessors, hunting skillfully, using fire regularly, and embellishing objects with colored paint. With their dead they buried stone implements carefully arranged in patterns and what seems to have been food, now consisting only of charred bones. Such ritual burials indicate that Neanderthals possessed some conception of a life after death, of a soul destined to make a long journey into another world. People with such beliefs could hardly be called brutes.

Modern Man. With our perspective of the history of living things, we could say that Modern Man arrived upon the scene only yesterday, because the forty thousand years that have elapsed since he appeared are an infinitesimal fraction of geologic time. But from the point of view of human history, Modern Man has already experienced a long life; forty times as long as the time elapsed since William the Conqueror invaded England, twenty times as long as the Christian Era, and nearly eight times as long as the time since the beginning of Egyptian civilization along the Nile. In 40,000 years at least 1600 generations lived their lives—not enough generations for evolution to have made a visible difference in the human skeleton, but enough generations for *Homo sapiens,* with the best brain yet produced, to make tremendous strides in technology and in less practical human wisdom. Almost certainly language had been invented long before Modern Man appeared. As a result, acquired experience (practical wisdom, if you like) could easily spread from one people to another and from parents to offspring. In such communication only one more step was needed—the invention of writing—so that experience could be written down and stored in libraries for the use of future generations. Modern Man took that step.

The earliest representatives of Modern Man were the Cro-Magnon people, named for a cavern in southwestern France where a number of their skeletons were found. Anatomically these people had developed well beyond the Neanderthals and were fully modern (Fig. 19-12). They were tall, taller indeed than the average European of today. Many of the men reached a stature of 6 feet and one skeleton measures 6 feet 5 inches. They were finely built, with higher, rounder, and narrower heads than the Neanderthals' (Fig. 19-7). There is little doubt that if we could meet a Cro-Magnon walking along a city street in clothing like ours, his appearance would attract little notice. Indeed some scientists believe that Cro-Magnons were the direct ancestors of modern Nordic people, that they were pushed gradually westward and northward by the later immigration of differently built people ("Mediterraneans") from the southeast, and that people of Cro-Magnon type are still to be found living today in the western parts of the British Isles and Norway. As for

Figure 19 - 12. Members of a band of Cro-Magnons at the ceremonial burial of one of their number, perhaps killed in a hunt. The body, dressed in furs and a shell necklace and accompanied by weapons and a mammoth tusk, is sprinkled with red ocher (limonite or hematite; Table 2 - A) before the grave is filled. The time is about 30,000 years ago. (Augusta and Burian, *Prehistoric man*, pl. 49.)

the origin of the Cro-Magnons themselves, all we know is that they showed up in Europe rather suddenly, and that apparently they came from the southeast. At about the same time, 35,000 to 40,000 years ago, other races of Modern Man began to show up in eastern and southeastern Asia, Australia, and Africa. Wide migrations seem to have been occurring throughout the Old World, although as far as we know the Americas were not yet occupied by man.

In this chapter we have traced the development of hominids (Fig. 19 - 13) and have sketched quickly what the successive groups were like. But although one of the unique characteristics of man among the other mammals is that he has learned to make tools and to create art, we have said almost nothing about the tools and the works of art themselves. These things form so important a part of the later history of man that they deserve a chapter to themselves, the chapter that now follows.

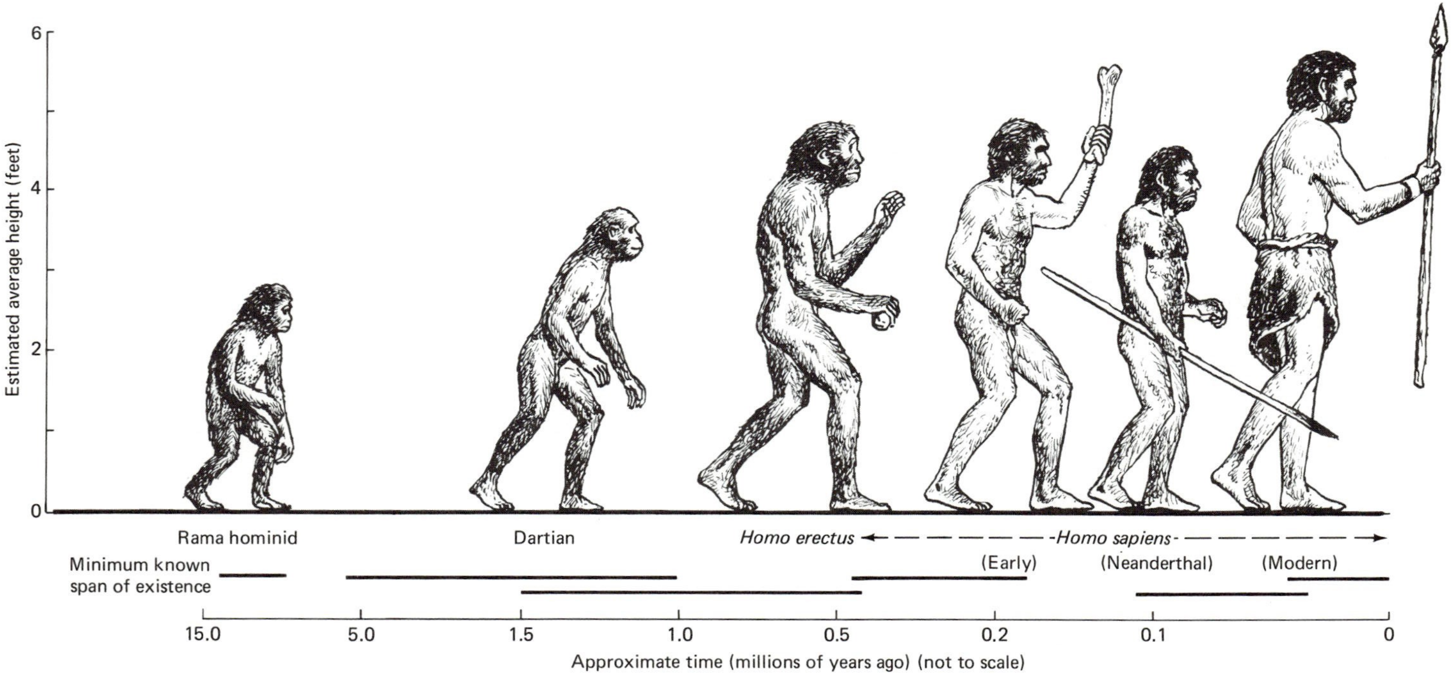

Figure 19 - 13. Time spans (thick bars), average height, and supposed appearance of six hominids.

References

Cornwall, I. W., 1968, Prehistoric animals and their hunters: Praeger, New York.

Howell, F. C., and Editors of *Life,* 1965, Early Man: Time, Inc., New York. (Life Nature Library.)

Howells, W. W., 1966, *Homo erectus:* Scientific American, vol. 215, p. 46–53.

Kurtén, Björn, 1972, Not from the apes: Pantheon Books, New York.

Pilbeam, David, 1970, The evolution of man: Funk and Wagnalls, New York.

Romer, A. S., 1964, The vertebrate story: Univ. of Chicago Press, p. 309–340, 351–411.

20

Prehistoric Evolution of Culture

The descent of man

Our review in Chapter 19 of the evolution of Modern Man from early hominids fifty million years old shows that the process was a series of transformations. These formed a continuous chain, even though the sequence of fossils is interrupted by wide gaps. Also the dates of the fossils, although few in number, have turned out to be in logical sequence, the most primitive fossils being the oldest and the most advanced fossils the youngest. Even though the fossil record is spotty, we accept as broadly correct the history it suggests. And yet we still find it hard to realize that the idea of man as a fossil, man as the maker of ancient stone tools, and man as related by blood to other mammals had not yet formed in the minds of any but a handful of scientists until about 1860, little more than a century ago. The general public did not catch up with this startling advance in knowledge of man's origin until much later.

Through long centuries stone implements had been dug up by accident here and there. Occasionally some astute person sensed something of their significance. Others thought of them as "thunderbolt stones" or some other kind of magic product rather than as the work of ancient men. Many more simply didn't think about the finds at all. In 1857, in a cave in southwestern England, stone weapons were found closely associated with the fossil bones of rhinoceros, hyena, and other animals long extinct. This find gave scientists a jolt, and other similar discoveries followed in rapid succession. In 1859 the publication of Charles Darwin's book *The Origin of Species,* containing the theory of organic evolution by natural selection, opened many eyes and put a gradual end to the belief that man had appeared on Earth only a few thousand years ago. The idea of the progressive evolution of man from early hominids slowly became established in spite of fierce opposition.

Artifacts and their occurrence

Having traced the evolution of human anatomy, we can now trace the evolution of *artifacts.* These are objects of any kind that have been made or at least modified by human work. An ape or monkey can be seen to pick up a stick and use it for some food-getting purpose. But the stick is not for that reason an artifact, because an artifact is something *made* in order to accomplish a purpose that has been thought out in advance. Having a well-defined purpose, the artifact is made to a pattern, a pattern that can be and is followed again and again. All hammers must possess characteristics of hardness, toughness, weight, and shape that make them suitable for pounding, whether they are made of stone or steel. The evolution of hammers through more than a million years shows that the pattern changes in detail through time, but it remains basically an implement designed for pounding. Neither apes nor monkeys, nor any other animals except one, can make artifacts. That one is man. With his flexible fingers and opposable thumb developed in an early history of grasping branches, his ability to walk erect developed by his later descent to the ground, and his big brain capable of thinking out a design that will satisfy a specific need, he alone has the ability to make tools and other artifacts.

Most ancient artifacts, particularly the older ones, are tools, weapons, or ornaments. They tell us at least as much about the history of people as human skeletons do. And because artifacts and fossil human bones are rather commonly found together in the same places, we have an immediate means of correlating artifacts with the people who made and used them.

Scientists collect and study artifacts much as they collect and study fossils. An artifact identifies the layer or stratum in which it is found, much as a fossil does. But the layer does not ordinarily consist of rock. More commonly it is a layer of refuse and rubbish left on the floor of a cave or rock shelter by people of an earlier time. Such layers are being made today and every day at the sites of town and city dumps. When scientists in the distant future come upon the huge rubbish dump that marks the waste from a large town in, say, New England, they will find upon excavation that it is built up of many successive strata. One layer, well down below what will by then have become the top, will include discarded television sets, radio sets, old tires, and broken electric-light bulbs. Underneath it will be a layer with radios, tires, and bulbs but no television sets. Farther down will be a layer with light bulbs but no radios. Still lower will be a stratum with broken whale-oil lamps and bits and pieces of horse harness. Beneath that, very likely old candle snuffers and home-made pottery will make up part of the mixture. Down to that depth the excavation will have represented at least 150 years of American industrial culture.

Many of the town dumps of prehistoric man occur in caverns and in rock shelters, slight recesses beneath overhanging rock ledges (Fig. 20-1). The floors of many such places are the upper surfaces of dumps, of piles of refuse several feet to several tens of feet thick. The refuse accumulated throughout periods some of which far exceeded 100,000 years and spanned the lives of far more than 5000 generations of people. Many kinds of things accumulated in these heaps of refuse that were tramped down and compacted daily by human feet. But the things that consisted of wood, other plant mat-

Figure 20 - 1. Rock shelter beneath an overhanging ledge of Jurassic limestone on the River Ain in southeastern France. The partly dug-out filling appears in the lower half of the view. It consists of refuse layers interspersed with sand that was washed in by the river in the latest glacial age, when the spreading ice cap of the Alps approached the place. At one time the glacier was only ten or twelve miles away, but it never reached the shelter. The filling contains at least three layers made by human occupation, roughly 20,000, 16,000, and 7000 years ago. The artifacts in each layer are distinctive. (R. F. Flint.)

ter, bits of meat, and skins disappeared through decay and oxidation long ago. What survived were stone and some of the bone, teeth, ivory, and shell embedded in a chaotic mixture of clay, silt, and sand brought in particle by particle on human feet and tramped down along with the rest. The accumulation was added to from time to time by rockfalls from cave ceilings. Sometimes bodies were buried in those places; the town dump was also the town cemetery. Now we begin to see why artifacts can often be correlated with skeletons and even scattered bones, why one can tell who made what, as the floor of a cave is carefully excavated and the strata in the refuse heap are examined.

The artifacts found in all but the uppermost layers are chiefly tools and weapons, most of them made of quartz (Table 2 - A, bottom). This mineral, especially a non-crystalline variety called flint, is well suited to the making of tools. It is easily chipped in shallow, concave chips like those one dislikes to see in the rim of a china plate, simply by a quick blow at the correct angle. By chipping one can fashion a lump of quartz into a variety of shapes, including shapes that have sharp edges. Being very hard, harder than steel, quartz retains a sharp edge very well. Being unusually durable chemically, quartz resists weathering. It is far more durable than bone, which consists mainly of calcium phosphate and calcium carbonate. Hence artifacts made from quartz have survived in many environments from which the bones that once lay buried alongside them have disappeared, destroyed by chemical weathering. We can say truthfully that all but the latest part of the history of man has been written in quartz.

Stratigraphy of artifacts

Artifacts are classified according to the methods by which they were made and the skill in workmanship they reflect. As we go upward through strata (some natural, some human rubbish) of later and later Cenozoic age, we find artifacts that have been made more and more cleverly. In a way this resembles the evolutionary change in fossils of all kinds as we go from lower to higher layers of sedimentary rock. But the resemblance is only partial. The evolution shown by fossils is the direct result of natural selection, which is purposeless. The evolution shown by artifacts is the result of deliberate planning by hominids. Because the physical and mental capacities of hominids were gradually improved by natural selection, hominids made better and better tools. So the evolution seen in fossils is direct, whereas the evolution of artifacts is indirect.

Clearly, then, artifacts show that human skills improved with time, at first very slowly indeed but later on at an increasing rate.

New skills in tool making were carried by people from one region to another and even from one continent to another. We find artifacts made with a particular degree of skill showing up in strata of the same general age in widely separated regions. As a result artifacts, like fossils, make it possible to match or correlate, at least roughly, layers of human refuse through long distances.

From the layers, each with its own kinds of artifacts, have been constructed charts that somewhat resemble the geologic column portrayed in Table 3 - A. They resemble that column because they have been built up and gradually improved in an almost identical way by the addition of new discoveries and new radiometric dates and by ever-closer correlations. Table 20 - A is a greatly simplified example of such a chart. Although it resembles the geologic column in style, it is based primarily on artifacts, whereas the geologic

Table 20 - A. Groups of artifacts, related cultural ages, and a few dates. (Greatly generalized.)

Top part of geologic column for comparison	
Quaternary	Recent or Holocene
Quaternary	Pleistocene
?	
Pliocene	

Artifacts	Cultural Ages	Approximate dates (years ago)	
		0	
Iron, copper, bronze	Age of Metals		Radiocarbon dates
		7,000	
Ground and polished stone; pottery	Neolithic Age ("New Stone Age")		
		12,000	
Pressure-flaked stone tools	Paleolithic Age ("Old Stone Age")		
		40,000	
Hand axes			(Gap)
		1,250,000	Potassium/argon dates
Pebble tools			
		3,000,000	

column is founded primarily on fossils. Nevertheless, the two charts have a common denominator in radiometric dates. This common denominator makes it possible to place the two side by side and compare them, provided both are drawn to the same time scale. Table 20 - A shows this comparison. Furthermore, the entire base of the table has been left open, because we do not yet know at what point in time artifacts were first made. The oldest one yet found is from East Africa and is about 2.6 million years old. Because the oldest Dartian fossil is more than twice that age, it is likely that simple artifacts of that greater age exist. But if they do, their discovery lies in the future.

Apart from the column of dates in our table are two other columns. One is for kinds of artifacts; the other contains the names of *cultural ages,* time spans during which the various kinds of artifacts were made. Cultural ages are analogous to the periods in the geologic column in which, for example, the Cretaceous Period represents the time span during which certain animals and plants found in Cretaceous rocks were alive.

Pebble tools; start of the Old Stone Age

As we see in Table 20 - A, the earliest known artifacts are *pebble tools.* Their name comes from the fact that they consist of smooth, rounded or flattened pebbles or small cobbles (Fig. 4 - 6) that hominids picked up from a stream bed or a beach and crudely chipped to form a sharp though ragged edge (Fig. 20 - 2). Probably the chipping was done at first with another stone held in the hand and used as a hammer. Perhaps later the leg bone of an animal was used as the hammer. A sharper and more accurate blow could be struck with a long bone because control was better. The resulting pebble tool, a sort of crude chopper, might have been used for such purposes as hacking through small sticks and flesh and for scraping raw skins.

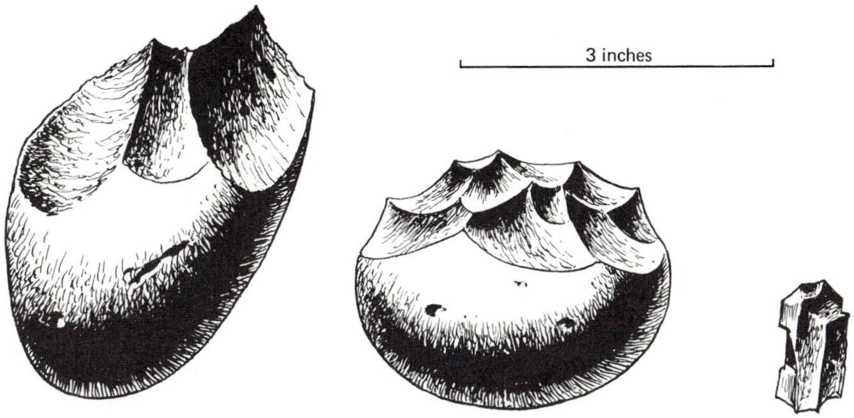

Figure 20 - 2. Two pebble tools, one very crude, the other made more skilfully. The by-product chip may have been used for skinning.

Here we may logically ask how a carcass could have been skinned. Certainly not with a pebble tool. But along with pebble tools have been found small, thin, sharp pieces of quartz that seem to be chips knocked off a stone in the making of a pebble tool (Fig. 20-2, c). Possibly these by-product chips or flakes were used in skinning. If so, such chips were man's earliest knives.

Although it was a clumsy tool that could do only clumsy work, the pebble tool was used through a period of time at least two million years long, improving a little during the process. The users were mainly Dartians and perhaps toward the end *Homo erectus.*

The making of the first pebble tool marked the beginning of the *Paleolithic Cultural Age* (which means Old Stone Age). That age embraced not only the time of pebble-tool making but also the evolution of pebble tools into implements of ever-greater refinement. But they were still made of stone and were still made by chipping and flaking. The Old Stone Age, the age of chipping and flaking of quartz, did not come to an end until little more than twelve thousand years ago. Its great length means that men lived with stone implements through a period more than 350 times as long as the period during which they have been fabricating metal. This in turn means that the learning process was very slow. Yet, through the long Old Stone Age technical education did not stagnate. Some improvements in working quartz were made.

Hand axes

The earliest improvement was the long, slow evolution of the pebble tool into the *hand axe.* Making a hand axe involved chipping a lump of quartz not merely along one side but all over its entire surface, creating a flattish shape that could be held firmly in the hand (Fig. 20-3). But edges could be made both straighter and sharper by striking off many little chips instead of a few big chips. For the small-scale chipping the maker used a hammer consisting of a small bone, a piece of antler, or a small stick.

For the men who lived in the Old Stone Age, making tools by this method was not a laborious job because they were skilled at it. You could say they were specialists. No doubt the children watched and imitated their parents and then learned by practice. A few modern anthropologists, also by practice, have made themselves spe-

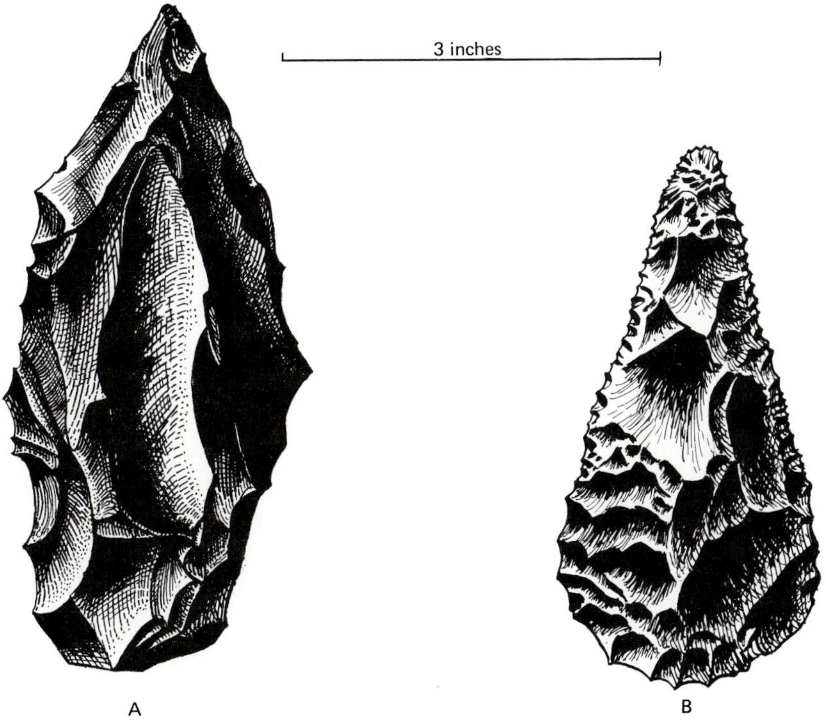

Figure 20 - 3. Hand axes. A. A crude type used about 300,000 years ago. B. An advanced kind used about 50,000 years ago.

cialists in the art. They have demonstrated clearly that a man can accurately fashion a lump of quartz into a perfect hand axe in just a few minutes, and make any of the more complex tools characteristic of the latest part of the Old Stone Age in scarcely more time. We do not therefore have to pity Stone Age men because making tools was a heavy task. It wasn't. They could do the job better and more quickly than all but a tiny handful of the men who are alive today.

Later artifacts

Those people, whom we may pity for other reasons if we like, included *Homo erectus* and the races of *Homo sapiens* from more than a million up to about 40,000 years ago. At about the latter time one of the later races made an important improvement in tool making by inventing the technique of *pressure flaking*. Against the edge of the tool-to-be the maker pressed, at a carefully calculated angle, a pointed implement of bone or wood rather like a very thick pencil. The pressure caused a small chip to fly off. With this method the worker could control both the size and position of each chip with great nicety, so that the shape of the resulting tool was determined very precisely.

With the close control provided by pressure flaking, it became possible to make finely shaped tools for a variety of purposes. For example, a small flat tool with a sharp point at one end was an awl for boring holes in skins or small pieces of wood, and double-edged, pointed implements were made for mounting at the ends of short poles to form spears.

A great aid in the making of thin implements developed when it was realized that the by-product flakes from the making of tools could themselves be finely worked. Toolmakers learned to strike off thin, narrow flakes several inches long from lumps of quartz, and then fashioned the flakes themselves into sharp knives, small saws, scrapers for dressing skins, and spearheads. Tool kits were becoming gradually more varied and more specialized. Between 25,000 and 12,000 years ago tools were brought to a final high degree of

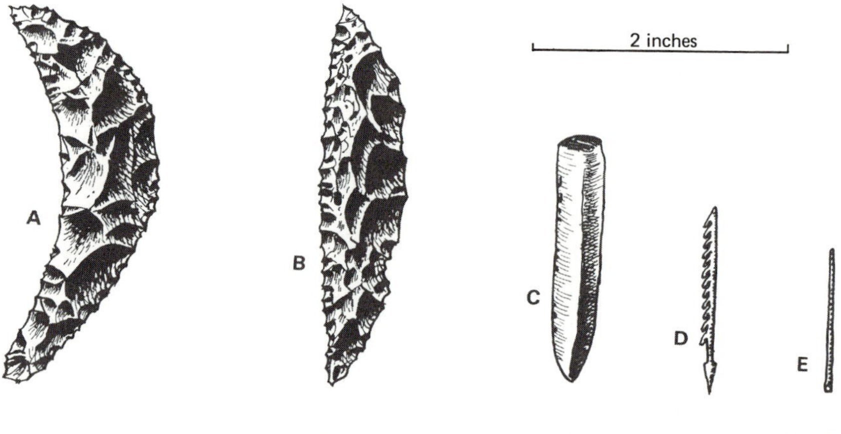

Figure 20 - 4. Samples from the Cro-Magnon people's tool kits. Specialized tools made by pressure flaking of quartz (A-C), and by carving and polishing of bone and ivory (D, E). Represented are a sickle, a saw, a knife, a harpoon head, and a needle with eye.

perfection by Cro-Magnon people. Before them, the latest of the Neanderthals had begun to work not only quartz but also bone and ivory, but the Cro-Magnons carried the process much farther (Fig. 20-4). The implements became so specialized that they even included fine ivory needles each pierced by an eye. From such finds and from others as well, it is evident that Cro-Magnon people wore clothing well made from skins. Many burials have been found, showing that at least the most important people were buried (Fig. 19-12) with food, implements, and ornaments, far more elaborately than were the Neanderthals. This shows that the Cro-Magnons had a strong belief in a life after death.

The beginning of art

Between 30,000 and 15,000 years ago, while the latest of the glacial ages came on and passed its climax, Cro-Magnon communities were increasing in size. In western and central Europe people lived in caves, many of which were closed at their mouths by stone walls. Where there were no caves, people lived in tents made of skins stretched over frameworks of bone, ivory, or wood. The economy was based almost entirely on hunting by large groups of the men (Fig. 19-9), while women probably gathered berries, fruit, nuts, seeds, and roots when and as they were available. But the diet was predominantly meat. This was provided by large glacial-age mammals such as the reindeer, horse, woolly rhinoceros, and woolly mammoth. These animals were likewise a source of clothing and of bone or ivory for making implements.

The dependence of these people on hunting may have been what led to the development of art. Not only did the Cro-Magnons paint splendid pictures on the walls and ceilings of large caves; they made sensitive engravings on bone and ivory and carved statuettes as well. Their pictures were mostly of food animals—not people, not carnivores, and not small mammals—and not uncommonly the figures were represented as wounded or dying in the hunt. Although the cave dwellings are floored with domestic refuse, they lack the paintings, which have been found only in large caves without domestic refuse. When these facts are put together it looks as though the large, painted caves were ceremonial places (corresponding in a way to temples or Masonic lodges) where a tribe gathered to promote success in the hunt. The animal they wanted to kill was painted on wall or ceiling, there was a secret rite or ritual, and possibly the hunt itself was enacted. Some of the paintings and engravings were done with such talent and skill that they must have been the work of individuals who specialized in art, who studied to become painters. This is easy to believe, for the Cro-Magnon brain and the capable Cro-Magnon hand were indistinguishable from ours today. Yet despite this specialization, it looks as though art first appeared in the texture of human life not as fine art created for the sake of beauty alone, but as a kind of industrial art, created for a very practical purpose: to increase the food supply by the practice of magic.

Nevertheless Cro-Magnon painting and drawing is of such high quality that it speaks to us as though it *were* fine art. It is the result of close and intelligent observation of the animals commonly hunted, and is full of strength and activity. These qualities are evident in the mammoth seen in Figure 9 - 11, and in the more elaborate works of art in Figures 20 - 5 and 20 - 6. Although magnificent studies of anatomy, most of the pictures display no sense of composition or arrangement. This is another reason for believing they were created for a ceremonial purpose rather than to be lived with and looked at every day.

Figure 20 - 5. A magnificent bison painted in colors on the wall of a cave in northern Spain about 15,000 years ago. (Courtesy of the American Museum of Natural History.)

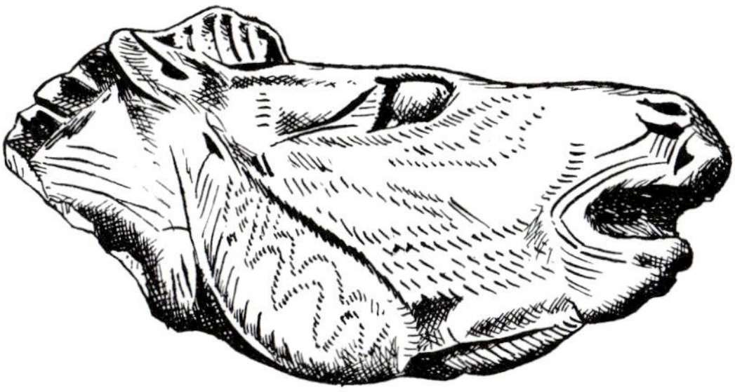

Figure 20 - 6. Head of a horse carved from reindeer horn by a very late Paleolithic artist about 10,500 years ago. (From Boule, M., 1921, Les hommes fossiles, fig. 163. By permission of Masson et Cie., Paris.)

Besides the splendid pictures, the Cro-Magnons created ornaments to be worn by people. At first these consisted of necklaces of shells, teeth, bears' claws, and carved ivory. It may be that in the beginning these were worn only by male leaders during ritual ceremonies. But much later they came to be worn by women as personal adornment.

The New Stone Age

Whether or not their magic helped them, the Cro-Magnons were skilled and efficient hunters. They gradually killed off, or contributed to the killing off, of the large mammals in Europe. The kinds of bones found in their refuse heaps, coupled with radiocarbon dates, show how their diet changed through a period of twenty or thirty thousand years. The large glacial-age mammals were all eaten, and by 12,000 years ago the diet consisted mainly of horses, deer, and smaller mammals. The food supply of a meat-eating population was diminishing, but people were increasing. According to careful estimates, right within the time span we are discussing the world's population curve began to bend upward toward the near-vertical climb that alarms us today (Fig. 20 - 7). In the 24,000-year period between 30,000 and 6000 years ago the estimated population increased more than 28 times. (And in the last 6000 years, a time span only one-fourth as long, it has increased 42 times again.)

With more mouths to feed and fewer large and even medium-size mammals to feed them, people turned to smaller and smaller mammals, birds, and fish. They used bows and arrows and hunted

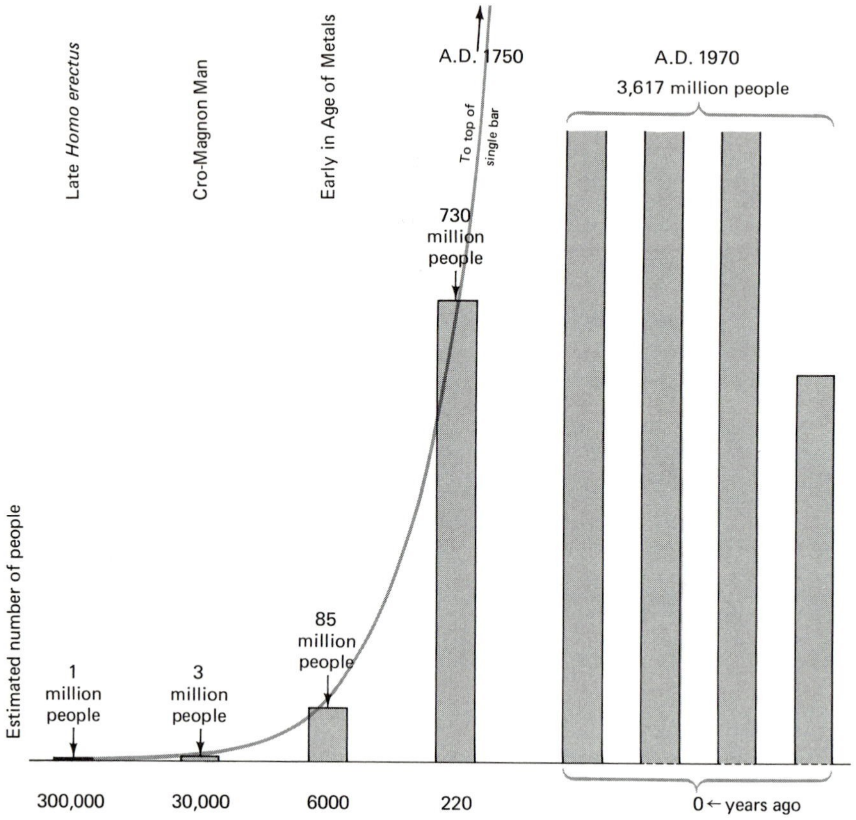

Figure 20 - 7. Growth of human population of the world through 300,000 years. The bundle of four bars at the right must be placed end to end to form a single tall bar, to be in proportion to the other bars. In 1970 there were five times as many people in the world as in 1750.

less in large groups. They gathered various sorts of vegetable food, including wild seeds. Before long they discovered that seeds not eaten but planted in the ground turned into plants bearing more seeds. That discovery made available a basic new source of food, grain, to replace the dwindling supply of meat. This in turn started a whole new way of living that began to take shape around 9000 years ago. Instead of continuing as hunting tribes living in caves, groups of people gradually became cultivators of grain, living in dwellings they built themselves right on the land they cultivated. Their lives became settled and villages began to appear.

With this change the character of the whole society changed. The transformation began in the comparatively dry Near East where forest land and the game it sheltered were scarce. Later it spread through the moister, densely forested lands of Europe, where at least some game was still present and where big trees had to be cut down and cleared away before grain could be planted.

The gradual progress of clearing the European forest is recorded by a change in the kinds of fossil pollen found in successive layers of sediment in small lakes and bogs (Fig. 17 - 7). In the samples of sediment, tree pollen gradually gives way to pollen of coarse weeds such as grow only in cleared lands. Even the earliest farmers had their troubles with weeds!

The transition from a purely hunting life to a more settled life of agriculture brought with it changes in implements. Hunting weapons became fewer and domestic implements increased. Clay was molded and baked to form pottery for the storage of water and grain and for other uses. Instead of being flaked, axe heads and other stone implements were ground and polished. It was with such axes that forests were cleared for primitive agriculture.

The clearing of forests and a more settled life led naturally to the domestication of animals. There is evidence that goats were domesticated as early as 8500 years ago, and sheep followed soon thereafter. Cattle had been hunted as wild game by both Neanderthals and Cro-Magnons, but now cattle were domesticated, and so were laid the basic foundations of our modern beef and dairy industries.

Figure 20 - 8. Neolithic dwellings built nearly 5000 years ago on pile-supported platforms over a lake in Switzerland. Restored from wooden ruins preserved against oxidation in saturated lake-floor mud. Note fishing nets and boat. (Courtesy of H. Müller-Beck.)

All these things—pottery, ground stone, permanent dwellings build of wood or mud brick, domestic animals, woven cloth, and basketry—are together characteristic (Figs. 20 - 8, 20 - 9) of the *Neolithic Cultural Age* (the New Stone Age), which lasted for several thousand years (Table 20 - A). Within Neolithic time was built the earliest city we know of. Jericho, not far north of Jerusalem, was a flourishing city well built of mud bricks 8200 years ago, and at that time had long been in existence.

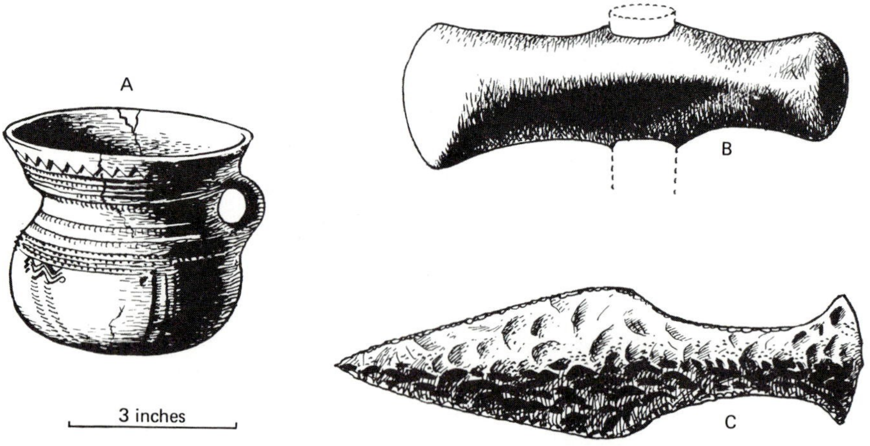

Figure 20 - 9. A group of Neolithic artifacts from western Europe. A. Pottery jar with handle. (Switzerland.) B. Axe head, made by flaking a piece of igneous rock and finished smooth by grinding. The hole through it accommodated a wooden handle (broken lines). (Austria.) C. Dagger with blade and a comfortable handle, made by flaking a piece of flint. (Scandinavia.) (Yale Peabody Museum.)

As the technical ingenuity of Neolithic men increased, a next logical step was the discovery of ores and the development of ways to smelt them and to fabricate the resulting metal into better implements than those made of stone and more durable vessels than those made of clay. The use of bronze, iron, and silver began as early as 7000 years ago, gradually replacing the New Stone Age with an Age of Metals in which we ourselves live. Metals are a very recent development in the evolution of industry. Our ancestors fabricated tools from stone during a period at least 370 times as long as the 7000 years during which they have used metals.

Prehistoric man in the Americas

Despite the spreading of Early Man through Africa and Eurasia at times late in the Cenozoic Era, no evidence has been found that the Americas were entered by men until very recently. Despite much searching, neither human fossils nor artifacts that date from more than about 14,000 years ago have been found.

The oldest North American artifacts are small stone points that formed the heads of short wooden darts or javelins. They are distinctive (Fig. 20 - 10), being finely pressure flaked around the edges, and with a wide groove or flute down the center of each flat side where the point was fitted between the halves of a split wooden shaft. Because of this form they are called *fluted projectile points.*

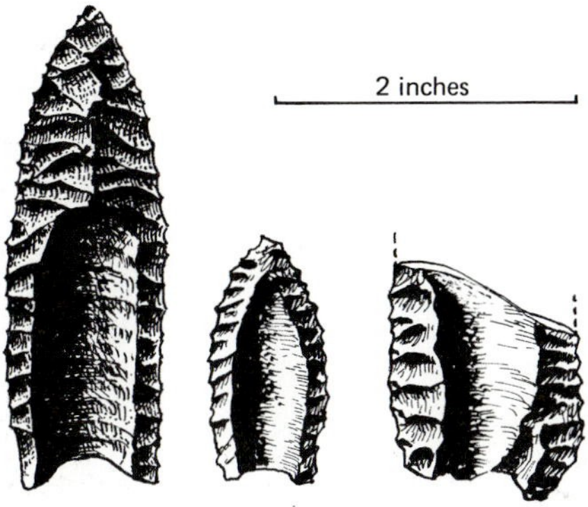

Figure 20 - 10. Two typical fluted dart points and a knife, made by Paleo-Indian big-game hunters in Colorado 10,000 or more years ago.

The finished darts were thrown at high velocity from the jointed launching sticks mentioned near the end of Chapter 18. The method used in making dart points was like that familiar to the Cro-Magnon people who hunted big game in Eurasia. It is thought likely that as those people spread across northern Asia following herds of large mammals, some of them—perhaps only a few hundreds—crossed over the land bridge (Chapter 16) that then connected Siberia with Alaska. Because the level of the glacial-age sea was then low, the wide bridge existed between more than 30,000 and about 12,000 years ago.

Probably these hunters, who belonged to the latest part of the Old Stone Age, did not cross over all at once. Probably they drifted across in groups of a few families over a rather long time. They may have lived in Alaska, hunting the abundant mammals there, for perhaps thousands of years before the ice sheet began to melt back, opening the corridor along the eastern base of the Rocky Mountains that we saw earlier on a map (Fig. 17 - 2). We suppose these very mobile people, living in temporary camps and skilled in hunting big game through thousands of years of practice in Siberia and Alaska, moved down the corridor and in Montana fanned out into the vast region south of the glaciers. The vegetation they would have encountered is sketched in Figure 18 - 1. The big game they hunted consisted mainly of woolly mammoths and other elephants, mastodons, bison, horses, camels, tapirs, and large ground sloths. These people may have been the chief cause of the killing off of big mammals within a short time, according to the hypothesis described in Chapter 18. Whether or not they were, people of this kind drifted southward and within a short time had reached the southern tip of South America. The two Americas had acquired their first human population.

These Stone Age hunters, who are often called *Paleo-Indians,* were followed during the next few thousand years by other races from northeastern Asia. The land bridge had been destroyed, drowned by glacial meltwater pouring back into the sea, but the later arrivals could easily have crossed the Bering Sea in primitive boats. They were people with Neolithic cultures brought from the Old World and are therefore often spoken of as *Neo-Indians.* Long before they arrived most of the big game had ceased to exist. Some of the immigrants hunted smaller animals, some who settled along the coasts ate seafood, some were primitive farmers, and some combined these ways of life.

No trace of the later history of the Paleo-Indians has been found; so we do not know to what extent they mixed with later immigrants. Whether there was mixing or not, the descendants of these groups lived in North America for thousands of years. They are the people encountered by the Norsemen, Columbus, John Smith, and the passengers on the *Mayflower* when they landed in America.

The past and future of *Homo sapiens*

Somewhere in the shadowy zone in which the Age of Metals overlaps the New Stone Age, the earliest known writing appeared.

With writing, knowledge could begin to be stored and handed down to later generations of men without having to pass perilously through each generation by word of mouth. This was the beginning of history based on written documents, in contrast to the far longer history based on fossils and other non-written documents that we have followed through many chapters.

We have devoted only one chapter to the history of culture because it is familiar territory; we see the effects of culture all about us every day. The path of culture from its first beginning has been two or three million years long, but the path of evolution, more than a thousand times as long, traces the process that eventually made culture possible. Its effects are less obvious and less widely understood. That is why we have devoted ten chapters to evolution. As members of a certain species of a certain genus of a certain group of vertebrate animals, we are not merely inhabitants of the Earth; we are a small part of its biosphere. The carbon, hydrogen, nitrogen, and oxygen with which our bodies are chiefly built were drawn from the Earth's chemistry. Our outward form and our inner mental ability are both products of the long chain of evolutionary change, guided always by successive environments through which our ancestors, recent and remote, have passed.

The chain has led from purely organic evolution into cultural evolution, with its Age of Metals subdividing itself with increasing speed into a Machine Age, an Electronic Age, a Nuclear Age. And through the last 30,000 years a human population growing larger with ever-increasing speed. Only very recently, within the lifetimes of people now living, have we begun to realize the danger that food production cannot continue to increase as fast as the multiplication of people. Only recently have we begun to think about the many evils that result when millions of people are crowded into limited patches of territory. We are only just beginning to understand that highly industrialized people are distorting the water cycle and the chemical cycle related to it. We are interfering with the biological cycle by polluting the atmosphere that supplies our oxygen and the hydrosphere that provides our water, two essential ingredients of the life process. We are creating waste products that destroy living things and that cannot be easily, if at all, converted biochemically back into their components. The problems posed by all this man-generated pollution and distortion are formidable. Nothing comparable with them, and on such a large scale, has ever existed before. Only men can solve the problems. No other animal is capable of dealing with them.

To speak of "man and other animals" emphasizes the fact that Modern Man, *Homo sapiens,* is indeed a zoological species. Although true, this does not imply that Modern Man is *nothing but* an animal species. Men are not like other animals because they have emerged from long competition with other species to occupy a dominant position among living things. The fossil record tells us that before *Homo sapiens* appeared, evolution was planless. It was a mere series of adjustments powered by variations in genes and shaped by environment. But in *Homo sapiens* the evolution of intelligence has gone so far that we possess a unique ability to learn, to stand apart and see ourselves in relation to our environment and the rest of the Universe, and to transmit to the succeeding generation a body of learned tradition that is the basis of culture. The content of that culture is determined by ourselves through our ability to choose and to make conscious decisions. This involves the planning of our own conduct, and in a sense represents a new, self-controlled evolution based on free choice, added to the old planless evolution controlled mainly by environment.

Clearly, men are no longer shaped wholly by environment. Possessing free choice, we have a considerable degree of control over it. This is a position of potential power and is shared by no other species. But the possession of power is always accompanied by something else: *responsibility.* Responsibility is a sense of duty, an obligation, in this case towards other men, towards other living creatures, and towards the Earth's natural processes—to maintain our environment free of distortion or deterioration. The capacity to be responsible, possessed only by men among all living creatures, is perhaps the main ingredient in what we call civilization. This unique endowment has a long evolutionary background, and from our concept of it we can derive an ethic, a philosophy based on the evolution of the Earth, the path we have traced in this book.

"Blind" evolution → evolution of the human brain → partial control over environment → responsibility.

Because of the mental ability that has come to us through evolution, we can look both backward and forward. We have looked backward through the record of the strata and the fossils they contain. In the record we have found an understanding of the Earth's cycles and have reconstructed the events of the Earth's past. We have built a history. That view backward, however incomplete, is a human triumph of which we have great reason to be proud. We were not yet evolved—we were not there to see those ancient events; we experience them only in our imagination. Yet we are sure they happened, because we possess the ability to perceive evidence, to reason from the evidence, and thereby to reconstruct the long historical procession that has gone before us.

With this same ability we can also look forward with responsibility to ourselves, our fellow creatures, and the environments we all share. We can visualize the dangerous consequences of man-made changes in the Earth's cycles, and firmly correct the changes that are not in our collective interest. If *Homo sapiens* fails to do this, no other species can do it.

References

Braidwood, R. J., 1961, Prehistoric men: Chicago Natural History Museum, Popular Series, Anthropology, No. 37. (Paperback.)

Howell, F. C., and Editors of *Life*, 1965, Early man: Time, Inc., New York. (Life Nature Library.)

Jennings, J. D., 1969, Prehistory of North America: McGraw-Hill Book Co., New York.

Pilbeam, David, 1970, The evolution of man: Funk and Wagnalls, New York.

Ucko, P. J., and Rosenfeld, Andree, 1967, Palaeolithic cave art: World University Library, McGraw-Hill Book Co., New York. (Paperback.)

Glossary

Names of rocks and minerals are not included in the Glossary, but are given in Tables 2-A, 2-B, 2-C, and 2-D.

Alluvium. Sediment deposited by streams in any land environment.

Alluvial fan. A fan-shaped body of alluvium built at the base of a steep slope.

Amnion. The innermost membrane of the sac that encloses the embryo of a mammal, reptile, or bird.

Artifact. An object of any kind that has been made or modified by human work.

Atmosphere. The gaseous envelope that surrounds the solid Earth.

Bedrock. Continuous solid rock, in contrast with regolith.

Biosphere. That part of the Earth which consists of organic matter, both living and nonliving.

Cerebrum. The upper, main part of the brain in vertebrate animals.

Conformable. The relationship between any two successive strata that were built up by the continuous, unbroken deposition of sediment.

Coprolite. Fossil feces of any kind of animal.

Correlation (geologic). Determination of equivalence of two strata in separated areas, as to geologic age or stratigraphic position.

Crust plates (also called *crustal plates*). Plate-like pieces of the Earth's crust, irregular in shape, that float on and move over the fluid-like heavy matter beneath them.

Culture. The ideas, customs, skills, arts, and so forth of a people at a particular time.

Delta. A body of sediment deposited by a stream that flows into standing water of a sea or lake.

Dissolution. Dissolving or being dissolved.

End moraine. A ridge-like accumulation of sediment built up along the margin of a glacier.

Erosion. The physical or chemical loosening of rock material and its removal from place to place.

Estuary. The wide mouth of a river that is affected by tides.

Fault. A fracture along which the opposite sides have been displaced relative to each other.

Foliation. A structure, common in many metamorphic rocks, that consists of parallel, thin, discontinuous layers of mica and other minerals and resembles the grain of a piece of wood.

Fossil. The naturally preserved remains or traces of an animal or plant.

Geologic column. A diagram combining in a single column the succession of all known strata, fitted together on the basis of their fossils and other evidences of relative age.

Geothermal gradient. The rate of increase of temperature with depth below the Earth's solid surface. Usually expressed in °C per unit of distance.

Glacial age. A period of time, apparently some tens of thousands of years long, when vast ice sheets spread over middle-latitude lands, chiefly in North America and Europe. Glacial ages were separated in time by milder conditions, in which the ice sheets melted away under the influence of warmer climates.

Ground water. Water contained in open spaces within bedrock and regolith.

Hogback. A sharp-crested ridge formed by a steeply inclined layer of sandstone or other hard rock, with weaker rock beneath it.

Hominids. The group of primates that includes man as well as all pre-human primates that are distinct from the line of apes.

Hydrosphere. The discontinuous water envelope that surrounds the lithosphere.

Igneous rock. Rock formed by solidification of molten material.

Lava. Molten rock material that has been forced up to the Earth's surface from a source within the Earth.

Lithosphere. The outer part of the solid Earth, including the crust and a zone immediately beneath it.

Magma. Molten material within the crust. It may include crystals and dissolved gases.

Mass-wasting. The gravitative movement of rock particles down a slope without the aid of a flowing medium such as water, glacier ice, or ordinary air.

Metamorphic rock. Rock formed within the crust, while in the solid state, by the response of a pre-existing rock to strong pressure, high temperature, or both.

Mineral. A naturally formed substance that has a well-defined chemical composition and a characteristic atomic structure generally expressed in definite physical properties.

Peneplain. A land surface worn down to low relief by streams and mass-wasting.

Petrifaction. The conversion of organic matter into a hard mineral substance.

Placenta. A flat organ filled with blood vessels that develops within the uterus in placental mammals. Connected with the growing embryo by the umbilical cord, it sends nourishment to and receives wastes from the embryo.

Plateau. A broad tract of high, rather flat land.

Regolith. The layer of loose, noncemented rock particles and mineral grains that generally overlies bedrock.

Rock. Any firm, coherent mass of mineral matter formed naturally.

Rock cycle. The continuous sequence of phases through which matter passes in the making and transformation of rock.

Savanna. Grassland dotted sparsely with trees.

Sediment. Particles or aggregates of particles that are derived from the erosion of rock or from the explosive eruptions of volcanoes, whether in transport or deposited by mass-wasting processes, water, air, or ice.

Sedimentary rock. Rock formed by cementation of sediment.

Solute. Dissolved substance.

Species. A group of individuals that possess characteristics in common and are capable of interbreeding. Their similarities result solely from their common ancestry.

Spore. A simple reproductive body, usually a single cell, that is capable of growth into a new individual, either nonsexually or through sexual union.

Stratigraphy. The systematic study of strata.

Stratum. A single well-defined layer of sedimentary or igneous rock formed on the Earth's surface.

Stream. A body of water, carrying solid and dissolved substances, flowing down a slope through a definite channel.

Topography. The physical features of an area, especially the relief and contour of the land.

Tree of life. A tree-like diagram in which fossil and living forms of life are arranged in order of similarity. It traces the principal paths of evolution.

Tundra. Treeless or nearly treeless vegetation of herbs, grasses, and shrubs of arctic character.

Unconformity. Lack of continuity between a stratum and the rock immediately beneath it.

Uniformity, Principle of. The concept that because the rock-making processes have always operated under the same (i.e., uniform) natural laws, today's rocks and processes form a valid basis for reading history from older rocks.

Vascular system. A system of vessels for the conveyance of a fluid, such as the sap in plants.

Vertebrate animals (or **vertebrates**). Animals that have backbones.

Vestigial structure. An anatomical structure that has no apparent function but is similar to a useful structure in near relatives.

Vivipary. Giving birth to living young as opposed to laying eggs.

Water cycle. The system of circulation of water substance from the ocean through the atmosphere to the lands and back to the ocean.

Weathering. The chemical alteration and mechanical breakdown of rock material exposed to air, moisture, and activities of organisms.

Index

Note: Names of groups of strata are omitted, but are given in Table 3–A, p. 42.

Asterisks indicate illustrations.

"... the possible relation of Stone-Age man to large mammals leads naturally to the absorbing story of the history of man himself ..."

(Page 351.)